T0177339

Progress Unchained

Progress Unchained reinterprets the history of the idea of progress using parallels between evolutionary biology and changing views of human history. Early concepts of progress in both areas saw it as the ascent of a linear scale of development towards a final goal. The 'chain of being' defined a hierarchy of living things with humans at the head, while social thinkers interpreted history as a development towards a final paradise or utopia. Darwinism reconfigured biological progress as a 'tree of life' with multiple lines of advance not necessarily leading to humans, each driven by the rare innovations that generate entirely new functions. Popular writers such as H. G. Wells used a similar model to depict human progress, with competing technological innovations producing ever more rapid changes in society. Bowler shows that, as the idea of progress has become open-ended and unpredictable, a variety of alternative futures have been imagined.

Peter J. Bowler is Emeritus Professor of the School of History, Anthropology, Politics and Philosophy at Queen's University Belfast.

Progress Unchained

Ideas of Evolution, Human History and the Future

Peter J. Bowler

Queen's University Belfast

CAMBRIDGE
UNIVERSITY PRESS

CAMBRIDGE
UNIVERSITY PRESS

University Printing House, Cambridge CB2 8BS, United Kingdom

One Liberty Plaza, 20th Floor, New York, NY 10006, USA

477 Williamstown Road, Port Melbourne, VIC 3207, Australia

314–321, 3rd Floor, Plot 3, Splendor Forum, Jasola District Centre, New Delhi –
110025, India

79 Anson Road, #06–04/06, Singapore 079906

Cambridge University Press is part of the University of Cambridge.

It furthers the University's mission by disseminating knowledge in the pursuit of
education, learning, and research at the highest international levels of excellence.

www.cambridge.org
Information on this title: www.cambridge.org/9781108842556
DOI: 10.1017/9781108909877

First published 2021

Printed in the United Kingdom by TJ Books Limited, Padstow Cornwall

A catalogue record for this publication is available from the British Library.

ISBN 978-1-108-84255-6 Hardback

Contents

Figures

Preface

This is a book about history in its broadest sense, from the history of life on earth to our ideas about the overall development of human cultures. More specifically it focuses on the role played by the idea of progress and the changing structure of that idea. As a historian of evolutionism I have long been interested in the idea of progress in both its biological and its social applications. I agree with Michael Ruse that there has always been a connection between the two areas: at least until recently, evolution was usually seen as biological progress analogous to and perhaps helping to justify faith in the advancement of humanity. In this book I explore the link between the two areas with a new emphasis, using the models and analogies provided by Darwinism and related theories in the life sciences to help us understand what I believe to be major changes in the way we think about human progress. If my readers think I spend too much time exploring the biological element, I can only plead that to me it seems necessary to explore this area in some detail so we can understand the distinctions I am trying to make.

The incentive to write this book arose from a new area of research which led me to think about how the idea of progress was projected into the future. I realized that there were parallels between the way people began to think about future progress and the insights that biologists were deriving from the latest developments in evolution theory. Here the Darwinism of the previous century was expanded to give a much more open-ended vision of progress, downplaying older models of progress which depicted it as the ascent of a ladder towards a final goal. The title *Progress Unchained* is meant to highlight the demise of a model of progress based on the old system of arranging living forms into a linear hierarchy, the 'chain of being'.

Given the scope of its topic this is a relatively short book, and this was made possible only by borrowing relentlessly from a host of historians who have helped me deal with areas of which I had no previous experience. Many areas of the book are shorter than they might otherwise have been because I have directed the reader to other publications to fill in details. As suggested above, this includes the whole swathe of my own research. I have thus ended up pulling together the whole sweep of my research career while at the same time learning about topics that were new to me, a nice combination at my time of life.

1 Introduction
Ladders and Trees

In 1771 the *philosophe* Louis-Sébastien Mercier published his *L'an deux mille quatre cent quarante* (*The Year 2440*), the first utopia set not on an imaginary island but in the future. He thus rates a whole chapter in J. B. Bury's classic survey of the history of the idea of progress. Mercier's imaginary time-traveller awakes to find a Paris transformed for the better. Thanks to the application of reason, society has been brought to a state of perfection in which everyone can live a comfortable life. There is a hint that the new state of affairs is not a static 'end of history' that will perpetuate itself without further change. Progress in science and technology will continue, and further discoveries will open up unimaginable potential for change. Mercier makes no attempt to guess what might be achieved, but his brief speculation highlights a crucial tension in the basic idea of progress.[1]

Predicting a future utopia certainly implies progress, but Mercier made no attempt to link this to a progressive trend that can be seen in the past history of civilization. The true idea of progress as a built-in historical trend that will continue into the future was consolidated a little later in the century by the marquis de Condorcet and others. From that point on the hope of further social and moral improvement was increasingly promoted as something to be expected precisely because history tells us that there is an inevitable trend in that direction. All too often, however, those who appealed to the idea disagreed over both the goal to be achieved and the driving force at work.

As Bury notes in his introduction, any definition of progress requires a value judgement as to the desirability of what is unfolding. The highly structured state proposed by socialists is anathema to liberals who see human liberty as crucial for happiness. Bury goes on to suggest that there are two versions of the idea: socialists know exactly what they want to achieve and thus imagine their future utopia as a final goal for social development, while libertarians are more likely to see progress as continuing indefinitely because free individuals can come up with new ideas. But even those looking for an expansion of liberty

---

[1] Mercier, *L'an deux mille quatre cent quarante*, chap. 31; see Bury, *The Idea of Progress*, chap. 10.

have a trajectory in mind which assumes that progress will produce the kind of benefits they value. Underlying Mercier's suggestion of unpredictable new developments in technology lies a very different model of progress, one that sees history itself as inherently unpredictable and open-ended. This makes progress a much more slippery concept because there can be no final goal to achieve and the past cannot be used to predict the future. The crucial insight is the recognition that there are always more ways than one of achieving greater complexity.

Rethinking the Shape of Progress

Imagining a utopia does not require a theory of historical progress: Sir Thomas More's vision placed utopia in some unknown part of the world, not in the future. Utopias are intended as critiques of the existing state of affairs. Even when an imaginary perfect society is described as emerging at some point in the future, if there is no attempt to explain how we get there, and no connection to trends perceived in past history, the suggestion doesn't function as a form of progressionism. A goal-directed view of progress does require the postulation of a perfect state towards which events are moving. But once the idea of an 'end of history' is abandoned, progress has a looser connections with utopianism: if there is no single goal in prospect, there is less inclination to describe any future state as 'perfect'. It is just supposed to be better than what went before according to certain criteria.[2]

The transition from a model of progress as a predictable ascent towards a predetermined goal to a more open-ended vision of history is the focus of this book. It's a distinction all too frequently ignored by historians of the idea; even Bury's division between the 'end of history' approach and the hope of indefinite progress doesn't quite capture it. The transition cuts across debates about the cause and purpose of progress because it involves two very different concepts of the 'shape' of historical development. It applies equally well to any theory of evolution, biological, social or technological, so we have to bring in changes in how we understand history in the broadest sense in order to appreciate its ramifications. In the end there is a crucial distinction that needs to be recognized: do you see development as the ascent of a linear scale, a ladder of perfection leading towards a predetermined goal, or as an open-ended process best represented by a branching tree?

In his account of science in the year 2440 Mercier claimed that naturalists had confirmed the validity of the ancient concept of the 'chain of being'. As described in Arthur O. Lovejoy's classic history, this was the belief that all

---

[2] On utopianism see for instance Claeys, *Searching for Utopia*, which does cover some areas that overlap with progressionism.

natural forms can be arranged into a linear sequence running from minerals through an ascending scale of organic complexity up to the human race. (In its medieval form it also continued up through a spiritual hierarchy to God Himself.) The notion that the world is based on a linear pattern established by the Creator had an enduring fascination, in part because it gave a rational structure to the natural world. In the late eighteenth century the chain was (in Lovejoy's terminology) 'temporalized' to make it a ladder of progress mounted by life in the course of the earth's history. When Condorcet and others created the idea of social progress, they defined the trend in terms of a similarly predetermined sequence of developmental stages. The parallel with the chain of being was noted at the time, and comparisons between biological and social progress continued to be made through into the modern era.[3]

The image of a ladder of creation could be related to the process of embryological development. The modern view (established by the early nineteenth century) is that the embryo starts as a single cell and gradually acquires the complex structures that turn it into an adult organism. The development of the embryo is goal-directed and takes place in a predetermined sequence of stages adding new levels of complexity. One way of understanding that sequence was to represent it as analogous to the chain of being, so that the 'lower' animals could be seen as immature stages of the human form. When exploration of the fossil record revealed a similar ascent of the scale through geological time, naturalists could argue that the historical process was goal-directed just like embryological development. Humanity was the goal of the whole process, prefigured from the very beginning.

This teleological vision was analogous to the foundation underlying the first ideas of social progress – the assumption that what will be achieved is the fulfilment of God's expectations for humankind. Traditional Christianity insisted that after the Fall, sinful humanity had no hope of redemption in this world. But a more liberal viewpoint gradually emerged offering the hope that – guided by Christ's example – we can gradually recover the state of felicity enjoyed by Adam and Eve immediately after their creation. Progress was goal-directed because the end-point had been specified from the beginning. Adding the development of life on earth into the sequence both extended the range of this model and allowed it to be seen as a parallel to the development of the embryo to maturity.

Bury notes that this 'genetic' world-view underpinned the linear and teleological interpretations of human social progress. For generations of late eighteenth-century and early nineteenth-century naturalists and historians, the world exhibited a predesigned ascent of a hierarchy of perfection designed by

---

[3] Lovejoy, *The Great Chain of Being*, esp. chap. 9.

God. The chain of being observed in God's creation of nature served as the blueprint for this linear model of development through time, both in nature and in society. My title *Progress Unchained* is meant to indicate that this model would soon be challenged by a far less structured view of historical development. This new approach was also mirrored in a model of change derived from nature itself, in this case the theory of evolution that came to be associated with the name of Charles Darwin. I shall argue that the new vision of organic evolution played a significant role in transforming ideas about progress in general.

Those committed to the linear model did not necessarily think that the same processes drove all phases of the ascent. This is why a focus on the 'shape' of progress is important: the chain of being represents a deeper foundation, based on a conviction that the world must have a simple hierarchical structure. Thinkers such as Condorcet who did not accept the religious faith in a divine purpose for the world nevertheless adopted the linear model, shifting the definition of the perfect society to be achieved into one based on material prosperity and happiness.

There was, however, a very different approach emerging, one which had less faith in the expectation that everything was governed by a divine order. The materialists who pioneered the modern idea of evolution saw that the laws of nature worked without reference to a predetermined plan. If there was progress, it was because those laws interacted in a way that at least sometimes allowed more complex structures to emerge. For Herbert Spencer and the Darwinians, biological and social progress were indeed the result of the same laws, but there was no reason to suppose that there was only one route leading to the most advanced state of animal or social evolution. Soon even thinkers who rejected materialism jumped onto the bandwagon. Henri Bergson's anti-mechanistic theory of 'creative evolution' helped to convince early twentieth-century thinkers that the direction of progress was not predetermined. The 'shape' of evolution could be uncoupled from the particular ideology which had given rise to its original manifestation.

The new approach encouraged the emergence of similar models of both biological and social progress. Darwinian evolution is a branching, ever-divergent and for all practical purposes unpredictable process driven by the complex interactions between organisms and their changing environments. When faced with a new environmental challenge species adapt by developing new characters that enable them to survive and flourish. If a species becomes exposed to two different environments it will evolve in two different directions, producing a node in the constantly branching 'tree of life' that became characteristic of Darwin's thinking about organic relationships. Progress towards more complex states is certainly possible, but it is not inevitable because most adaptations have only local significance, and there can be no one goal towards which evolution as a whole is directed. The tree replaces the ladder as the 'shape' of evolution, and progress has to be defined as a general but haphazard increase in complexity or sophistication rather than an advance towards a predetermined goal.

The development of a new adaptation by a species is in some respects analogous to the invention of a new technology (physical or social) by a human population. This parallel was explicitly recognized by Darwin himself and by twentieth-century thinkers such as H. G. Wells, which suggests that it was no accident that the idea of open-ended progress emerged just as the modern Darwinian synthesis was being consolidated in biology. In a few cases, an adaptive innovation has more than local value: the new structure has entirely new functions that can trigger a phase of rapid expansion, as when a 'higher' class such as the mammals or the birds enters onto the scene. Biological progress in the Darwinian world is opportunistic, unpredictable and to some extent episodic – as was the new understanding of human progress. New ideas about the emergence of humanity from an ape ancestry stressed that it was impossible to treat our species as the predictable goal of evolution. Humans were a contingent outcome of a complex process that had produced all the other species. Small wonder, then, that our own efforts to improve ourselves showed a similar lack of direction.

Social progress may not be driven by the same mechanism as its biological equivalent, but if there are many ways of developing a more complex society or culture, no one species or culture (not even one's own) can be seen as the high point towards which all are progressing. Cultural relativism is the social science equivalent of the evolutionary biologists' tree of life, and there are many examples of analogies being drawn between the two areas. And if the past can be seen an ever-branching tree, the future for any society becomes unpredictable because we cannot be sure what new inventions will be devised that will impinge on the way people live. Even in today's globalized world, it is by no means certain that all nations are evolving in the same direction.

As the implications of the new visions of progress became apparent, new impetus was given to those who had always doubted whether there was any realistic hope of the human race achieving universal happiness. There have always been pessimists, not all of them evangelical Christians, who think that human nature is so imperfect that any effort to improve things is doomed to fail. The genetic model of history itself was open to a less optimistic view of the future: a society might develop towards maturity, but in the natural course of things that state is always followed by a decline to senility and death. This way of understanding historical development drove the early twentieth century's most widely recognized critic of progress, Oswald Spengler.

Spengler's concerns about the future were echoed by numerous critics of modern culture's increasing dependence on technological innovation. In some cases a form of technological determinism created a new and nightmarish vision of a final 'end of history'. Aldous Huxley's *Brave New World* parodied the rationally ordered World State proposed by H. G. Wells and others. His future's nightmarish qualities derived from the applications of biological technology predicted in J. B. S. Haldane's *Daedelus*, published eight years earlier in 1924. Like Wells, Haldane was no simple-minded optimist, and he foresaw reproductive technologies that could change society for ever and perhaps even create new versions of the human race. Here was a more optimistic vision of progress in which innovation continued to offer new opportunities: there could be no static future state, whether utopian or otherwise. The problem for the optimists was that no one could be sure whether the results would be beneficial or harmful to our aspirations. The developments that Wells and Haldane welcomed were seen by their critics as destructive of all traditional human values. The optimists' utopias were the pessimists' nightmares, all the more worrying because no one could be sure what new innovation might catch on next.

The new visions of the future depended not only on a more complex model of social development but also on a new definition of what constitutes progress. In the linear view of history the goal of progress was a society that maximized human happiness. There were ideological differences over how this would best be achieved, but all could agree on the moral value of what they were aiming at. Once achieved, the utopian society would be static or it would no longer be utopian. If utopia is the mature phase of social evolution, no one would want to go beyond it. There was no expectation that social organization might be upset by further developments in technology, once an adequate physical environment could be ensured for all.

When progress was reconfigured as a branching tree of opportunities, progress had to be defined in utilitarian terms. Better control of the environment was the key – as in biological evolution – and there was much more room for disagreement over the moral value of what would emerge. Enthusiasts proclaimed the richness of the opportunities for personal fulfilment that would become available, but pessimists foresaw that the search for purely material satisfactions might have disastrous consequences for everything they valued. The best the technophiles had to offer was the prospect of a transformation of humanity, perhaps into forms that would allow an expansion out into the cosmos by space travel. Science fiction predicted that even that might not really ensure the elimination of conflict and hardship.

In my *A History of the Future* I outlined many of the forecasts made by scientists and science-fiction writers during the early twentieth century. By the 1960s the issue was becoming crucial as the pace of innovation increased, allowing futurology to become a central feature of public concern. The RAND Corporation was but one organization seeking to predict and direct what might happen, while books such as Alvin Toffler's *Future Shock* became bestsellers. In our own time scientists compete to predict what might happen, and Yuval Noah Harari's *Homo Deus* has brought home to many the enormous potential for technology to transform not just society but human nature itself. Significantly, Harari's book was a follow-up to an earlier survey of how technology has changed our world in the course of history.[4]

We have come to realize that progress is not a step-by-step ascent towards utopia but an open-ended process that can produce any number of potentially more sophisticated futures. The hope that globalization might bring our cultures together has withered away as tribal, ethnic and religious differences prove as divisive as ever. Even if there might come to be a future global civilization, the West can no longer assume that anything other than its invention of the scientific method will lie at its heart. Critics who worry about the dangerous implications of the new technologies on offer are emboldened by the very fact that the enthusiasts cannot agree on which new gadget or technique will be most influential in shaping the more exciting world they expect to emerge. Defining progress in terms of complexity or sophistication makes more sense in the real world, but has undermined any hope of agreement over the moral implications of what might be produced.

Since the first drafts of this book were completed our society has been rocked by the impact of the global heath pandemic, leading to dire predictions of economic catastrophe. Yet optimists still think that it is technological innovation that will deal with both the medical issues and a much wider range of problems that we were already facing before the outbreak. This new situation perhaps helps to drive home the Darwinian aspects of how we can think about change: we seek to control our environment for our own benefit but the environment itself is unpredictable, and opportunities and innovations have often emerged in response to external challenges.

Historians and the Idea of Progress

The transition from the linear to the open-ended vision of progress was not a simple replacement of one idea by another. The linear, teleological model is certainly the original, yet it survived in one form or another through into the late twentieth-century expectations that liberal capitalism might represent the end of history. Recognition of diversity was a later development, routinely subverted by efforts to give the branching tree of evolution a central trunk representing the main line of progress. Small wonder that historians of the idea

---

[4] Bowler, *A History of the Future*; see the epilogue on the intensification of debate in the 1960s. For a modern example of scientists' predictions see Al-Khalili, ed., *What's Next?* Harari's *Homo Deus* was preceded by his *Sapiens: A Brief History of Humankind*.

of progress have not been able to offer a coherent account of the debates it has engendered. They have recognized many different versions of the idea, but have seldom understood that beneath the disagreements lies a more fundamental transition from a linear to a less-structured vision of how the world might become more complex.

For as long as I can remember my library has contained an (increasingly battered) copy of the 1955 reprint of Bury's *The Idea of Progress*. While noting the links to theology, Bury argued that a fully fledged idea of progress could not emerge until Enlightenment thinkers became convinced that Western civilization had advanced beyond the achievements of antiquity. He saw the many different versions of what the optimists thought might be the goal of human progress. Most wanted more happiness but disagreed over whether this could be brought about by freedom, a more orderly state or enforced economic equality. Bury did see a difference between those who knew exactly what the final goal should be and those who thought there might be ongoing progress towards ever-greater felicity. Yet he ended his survey with the late nineteenth century, when the idea had become an 'article of faith' – as though there was by then a unified vision of what to expect. His epilogue notes only a growing lack of enthusiasm for the idea: it was originally published in 1920, when war and economic depression had undermined the enthusiasm of the late Victorian era. Bury himself had by this time given up on any hope of seeing a pattern in history, comparing it to Darwinian evolution on the grounds that both areas had to allow for chance events brought about by the intersection of independent causal chains. Curiously, he suggested that the increasing role played by science would limit the opportunity for such chance events to affect the course of history.[5]

This last point is certainly not valid for science-based technology. The later edition of Bury's book has an extensive introduction by another historian, Charles A. Beard, who had edited two volumes seeking to predict the future. *Whither Mankind?* of 1928 had articulated the concerns of literary figures and moralists who shared Bury's pessimism. But *Towards Civilization* two years later brought out the hopes of scientists and inventors, including Lee De Forest, who hailed radio as a vehicle for worldwide cooperation. Bury had made limited references to the role of technology and industrial innovation as components of progress, but his remarks about science seem to imply that the direction of change is increasingly restricted by this factor. Beard appreciated that while invention exploits scientific information, it is by no means constrained by it because so many opportunities are opened up by the increasing

---

[5] Bury's *Idea of Progress* originally appeared in 1920; the Dover reprint is of a later edition published in 1932. His 'Darwinism in History' and 'Cleopatra's Nose' are reprinted in his *Selected Essays*, pp. 23–42 and 60–9.

breadth of research. His introduction to Bury's book builds on the insights in his edited volume to acknowledge the ever-increasing diversity and impact of new inventions.

The 1955 edition of Bury's book fits into a wave of publications on the topic in the mid- to late twentieth century. A short account by Morris Ginsberg in 1953 stresses how the goal-directed visions of progress that treated Western civilization as the goal were challenged by the growing willingness to admit that other civilizations were unique products shaped by different circumstances. R. V. Sampson's *Progress in the Age of Reason* stressed that all theories based on the hope of perfecting human nature or society must include some notion of a goal, or at least of the way forward, and are hence teleological.[6]

Two later accounts by Sidney Pollard and Robert Nisbet included chapters on the rejection of the idea of progress in the twentieth century, Pollard's entitled 'Doubters and Pessimists' and Nisbet's 'Progress at Bay'. There was an increased willingness to accept that where progress was endorsed by twentieth-century thinkers it was in a context that made it less obvious what the future might bring. Nisbet included discussions of futurological predictions by figures such as Herman Kahn and Julian Huxley. He also noted the cosmic vision of human spiritual progress popularized by Pierre Teilhard de Chardin, a late manifestation of the view that the goal is marked out in advance by our Creator. Against this revival of teleology Nisbet notes a growing focus on the uncertainty of the future, but his commentary does not suggest a transformation in the very idea of progress in history. A more perceptive analysis by Charles Van Doren anticipated Nisbet by showing an awareness of the complexity of modern ideas of progress, including the hopes for a technologically enhanced future expressed by science-fiction writers such as Arthur C. Clarke.

More seriously, Van Doren invoked Karl Popper's attack on historicism in which he argued that our inability to predict technological inventions made it impossible to see how society might evolve in the future. Here we see emerging something like the approach I want to take: we cannot anticipate the future, which in turn means that we need to appreciate the contingent nature of our present situation. If there is no pattern in history there can be no linear sequence of progressive states in social evolution, and the uncertainty of technological invention suggests at least one reason why that is so. My argument is that we need to generalize this point by recognizing that the linear, teleological model of progress has increasingly been challenged not by a complete loss of faith but by a redefinition of progress to allow for an open-

---

[6] Ginsberg, *The Idea of Progress*; Pollard, *The Idea of Progress*; Van Doren, *The Idea of Progress*; Nisbet, *History of the Idea of Progress*.

ended and unpredictable advance towards a more complex situation. The emergence of the modern Darwinian synthesis and new ideas about human origins paralleled the redefinition of human progress. Thinkers in both areas routinely interchanged ideas, metaphors and analogies.

The most detailed study of later ideas of progress is that of W. Warren Wagar. He sees the idea as a 'thought form' or what Lovejoy called a 'unit idea' – a concept so basic that it can be exploited by many different ideologies and belief systems. Wagar outlines an even wider range of applications than those noted by Bury, along with the reasons advanced by critics for rejecting any form of the model. More than the other authors he forces us to appreciate just how slippery the faith in progress became as it was expressed by idealists, materialists, rationalists and socialists – to say nothing of those who still saw it as the unfolding of a divine plan. He also asks whether progress must be seen as unilinear, or whether it might be discontinuous or spiraliform (I prefer the term 'cyclical'). There is a hint that it might even be irregular, but despite the inclusion of Darwin in the subtitle his analysis of the early twentieth century doesn't bring out the possibility that a new, open-ended vision of progress has emerged.[7]

Wagar shares Bury's view that the true idea of progress is a modern invention, with only limited connections to earlier theological visions. This position is shared by several detailed studies of the Enlightenment manifestations of the idea, although early modernists are more inclined to note the role of liberal religion in the thoughts of Francis Bacon and other seventeenth-century precursors. Carl Becker's classic (although highly controversial) *Heavenly City of the Eighteenth-Century Philosophers* also reminds us that it is still possible to see relics of the theological viewpoint even here.[8]

Detailed studies of nineteenth-century interpretations of progress tend to focus on the key figures: Comte, Hegel, Marx and Spencer. National studies include Arthur A. Ekirch on America and my own *The Invention of Progress*, which demonstrated parallels between Victorian ideas on the development of life on earth and the progress of human societies. I was aware of the continued influence of the linear, developmental model (even if in the form of a series of discrete steps) long after the Darwinian 'branching tree' was introduced, but I did not follow this insight through to examine the eventual proliferation of open-ended models of social progress. There are numerous studies of social Darwinism which assume that the ideology was associated with progress, but

---

[7] Wagar, *Good Tidings*. On pessimistic visions of the future see the same author's *Terminal Visions*.

[8] In addition to Becker's *The Heavenly City of the Eighteenth-Century Philosophers*, see Tuveson, *Millenianism and Utopia*; Frankel, *The Faith of Reason*; Sampson, *Progress in the Age of Reason*; and Manuel, *The Prophets of Paris*.

these often show limited awareness of the complexities inherent in the efforts to link biological and social evolutionism.[9]

Biographical accounts of thinkers associated with the idea of progress are of limited value in helping us understand how the 'shape' of the idea has changed. Hegel and Marx are usually presented as key figures in the creation of a model based on a sequence of developmental stages – although their ideas on what constituted the goal and the driving force were antithetical. Both attracted numerous followers, a huge international movement in the case of Marx, but the result in each case was a proliferation of conflicting interpretations of their teachings. Most of these were based on the assumption of a linear (if discontinuous) pattern of change. Recent studies suggest that the disciples of both Hegel and Marx may have over-simplified their messages to endorse the linear model of social evolution.[10]

Another important figure with whom I am better acquainted is Herbert Spencer. His progressionist philosophy fell from favour so dramatically that until recently he was little studied by historians. Here again, though, closer studies now take us only a limited way towards understanding what he actually stood for on the question of the shape of evolution. His law of progress stressed the increasing diversity as well as complexity of living forms, apparently endorsing the 'tree of life' image. Yet his views on human social evolution – which he supposed to be driven by the same mechanism – were usually associated by his followers with a linear model based on development towards a final goal of unfettered individualism. This vision closely follows that of the philosophical historians, anthropologists and the Victorian archaeologists who saw social evolution as a step-by-step ascent from savagery to the Industrial Revolution. In fact, Spencer's vision of social evolution was not as simple as many of his disciples assumed.

Wider debates among modern historians have only occasionally thrown light on the question of progress. The issue that comes closest to the theme of the present study is that focused on the roles of chance and necessity. Historians who are suspicious of the claim that chance occurrences might deflect the course of events are more likely to identify laws or trends that appear to be in control. They are not necessarily drawn to the idea of progress, but the possibility that the trends they seek might have a progressive element is always there. One contribution is that of E. H. Carr, whose *What Is History?* is routinely cited as an example of a historian denying a role for chance. Carr was

---

[9] Bowler, *The Invention of Progress* and Ekirch, *The Idea of Progress in America, 1815–1860*. Classic studies of social Darwinism include Hofstadter, *Social Darwinism in American Thought*; Bannister, *Social Darwinism*; and Jones, *Social Darwinism and English Thought*.

[10] A good example is Dale, *Hegel, the End of History, and the Future*. For more details see the analysis in Chapters 4 and 5 below.

a Marxist, and he did see history as a record of progress, including an explicit chapter on this theme in his book. It would be easy to see him as an advocate of the linear model of progress usually linked to Marxism. Yet his final chapter recognizes both the bewildering acceleration of technological invention and the need for history to take on a global dimension that will deflect attention from the rise of the West's industrialized society. Neither of these insights seems to have encouraged him to trace the element of diversity implicit in these requirements back into the past.

The same failure of imagination characterizes some of the historians whom Carr castigates for seeing history as a record of events whose outcome is dictated by chance rather than some overarching trend. Curiously, one target is J. B. Bury, whose essays on the role of chance in both biological evolution and human history were noted above. Carr also engaged with other exponents of the role of chance, including Isaiah Berlin and Karl Popper. The latter's attacks on what he called 'historicism' – by which he meant the goal-directed theories attributed to Hegel and Marx – were based on the claim that each had in its own way led to totalitarian dictatorships. Carr noted that Popper's own philosophy of the scientific method stressed the role of creativity in the formulation of new hypotheses. The implication that scientific and hence technological innovation are unpredictable was one of the reasons why Popper thought efforts to see a pattern in history were doomed to failure. He also saw the parallel with Darwinism (which makes the idea of a 'law' determining the direction of evolution untenable). Popper claimed to be an optimist about the future, but distinguished this from belief in progress.[11]

These debates from the mid-twentieth century remain of interest, but things have moved on and the role of contingency is now increasingly recognized. One indication of this growing interest is the willingness of historians to take counterfactualism seriously. Counterfactual history is based on the assumption that there are key events in the past that might very easily have turned out differently and led to a very different present from the one we are living in. I have tried my own hand at this, imagining a world in which Darwin did not return from the voyage of the *Beagle* in order to emphasize that non-Darwinian theories had the potential to promote evolutionism in the absence of the idea of natural selection. There have been numerous efforts by historians to identify other potential turning points, collected in volumes such as Robert Cowley's *What If?* and Niall Ferguson's *Virtual History*. Counterfactualism makes sense

---

[11] Popper, *The Poverty of Historicism* and *The Open Society and Its Enemies*. For Carr's response to Bury see *What Is History?*, pp. 94–5, and to Popper pp. 85–7 and 149–51. On the debates between Carr and his opponents see Richard Evans's introduction to the edition of *What Is History?* cited in the bibliography, pp. ix–xlvi, and R. W. Davies's notes on Carr's plans for a second edition, ibid., pp. lv–lxxxiv.

only if one rejects the idea of historical inevitability and allows a role for chance. Ferguson's introduction to his volume is a detailed account of the debate between the two positions.[12] The inevitability of historical development does not always equate with progress, but those who see a significant role for chance are even less willing to see events as generally progressive. The interpretation of twentieth-century thought I propose assumes that there *can* be a version of the idea of progress that allows a role for chance, so that progress does not occur in a predetermined direction. The transition from goal-directed progress to this more open-ended vision is the key to understanding how the idea survived – and why it is so often attacked.

Science and the Idea of Progress

Lest my repeated comparisons between models of social history and those proposed by evolutionary biologists seem out of place, I can point to Ferguson's defence of counterfactualism, which invokes Stephen Jay Gould's well-known argument that if the 'tape of evolution' were replayed from the origin of life onwards it would be most unlikely to produce anything resembling the human species.[13] Gould was no orthodox Darwinian, but he did appreciate that the evolution of life on earth involves a complex interplay between a bewildering array of processes (ranging from asteroid impacts to genetic mutations), so that trivial events can sometimes have major implications. Evolution is a process by which populations adapt to changes in their environment, but since this includes new environments encountered by migration there is a constant tendency for species to divide and branch out in separate directions. Slight changes in the circumstances of migration can decide whether and how branching will occur. The branching is thus inevitably haphazard and to all intents and purposes unpredictable. If we could replay the tape, somewhere along the line a key event founding a new species ancestral to ourselves might not occur and we would simply be written out of history.

Gould's argument doesn't make everything the product of chance in the sense understood by philosophers who worry about such things. All the processes involved are governed by natural law and are thus in principle predictable. But the interplay of the factors involved is so complex that any hope of predicting the outcome is out of the question for any intellect short of the Almighty. Ferguson's point is that the same complexity is involved in human history. Here too 'chance' events can have major consequences, as

---

[12] See Ferguson's introduction to Ferguson, ed., *Virtual History*, pp. 1–90. See also Cowley, ed., *What If?* and my own *Darwin Deleted*.

[13] Gould, *Wonderful Life*, pp. 45–52; see Ferguson, ed., *Virtual History*, pp. 75–7. Gould's arguments are explored further in Chapters 6 and 7 below.

I imagined when I tried to reconstruct what might have happened if Darwin had fallen overboard from HMS *Beagle* during a storm. Translate Gould's argument for the array of species produced by biological evolution into the array of societies and cultures produced in human history and it is easy to see why a process governed by natural laws can still allow for unpredictable outcomes. There can be no predetermined goal of evolution or of history, because the diversity of species and cultures is the result of a branching process that is, to some extent at least, haphazard rather than deterministic. Technological innovation is unpredictable too, because we cannot foresee what inventions will be made – nor which will succeed in the real world.

Gould knew that biological evolution is mostly a process by which populations adapt to their environment, and he roundly condemned those who claimed to see a general progression towards 'higher' levels of organization; indeed he questioned the very idea that we could define one organism as more advanced than another. He was partly responsible for the view that Darwin himself refused to see evolution as progressive. Few historians now accept this interpretation of Darwin's position, suspecting it to be an attempt to bring him more into line with the most radical modern thinking. Michael Ruse has written a substantial account of the development of evolutionism showing the extent to which it has always been associated with a belief in social progress, Darwin being no exception.[14]

Ruse does not, however, bring out what I take to be the important transition in the definition of progress implied in Darwin's approach. Pre-Darwinian ideas mostly centred on a linear model ascent towards the human form and a future social utopia, and there were many attempts to defend this position against the more complex viewpoint that now began to emerge. But for Darwin himself there could be no inherent progressive trend and no preordained goal towards which life was advancing (Gould was right to this extent). Adaptive evolution often produces specialization, but this usually involves no advance in the level of organization and may even result in degeneration. Over-specialized organisms also leave themselves vulnerable to extinction in the case of rapid environmental change. Being better adapted means just that: it does not imply that the organism is more advanced than any other.

Darwin and his followers realized, however, that every now and again the evolutionary process does seem to have come up with a genuine novelty, a new adaptation that also has the potential to be exploited in ways hitherto unavailable to any living thing. By offering multiple opportunities for further development it sets the stage for an explosion of diversity and sometimes the conquest of new environments. Life can sometimes invent a new function that

---

[14] Ruse, *Monad to Man.*

is full of potential, and it can still make sense to define this as a progressive step. The fossil record suggests that the resulting steps forward often occur only in response to a major environmental challenge, and that sometimes an innovation with real potential can be blocked by well-entrenched rivals from the previous era.

I use the term 'invention' here because I think there is a useful parallel between the production of new structures in biological organisms and the process by which we ourselves invent new technologies. There is of course a crucial difference in that human inventors use their ingenuity to create new designs to achieve their goals, while natural selection is a purely mechanistic process. Nevertheless it does seem that the complex interplay of biological activities – mutation, development, selection, etc. – can innovate in the sense that every now and again something more sophisticated or capable of new functions is produced. And once it is produced it is maintained in later populations to become the basis for a range of new applications. So although most evolution does not result in progress, in the long run the potential of living things to exploit and experience their world is ratcheted up by the occasional significant innovation and life as a whole advances to new levels.

Here is progress in a new and unpredictable format. The progressive steps may result from events as unlikely as a few castaways arriving on an island lost in the ocean, but they open up new pathways into the future. Ferguson was right to highlight Gould's argument that contingency plays a role in history, but we need to feed this into Ruse's survey of the link between evolution and progress to see that Darwinism points the way towards a redefinition of the idea of progress itself. The new way of thinking goes far beyond Darwin's theory of natural selection, although that did drive home the fear that life evolves by 'chance' events. The same need to redefine the idea was popularized in the early twentieth century by Henri Bergson's apparently non-Darwinian philosophy of 'creative evolution'. What Darwin and Bergson realized was that contingency doesn't make progress impossible: it just means we have to rethink what counts as progress. We need to recognize that it cannot be a predetermined process and cannot be aimed at a particular goal, because it is only a by-product of the complex web of interactions that drive historical change. Progress must be seen as the introduction of occasional novelties that are useful in giving life – including humans – better ways of responding to or using the resources available. Technologies can be lost, resulting in actual degeneration, but this only occurs in unusual circumstances such as extreme isolation, as Darwin saw in the natives of Tierra del Fuego.

Using terms such as 'invention' and 'technology' in the areas of both evolutionary biology and human progress brings out the sense that what is crucial in the new definition is the element of innovation and utility. Things are better because they work better, not because they serve some higher moral good. That is why progress becomes a much more contested idea: most new technologies have harmful as well as useful applications, and whether they are beneficial overall depends on your point of view. We all want to live more comfortable, more rewarding lives, but all too often things that offer new opportunities turn out to have a darker side. Innovation and creativity are needed to get on in the world, but they can have unpredictable consequences and are seldom of benefit to all.

The move towards a more open-ended view of progress took place against a wider background of secularization in which the hope of spiritual fulfilment was replaced by a concern that everyone should be happy in this world. The rise of utilitarianism in moral theory did not by itself prompt the reassessment of progress. After all, the linear version could be understood as either an advance towards a state where all were spiritually transformed or as a drive to increase the greatest happiness of the greatest number, with happiness being seen as a state in which pleasure far exceeded pain (in the language of Jeremy Bentham's utilitarianism). But the more materialistic world-view helped to create the cultural environment in which technology was recognized as a major force for creating happiness. It also increased the awareness that the invention of new ways of mastering nature for profit could generate greater material comfort. It is no accident that Darwin's theory of evolution had a utilitarian basis, seeing the adaptation of species to their environment as a parallel to human efforts to improve ourselves. In making this move, however, Darwin focused attention onto the diverse ways in which adaptation occurred, and by implication onto the diverse ways in which human civilizations seek to achieve material wealth and power. It became obvious that there was no single route towards that goal, both in the past and in the future. Wider recognition of the true diversity of biological and cultural evolution coincided with a revolution that made it obvious to all that technological innovation was opening up a cornucopia –or a Pandora's box – of unpredictable opportunities.

The appearance of new technologies has not always been understood as the result of creativity even in the world of social evolution. Even when the pace of the Industrial Revolution quickened it was still possible to see invention as a more or less automatic response to a perceived blockage in the means of production (this seems to have been Marx's view). Bury's suggestion that the advance of science is inexorable and will thus limit the role of chance in history offers another example of this way of thinking, in this case based on the assumption that invention can only be derivative from discovery. The fear that a technocratic society must result in a totalitarian 'brave new world' is a late by-product of the same way of thinking.

Historians note, however, that by the start of the twentieth century there was a growing awareness of the multiplicity of innovation. Inventors were now coming up with a bewildering array of new devices, each with the potential to open up new opportunities and transform the way we live. It was suggested by thinkers as diverse as A. N. Whitehead and H. G. Wells that the creation of a link between scientific discovery and technological innovation was a cultural event of crucial significance that paved the way for a new phase of history. It was equivalent to one of the great breakthroughs in biological evolution. Originating in the particular circumstances of early modern Europe, this interaction was now having increasingly obvious effects on society.[15]

By the start of the twentieth century the effects of the new approach to invention, already apparent in the original industrial revolution, were becoming ever more apparent. In his *The Vertigo Years* Philip Blom shows how the period from 1900 to the outbreak of the Great War saw the appearance of a series of major inventions that were expected to improve everyday life, including radio, cinema, the automobile and aviation. Writers speculating about new developments began to realize that they had the potential to drive social evolution in different directions. In *The Sleeper Awakes* Wells drew on the inventions that made the skyscraper possible (steel-framed building and the elevator) to imagine a future world in which everyone lived in gigantic roofed cities – but almost immediately realized that new developments in transport such as electric railways and the automobile would allow suburbs to spread ever further outwards. The promoters of new technologies had to compete for the opportunity to transform society, and who succeeded would determine the future direction of social evolution.

This competitive, and hence Darwinian, element is now increasingly recognized by historians of technology. Where once the story of industrial development was written as though the sequence of innovations was inevitable, each new technology being obviously superior to what went before, we now see that the outcome of the clashes between rival inventors was seldom predictable at the time. Even in the nineteenth century, those involved found it hard to be sure which technologies would prevail. To begin with, no one could be sure that electric lighting would turn out to be superior to gas. Marconi saw radio solely as a new means of personal communication equivalent to the telegraph: it was amateur radio hams who first realized the possibility of broadcasting to a wide audience. The enthusiasts for airships and aeroplanes fought a running battle through the inter-war years. Sometimes the public could see what they wanted – if not how it would be achieved – but in other cases innovations seemed to come out of the blue to transform society. No one foresaw the computer or the internet.[16]

---

[15] On the insights of Whitehead and Wells in this area see Chapter 7. Joel Mokyr's *The Gifts of Athena* is a modern exposition of the same point.
[16] On the unpredictability of invention see for instance Marsden and Smith, *Engineering Empires* and my own *A History of the Future*.

It was surely no coincidence that an appreciation of the role played by competing innovations should emerge in both evolutionary biology and people's thinking about technological invention simultaneously in the decades around 1900. We talk about 'social Darwinism', but something more than a worship of the struggle for existence was emerging here. This was not merely the growing awareness of the principle of natural selection, although that surely played a role. The popularity of 'creative evolution' suggests the emergence of a more general recognition that the world could change through the emergence of multiple competing and unpredictable innovations. The old certainties were collapsing, including the goal-directed model of progress. Change was haphazard and open-ended, which for some meant the collapse of any sense that there was a meaning to human history or a purpose to modern civilization. All too often the history of the twentieth century has been seen as a record of their disillusionment. Yet a version of progress could be salvaged by incorporating it into the less certain model of change offered by Darwinism and its equivalents in the study of human society.

Ranking cultures and civilizations into a hierarchy defined by the historical development of the West was no longer acceptable. Prehistory and the origin of races had always represented the point at which biological evolution intersects with social, cultural and political history. The chain of being had inspired many attempts to rank the human races into a hierarchy with Europeans at the top, and the first evolutionists had all too often succumbed to the temptation to use their theories to underpin this imperialist ideology. The anthropologists and sociologists of the early twentieth century who promoted cultural relativism did not see their position as a version of progressive evolution precisely because they identified the term 'evolution' with the old linear model. Once the various races and cultures were accorded equal value, study of prehistory would show that all the great recent civilizations must have emerged from simpler origins if one traced them far enough back. The history of species, races and cultures all showed the same pattern of divergent and irregular progress. Putting all the pieces of this insight together is not easy, especially in a world where the fruits of technological progress are increasingly seen as tainted. But there are still enthusiasts who hope that the future will be even more exciting than the past.

Pattern, Process and Purpose

The newer version of progressionism transformed far more than the 'shape' of the process. We have moved from the traditional linear hierarchy to a model in which there can be no preconceived goal because progress can occur in different directions. The driving force is no longer a divinely implanted or naturally inevitable ascent towards an anticipated utopia but the invention of more complex and sophisticated ways of dealing with the material world. This is a utilitarian definition – and not everyone will agree that what seems useful is actually desirable. Enthusiasts tout their rival visions of what their inventions can do to improve our lives and open up new opportunities for wider horizons. They see the 'obvious' advantages but never the potential dangers. They seldom appreciate that what seems beneficial to them may seem harmful or even pointless to others with different values. There is no longer a clear sense of moral purpose underlying the process. It is simply assumed that 'doing things better' confers benefits on society.

The exponents of goal-directed philosophies of progress did not agree on the exact nature of the utopia they expected to come about, but they were at least clear that the human race would become morally better as well as happier in the future state. Technophiles assume that we shall be in some sense happier with what they hope to produce, but they seldom worry about the deeper moral and spiritual issues that used to motivate the utopians. Religious thinkers and moralists who could tolerate or even approve of the old idea of progress now routinely complain that Darwinism has corrupted our values by suggesting that the world is driven solely by chance and material success. This cannot guarantee meaningful progress, they argue, and their cautions certainly have some justification. Technophiles respond by arguing that we live better lives now than our ape-like ancestors did on the African plains, and it is not unreasonable to see the advance of life from its primitive origins to the diverse modern world – ourselves included – as something worthwhile. Whatever the irregularities and disadvantages, the ratcheting-up of new ways of understanding and controlling the material world has achieved something of significance.

Historians disagree over the extent to which the linear model of progress towards a predetermined utopia derived an impetus from the liberal Christian vision of God's plan for the human race. Nevertheless, the hope of achieving a morally – and perhaps spiritually – perfect society retained the teleological expectation of an ascent towards a goal that could already be imagined. The most popular interpretation of Hegel's philosophy of history retained this sense of an inbuilt logic that would guarantee progress towards a goal of spiritual value. At the same time the palaeontologists who made the first efforts to understand the ascent of life revealed by the fossil record also appealed to a divine plan of creation. Teilhard de Chardin's vision of spiritual evolution was a twentieth-century version of the same way of thinking.

The more humanistic versions of linear progressionism that emerged towards the end of the eighteenth century may have paid less attention to spiritual perfection, but they still foresaw moral as well as social improvement. Education could be used to transform human nature. Moral improvement could be achieved by the application of reason to human affairs, devising the most effective way of running a society. This approach also had the advantage of linking the process to the increasingly obvious developments in technology that were kick-starting the Industrial Revolution.

The problem facing those who wanted to see reason as the driving force of progress was that history revealed periods when that faculty seemed to have been little used. Comte eventually tried to argue that rationality would inevitably transform the way we make sense of the world and hence human nature, but this merely evaded the problem of the so-called Dark Ages. The impression was that after a number of false starts rationality had only now got the bit between its teeth and started to gain momentum. To find a genuinely progressive force in history the rationalists had to borrow the Hegel's technique, modifying his idealist vision of the Absolute seeking to achieve its goal through the successive interactions of the dialectic. Marx borrowed this directly, while the parallel between embryological development and evolution was still routinely invoked in the post-Darwinian era to give the impression that the advance of life towards higher states was inevitable. The branching tree of evolution retained a main trunk leading through to humanity, and even the side-branches were sometimes seen as unfolding through linear trends.

These theories frequently depicted progress as occurring in a series of discrete steps or stages. Here the legacy of the chain of being became ambiguous; it certainly provided the template for a hierarchical model of progress, but the chain had traditionally been seen as a continuous sequence of forms stretching up from the simplest to the highest. There could be no distinct species of plants and animals because every intermediate form must actually exist. Our almost instinctive tendency to identify and name distinct species was made possible only by the fact that some forms were extremely rare. When temporalized, the chain ought thus to have encouraged the idea of steady, continuous progress even though this hardly corresponded to our equally inbuilt tendency to identify and name key episodes in the history of civilizations. So to preserve this sense of discontinuity, the hierarchies of biological and social evolution were routinely envisaged as series of discrete stages of development. The dialectic did this automatically for Hegel and Marx, encouraging the latter's obsession with revolution.

Discrete stages of development were also invoked in early studies of the fossil record. Martin Rudwick notes the influence of the biblical story of the days of creation in encouraging geologists to think of earth history in terms of distinct periods. The naturalists who compared evolution with embryological development could also appeal to the fact that many forms exhibit metamorphoses defining distinct stages in the life-cycle. The eventual decline of the individual towards senility and death also seemed to argue for periodicity within the successive stages. Thus the rise and fall of successive empires in the course of human progress could be compared with the rise and fall of the major forms of life revealed by the fossil record. Progress occurred in the long run, but in a series of distinct steps.[17]

The role of continuity in the open-ended view of progress is more complex. Darwinism – or indeed any mechanism of adaptive change – is almost by necessity a process of gradual development. Small adaptive modifications add up over many generations to give something entirely new. The possibility of a significant new adaptive structure emerging in a single step by chance alone is remote (although there have been efforts to modify Darwinism by postulating small but discrete steps). Adaptive evolution is continuous *in time*, but not in terms of its outcomes. Species eventually become distinct because once a lineage divides, the separate branches move out in different directions, leaving gaps in between. There are several parallels here with technological and social innovation in human history. Few major innovations spring fully formed from their inventor's mind. Almost invariably a significant process of development is required to make the new process effective, often involving many contributors. The creation of a new technology or social institution has major consequences for the future development of a society, not least in that it drives a wedge between it and others that do not adopt the innovation or adopt another one instead.

There is still room for a limited element of discontinuity, however. Although most biological adaptations, like most human innovations, are useful only in local circumstances, there are occasional breakthroughs that produce something with wider applications. Major innovations, biological, technical or social, can – once established – initiate or make possible a rapid expansion of the entity involved, often at the expense of its rivals. This is not an automatic consequence, since the effects of the innovation may be suppressed for a while by a dominant existing form. The mammals evolved the powerful adaptation of warm-bloodedness but they did not get the chance to dominate the earth until an environmental catastrophe (possibly an asteroid impact) eliminated most of the great reptiles. Once a new technique does get its chance, though, its rise to dominance can be quite rapid and the decline of its rivals occurs in tandem. The rise and fall of biological classes and human empires is not driven by some mysterious fund of energy that eventually declines into senility. It is a natural consequence of progress by innovation in a system where there are many players in the game.

Natural selection is a continuous process because only small adaptive improvements are likely to appear in any one generation, so selection is needed to allow them to accumulate and ultimately change the whole population. Darwin's theory has been associated with a materialistic, anti-teleological view

---

[17] Rudwick, *Earth's Deep History*, chap. 1. The cyclic form of progressionism is explored in my own *The Invention of Progress*.

of the world because the production of the variations that are its raw material is not driven by the needs of the organisms. To its critics, this reduces everything to chance (which in modern biology means genetic mutations, very few of which have any adaptive value). The process of biological evolution is thus fundamentally different from the production of innovations among humans, which we naturally assume to be the result of individual creativity and forethought.

There were alternatives to the Darwinian theory of natural selection, plausible at least until the synthesis with the new science of genetics in the early twentieth century. The so-called Lamarckian theory of the inheritance of acquired characteristics and Bergson's philosophy of 'creative evolution' driven by an *élan vital* (both described in the next chapter) were seen as alternatives to the materialism implied by a selection mechanism dependent on random variations. They implied that biological evolution produced innovations by something much closer to human inventiveness because they allowed the animals' purposeful response to environmental challenge to shape their future evolution. In fact neither was unambiguously linked to the vitalist notion of a non-material force enlivening biological organisms. Darwin and Herbert Spencer both accepted a role for Lamarckism, and Bergson himself did not see his *élan* as a non-material entity. In terms of the overall shape of the tree of life, it makes little difference how adaptive evolution works: if it is the main driving force of change there can be no predetermined goal for the whole process.

This ambiguity in the theory's ideological background is relevant because the notion of biological innovation played a central role for the pioneers of the 'Modern Synthesis' of Darwinism and genetics. Key figures such as Julian Huxley were inspired by Bergson's vision and were able to think of natural selection as a process that was also genuinely creative, producing a wide range of innovations that have allowed living things to advance in many different directions. The Darwinian synthesis that Huxley promoted was based on the belief that the divergent, open-ended tree of life involved innovation and creativity just as did human social and technological progress – of which Huxley was also an ardent advocate.

At this point the link between the idea of progress and the assumption that there must be a fixed goal towards which the world is progressing began to fragment. Bergson had insisted on the open-endedness of evolution even though his philosophy appealed to many who wanted to see the process as having some ultimate purpose. The purpose was defined by an unstructured push from below (so to speak) generated by the day-to-day activity of the organism, not by a pull from above towards some predetermined goal. The open-ended vision of evolutionary progress that Darwin himself had pointed towards now began to emerge. As Bergson's philosophy lost its influence, the more flexible vision of progress it had helped to inspire retained its hold, thanks in part to its links with the increasingly dominant Darwinian theory in biology.

The triumph of Darwinism came just as the pace of invention quickened, encouraging the hope that innovations would open up new avenues of social change. As commentators speculated about the future it became obvious that here, just as in biological evolution, new developments were essentially unpredictable because of the many new techniques being pioneered only some could succeed and thus shape the future. Optimists could still expect exciting and potentially beneficial effects in ever-increasing proportions, so the trajectory of development could be assumed to be upward even though the exact outcomes were unknown. Predicting a single goal towards which society would progress seemed increasingly simple-minded. The open-endedness of technical innovation meshed neatly with the assumptions that humanity was not the intended goal of creation and the West not the only viable route to an effectively functioning society. Recognizing the diversity of life and the diversity of cultures meant that the only way of retaining a sense that history moves in a meaningful direction was to accept that there was no single direction that counted.

Biological evolutionism, palaeoanthropology, prehistoric archaeology, and cultural and social history all suggested that becoming more complex and more sophisticated wasn't inevitable, but it did happen from time to time and the mechanisms at work usually ensured that each new step would be preserved until the next one occurred. The pioneers of modern palaeontology had recognized that there were major discontinuities in the development of life on earth. This insight was ignored by Darwin and his immediate followers because of their commitment to the principle of continuity, but in the twentieth century it was revived to help create a new Darwinism that had room for the possibility of major new innovations that introduced new and higher classes of living things. At the same time, students of history recognized that what Whitehead called the 'invention of invention' was a crucial step forward in cultural history equivalent to the occasional breakthroughs made in organic evolution. Here was yet another parallel between the two levels of progress. If the West had contributed something of real significance it was the scientific method and the drive to apply it to practical ends – a force that accelerated the pace of innovation and is now being taken up by other cultures with different origins and different agendas.

The new and more utilitarian view of what constitutes progress coincided with and almost certainly contributed towards a growing scepticism in many quarters. Everyone could see that the triumphs of technology often had as many harmful as beneficial consequences, including military applications, economic disruption and environmental degradation. It was all too easy to see the whole project of Western civilization as inherently flawed, the idea of progress itself as a product of the arrogance that had led Europeans to see themselves and their culture as the goals towards which the universal evolution was aimed. Pessimism made a comeback, but it was a pessimism fuelled by doubts not just about inevitability of 'improvement' but about the value of complexity for complexity's sake. The fact that any perceptive critic could see a downside to every one of the advances hailed by the enthusiasts made is possible to challenge the moral foundations of the new progressionism. Spengler saw the decline of the West as a consequence of an inverted application of the principle of inherent development, assuming that a decline towards senility and death invariably followed the rise to maturity. Most historians came to doubt that it was possible to see any meaningful pattern in the chain of events, the only exceptions perhaps being the historians of science and technology.

In the twenty-first century there is still no shortage of amazing new developments in technology and society promoted by the enthusiasts of Silicon Valley and the other great research centres opening up around the world. All have the potential to transform our lives in ways almost impossible to conceive – but no one can be sure which will actually have the greatest impact, just as no one could have predicted the emergence of humankind from the earlier steps in the history of life or the appearance of a technologically driven world from the pageant of cultural history. We are now beginning to suspect that what the Darwinians who were inspired by Bergson worked out years ago might be true: artificial intelligences may demonstrate that creativity can indeed be a feature of purely material systems. Who knows what that might mean for humankind? The future is now as open-ended as the past, an insight now driven home by the challenge of a global pandemic. Even in the face of this crisis, however, there are some who see technological innovation as the answer not only to the immediate challenge but also to the problem of building a better world in the future.

To chart the emergence of this new way of thinking about the evolution of life and humanity this book has to range across a wide array of activities in the natural and social sciences, in philosophy and history, and in thinking about the future. Some developments will require the conventional historians' technique of following the origin and development of an idea in a more or less sequential form. But the real meat of the argument comes from trying to uncover parallels and interactions between strands of thought in a range of disciplines and, more importantly, the emergence of similar new approaches in them all. That innovations in one area were inspired, directly or indirectly, by developments in another is a key aspect of the story. Tracing the emergence of a new world-view is inevitably an exercise in interdisciplinarity.

Part I

The Ladder of Progress and the End of History

2 From the Chain of Being to the Ladder of Creation

We begin with the emergence of a vision of the history of life on earth which sees it as a preordained advance towards the goal represented by the human race. This chapter charts the emergence of this version of what became known as evolutionism and shows how it initially defended itself against the encroachment of evidence suggesting that the process has no built-in direction. Illustrating the popularity and flexibility of the linear model of development in the life sciences provides an example we can then use to throw light on how that model also operated across a range of disciplines emerging to study human history and prehistory. In some cases, there were direct interactions between the various strands in which the model was applied, while in others there seems to have been a more general conceptual pressure at work encouraging thinkers to interpret their fields in terms of a goal-directed process of development.

Decades before what we today call 'evolution' became acceptable, naturalists began to challenge the traditional story that the world was created only a few thousand years ago. By around 1800 it was accepted that the earth had actually been in existence for a vast period of time, prompting various attempts to explain the succession of living forms being revealed by the fossil record. Not all focused on the idea of progress, but the record itself indicated a succession in the appearance of the higher forms of life, and for many naturalists this seemed to be evidence for a progressive trend in creation. At first there appeared to be no human fossils, so it could be assumed that we had appeared at the very end of the process. The Christian view that the creation of humankind was the key point of God's decision to create the cosmos encouraged many commentators to see the history of life as a progression aimed at this goal.

A template for this teleological approach already existed in the *scala naturae* or great chain of being. As Arthur O. Lovejoy's classic history shows, this apparently simple but actually quite complex model of the world had its origin in various strands of ancient thought. By the eighteenth century it had become a commonplace, widely invoked by moralists, religious thinkers, literary figures and naturalists to impose order on the apparently bewildering complexity of nature. It was not simply a description of the relationships between natural forms: in its original conception there was also a spiritual hierarchy stretching up to God. Humans occupied, in Alexander Pope's famous words, the crucial 'middle state', the point where the material scale below intersected with the spiritual scale above.

Here we are concerned primarily with the lower section of the chain, the scale of natural forms stretching from the simplest minerals through plants and animals up to humanity. This drew both on our sense that humans are somehow more significant than any other species and also on the common-sense appreciation that some animals seem 'higher' or more developed than others, and animals as a whole more complex than plants. This rough sense of a hierarchy based largely on assumed levels of consciousness was codified by the chain concept into a clearly defined structure, a unilinear pattern of relationships based on structural resemblances. Some of these resemblances are very superficial by the standards of modern biology, where the science of form (morphology) focuses on internal structure rather than external appearances.

The *scala naturae* (literally the 'ladder of nature') had obvious potential to be exploited as the basis for a temporal progression in the course of the earth's history. A ladder is meant to be climbed, although the metaphor of a chain carries no such implication. In Lovejoy's terminology, the chain or ladder was 'temporalized' in the late eighteenth century to give the first theories seeking to explain the history of life as a progression towards the human form. As William Bynum later noted, however, Lovejoy's focus on ideas led him to miss one of the darker applications of the chain, its use to justify the claim that the human races should be ranked into a hierarchy with whites at the top.[1]

As originally conceived, though, the chain was a purely static hierarchy. Its regular structure was seen as evidence that the universe was designed by the Creator to a rational plan, a plan that also identified the crucial status of humankind. To modern eyes it seems ridiculous to imagine that anyone could have seriously believed that all the species could be shoehorned into a single continuous sequence of forms. We have to bear in mind, however, that the idea was conceived long before Europeans had begun to explore the world geo-graphically, let alone the fossil evidence for past life. The situation would change as the true diversity of living forms became apparent, but in the meantime any practical disadvantages for naturalists seeking to classify species were outweighed by the chain's philosophical, religious and moral advantages as a world-view. The technical difficulties that arose as ever more species had to be crammed into the sequence will emerge in Chapter 7.

---

[1] Bynum, 'The Great Chain of Being after Forty Years'. On the human implications of the chain see Chapter 3 below.

To the eighteenth-century naturalists who first temporalized the chain it offered a way of showing how the Creator might have arranged things so that His plan for the eventual appearance of humankind would unfold in an orderly fashion. This seemed plausible when the complexity of the geological record was only just being recognized, and their ideas bear only a superficial resemblance to later efforts to reconstruct the history of life on earth. Their thinking was also limited in another, less obvious way. Like the naturalists of a later generation, they sought to establish a link between the history of life and the process by which the individual organism is produced, but their ideas about what was called the process of 'generation' were essentially materialistic. Conservative thinkers saw the growth of the embryo as a mechanical expansion of a pre-existing miniature or 'germ' originally created by God. The sequence of germs might embody a divine plan, but there was no significance in the actual process of growth.

It was the new philosophy of life associated loosely with the Romantic movement of the early nineteenth century that introduced a more dynamic view of generation, seeing it as an active process by which the unformed material of the fertilized ovum gradually develops the more complex structures of the mature form. Here was a concept that could turn the linear hierarchy of the chain into a model for a progressive trend in earth history. The ascent of a single hierarchy towards a final goal which we see in the embryo could be seen as a parallel to the development of life on earth, introducing what became the concept of 'recapitulation'. Focusing on the human embryo allowed its development to be seen as a speeded-up image of the whole history of life up to its intended goal. This link appeared in some of the early efforts to understand the fossil record in terms of a sequence of divine creations, but then exerted an even greater influence on theories of evolution.

There were other reasons why the linear version of the hierarchy of organic forms proved so resilient. Michel Foucault suggests that the 'classical' way of looking at things dominant in the eighteenth century encouraged the creation of ordering systems and assumed that nature must conform to our representations of it. The *scala naturae* was an obvious candidate for such a system.[2] This may be an expression of an even deeper mind-set. Naturalists with no religious axe to grind were nevertheless convinced that there must be a regular order or pattern underlying the diversity of living forms. This viewpoint emerges from a personality type that has underlying psychological foundations. Some people just cannot bear to live with the possibility that the world might be nothing more than the product of haphazard events. The opposite personality type is willing to live with a certain level of uncertainty and

---

[2] Foucault, *Les mots et les choses*, trans. as *The Order of Things*.

unpredictability.[3] The patterns sought by those who crave order at any cost are not necessarily linear. But a linear pattern is the simplest and most obvious, and it became a favourite among evolutionists who saw the development of life as the unfolding of a divine plan.

The Scale of Nature

By the early eighteenth century the chain of being had become the conventional image used by religious and literary figures seeking to convince their readers that the diversity of nature was based on a rational plan established by the Creator. Lovejoy's survey quotes numerous examples, of which Alexander Pope's lines from his *Essay on Man* (published 1733–4) are the best-known.

> Vast chain of being! which from God began,
> Nature aetherial, human, angel, man,
> Beast, bird, fish, insect, what no eye can see,
> No glass can reach; from Infinite to thee,
> From thee to nothing. – On superior pow'rs
> Were we to press, inferior might on ours;
> Or in the full creation leave a void,
> Where, one step broken, the great scale's destroy'd;
> From Nature's chain whatever link you strike,
> Tenth or ten thousandth, breaks the chain alike.[4]

Pope's emphasis on the impossibility of any form being destroyed lest the chain be broken illustrates the viewpoint that made it so hard for naturalists to accept the possibility of extinction. Continuity must be complete to preserve the coherence of the whole. This leads in turn to what Lovejoy calls the principle of plentitude: if God is infinite then everything that could exist must actually exist or He would not have applied His powers to the full. By the same token there can be no distinct species: if we can imagine the intermediates between two known forms, then they must exist somewhere even if we have not found them yet. The implication is that the chain is really more like a continuous ribbon or rope.

Pope's identification of humanity's crucial position in the chain is equally well known. We are placed in the 'isthmus of a middle state' unable to decide whether we are human or beast, 'created half to rise and half to fall'.[5] The implication that we are in a 'middle state' threatened the traditional belief that the spiritual part of our nature lifts us completely above the level of the 'beasts

---

[3] Bowler, 'Philosophy, Instinct, Intuition'.
[4] Pope, *Essay on Man*, epistle 1, lines 237–46, in *The Works of Alexander Pope*, vol. 7, p. 18.
[5] Pope, *Essay on Man*, epistle 2, lines 1–10 and 15–18, in *The Works of Alexander Pope*, vol. 7, pp. 25–8.

that perish'. Even before the chain was temporalized, religious thinkers and moralists had to confront the question of how this unique status could be maintained if there was complete continuity between the highest animals, usually presumed to be the apes, and ourselves.

To what extent was the chain accepted as a useful model of natural relationships by those who wanted to classify the increasing diversity of known living forms? It wasn't just the new species being discovered around the globe that made the chain seem increasingly like a procrustean bed into which a more complex world was being forced. Fossils too were becoming a problem and – as Pope reminds us – the microscope was revealing a vast array of minute creatures at the bottom end of the scale.

In the Renaissance there had been little interest in constructing a coherent ordering of species, most observers being more interested in mere description.[6] Many of the innovations in biological taxonomy from the late seventeenth century onwards either ignored the possibility of a linear arrangement or openly challenged it. Nevertheless the chain model did exert some influence, and the detailed survey by Henri Daudin is still worth reading as evidence of the ongoing conviction of some naturalists that they could recognize a linear pattern among the species they worked with. The botanist Anton Laurent de Jussieu took this approach seriously, as did Jean-Baptiste Lamarck – who figures prominently among those who sought to temporalize the chain.[7]

Other eighteenth-century naturalists have been associated with the chain by historians, but some caution is needed here. Concepts that originated as components of the chain model could take on a life of their own and could be used by those who had little interest in the possibility of a linear scale. This is most obvious with the principle of continuity, which inspired a wide range of naturalists to question the reality of species. Extracts from the comte de Buffon's classic *Histoire naturelle*, for instance, can be quoted to show that he sometimes argued for continuity. But this does not mean that he was interested in the idea of creating a linear scale. The same point can be made for Oliver Goldsmith's *A History of Animated Nature* of 1774, a successful popularization that drew much of its information from Buffon. Although hailed in one study for its use of the chain as an organizing principle, it actually described the animal kingdom in terms of multiple families more reminiscent of the branching tree of life. While always on the lookout for intermediates, Goldsmith was willing to admit that there were anomalies that did not fit into any of his families.[8]

---

[6] See Ogilvie, *The Science of Describing*.
[7] Daudin, *Études d'histoire des sciences naturelles*; on Jussieu and the chain see vol. 1, p. 207, and the introduction by Frans A. Stafleu to the reprint of Jussieu's *Genera plantarum* of 1789.
[8] For the original claim see Lynskey, 'Goldsmith and the Chain of Being'.

Those who studied nature from a more philosophical perspective were more inclined to take the idea of a regular pattern such as the chain seriously. Of these the two best-remembered are Jean-Baptiste Robinet and Charles Bonnet. Both show how more conservative thinkers linked the idea of a rational plan of creation with a theory of reproduction that became popular immediately before the emergence of the modern view of embryological development. This was the theory of pre-existing germs, often called the preformation theory, according to which all living things originate from a miniature structure within the egg or sperm that already contains all the structures of the mature organism. The principle of *emboîtement* or encapsulement held that all reproduction is based on the successive expansion of these miniatures, which had been originally created by God within the first members of each species, stored one within the other like a series of Russian dolls.[9]

This theory was introduced in the seventeenth century to reconcile the new mechanical philosophy inspired by developments in the physical sciences with the traditional belief that the universe was a stable divine creation. If – as Descartes and others claimed – the animal body was nothing more than a complex machine, it was hard to see how it could manufacture another equally complex structure in its own likeness. Even if this were possible, there could be no guarantee that successive generations would be exact enough copies to preserve the structure of the species and guarantee the stability of the organic world. For thinkers such as Robinet and Bonnet, the pre-existence of germs upheld their conviction that the Creator had provided the material universe with a mechanism capable of maintaining the order represented by the chain of being.

Lovejoy devotes fourteen pages to Robinet despite conceding that he tarnished his reputation by endorsing sailors' tall stories about the existence of mermen and mermaids. He certainly committed himself to the chain, devoting the seventh chapter of the first volume of his *De la nature* (1761–6) explicitly to the topic. But his later fascination with unlikely intermediates suggests that the principle of continuity was far more important to him. He rejected a suggestion made by P. L. M. de Maupertuis that there are now gaps in the sequence of creation opened up by catastrophic extinctions in the past.[10]

---

[9] On the theory of pre-existing germs see Pinto-Correia, *The Ovary of Eve* and Roger, *Les sciences de la vie dans la pensée française du XVIIIe siècle* (partial trans. as *The Life Sciences in Eighteenth-Century French Thought*); also Bowler, 'Preformation and Pre-existence in the Seventeenth Century'.

[10] Robinet, *De la nature*, vol. 4, pp. 7–9 on Leibnitz; vol. 1, pp. 20 and 64 on the fixity of species. The reference to Maupertuis's idea is in vol. 4, pp. 9–11; see the *Essai de cosmologie* in Maupertuis's *Oeuvres* (in the Lyons, 1766 ed.), vol. 1, pp. 72–3. See Murphy, 'Jean-Baptiste Robinet'.

As a consequence of his commitment to continuity Robinet found it impossible to admit the existence of discrete species: every point on the scale exists and is perpetuated by the mechanism of reproduction by pre-existing germs. In his first volume he argued that even minerals replenished themselves in this way, and in the fourth he suggested that all matter is, in fact, composed of germs with the potential to develop. At this point Robinet still saw the diversity of creation as a static system designed by God, with the linear pattern as only a limited guide to its structure.[11]

Charles Bonnet made his name with the discovery that aphids could reproduce by parthenogenesis, the process by which the females can produce offspring without the intervention of a male. Research with the microscope damaged his eyesight, and he spent the rest of his career reflecting on the relationships between God, nature and the human soul. He became the most eloquent proponent of the theory of pre-existing germs, assuming these were contained within the female ovum, so that in effect the whole human race had been enclosed within the ovaries of Eve. In the aphids the germs could evidently develop without the stimulus of the male semen. His *Considérations sur les corps organizés* of 1762 argued that the sequence of germs was originally formed by God within the first females of each species to ensure that the plan of creation would be preserved through succeeding generations. His system has been described by some historians as an effort to preserve the impression of absolute stability in the world.[12] In fact, Bonnet conceded that the germ defined only the species to which the organism would belong, allowing limited variability because of environmental influences. Although generally supportive of the principle of continuity he accepted that there were distinct species in nature – although the gaps between then were very small.[13]

Bonnet's *Contemplation de la nature* of 1764 develops his overall philosophy of nature. He reiterates his view that the germs guarantee that creation is a stable and coherent whole.[14] In part 3 of the work, entitled 'Vue générale de la progression graduelle des êtres', he identifies the plan of creation with a linear chain of being. The sequence runs down the scale from humans through the various forms of vertebrates. Snakes lead to slugs and then to the rest of the

[11] On continuity see Robinet, *De la nature*, vol. 4, p. 123; on mineral reproduction and all matter being germs see vol. 1, pp. 210–17, and vol. 4, p. 113.
[12] Whitman, 'Bonnet's Theory of Evolution' and 'The Palingenesia and the Germ Doctrine of Bonnet'; also Glass, 'Heredity and Variation in the Eighteenth-Century Concept of a Species'. For a more recent evaluation of Bonnet's philosophy see Anderson, *Charles Bonnet and the Order of the Known*.
[13] Bonnet, *Considérations sur les corps organizés*, vol. 2, p. 314 (*Oeuvres d'histoire naturelle et de philosophie*, vol. 6, pp. 392–3) and on species ibid., vol. 1, p. 123 (*Oeuvres*, vol. 5, pp. 231–3).
[14] Bonnet, *Contemplation de la nature*, vol. 1, pp. 154–64 (*Oeuvres*, vol. 7, p. 4).

```
Humans
Monkeys
Quadrupeds (mammals)
Bats
Ostriches
Birds
Aquatic birds
Flying fish
Fish
Eels
Sea serpents
Reptiles
Slugs
Shellfish
Insects
Worms
Polyps (Hydras)
Sensitive plants
Trees
Shrubs
Herbs
Lichens
Mould
Minerals
Earth
Water
Air
Etherial matter
```

Figure 2.1 The chain of being according to Bonnet's *Contemplation de la nature* (1764). Bonnet does not provide such a representation himself; this is assembled according to the sequence described in his text.

invertebrates (the notion that any 'creepy-crawly' could have a structure as complex as a vertebrate was alien to the hierarchical vision of the time). Bonnet uses the polyp or freshwater hydra, whose regenerative powers had been discovered by Abraham Trembley, to bridge the gap between animals and plants and then creates a sequence down through the vegetable kingdom to moulds and minerals (see Figure 2.1).

Some of the intermediates used to preserve an impression of continuity seem incongruous today, although they would not have appeared ridiculous in an age that did not judge affinities on internal structure. Bonnet himself recognized that there were practical problems inherent in the effort to shoehorn nature into a unilinear sequence. He conceded that insects and shellfish were difficult to fit in, and seemed puzzled by the status of sea-lions and whales.[15] Nevertheless his preference for a more or less linear hierarchy stretching down from humanity through the whole of the animal and vegetable kingdoms is apparent. If there were branches off the main stem, they were relatively insignificant.

The universe depicted in the *Contemplation de la nature* was essentially stable, but there was one suggestion of an in-built potential for future change. Bonnet had written extensively on the nature of the soul, dealing with what we call psychology today, and as a Christian he was convinced that every human soul would eventually be resurrected in a more perfect body. He suggested that there might be a second type of germ in addition to those encapsulated in women and intended for normal reproduction. Everyone, male and female, would contain a miniature destined to unfold into the better-endowed body to be enjoyed after resurrection; it would survive the death of the original body and lie dormant, waiting for the last trump.[16]

The Chain Temporalized

Robinet and Bonnet were cited by Lovejoy as examples of eighteenth-century thinkers who temporalized the chain of being. In their later works, both saw the possibility that the development of the germs could be programmed (to use a modern term) to unfold in a sequence that produced successively higher beings in the course of the earth's history. These ideas do not correspond to theories of transmutation or evolution as we understand it today because they do not involve the appearance of genuine novelties. Absolute stability is preserved in the sense that all the germs were still thought to have been created by God at the beginning of the universe. What unfolds is what was predestined to unfold.

Nevertheless, their ideas do recognize that the earth's populations have changed their appearance in the course of geological time. By the 1760s naturalists studying the earth's crust had demonstrated that the planet must be very old and had gone through major transformations in the course of its history. Robinet and Bonnet both appreciated the need to take this evidence into account. To this extent their ideas serve as a link between the original,

---

[15] *Contemplation de la nature*, vol. 1, p. 59 on insects and shellfish (*Oeuvres*, vol. 7, pp. 133–4). On the chain see Anderson, *Charles Bonnet and the Order of the Known*, chap. 2, esp. pp. 39–40.

[16] This is in part 4, chap. 13; see *Contemplation de la nature*, vol. 1, pp. 85–90 (*Oeuvres*, vol. 7, pp. 207–12).

static conception of the chain and the theories of evolution proposed by J. B. Lamarck and others that would influence nineteenth-century evolutionism.

Robinet's later views are somewhat incoherent since they involve both speculations about progress and the idea of an organic 'prototype', a basic design from which many different forms can be developed. The latter emerged in a separate work and seems more like an anticipation of the branching model of evolution.[17] The temporalization of the chain occurs in the fourth volume of *De la nature*, where he imagines that the germs of successively higher forms might unfold into real organisms in sequence over the course of ages. He hints that the first examples of a particular type of organism can unfold only in the body of the form immediately below on the scale. From this initial insight he later develops a vision of the natural world ascending a preordained hierarchy of organization towards the human form.

As for the order of the developments, I am sure that nature has always proceeded from the least composite to the more composite. The most complex organization that we know, and that which produces the most phenomena, is that of man. There would thus have been times when no human germ had developed. But how many millions of years or centuries were necessary to bring the human seed to maturity? How many manifestations to bring it forth? We are no longer in a position to say. There are no intervals, small or large, between the successive and neighbouring development. Nature passes from one to the other without discontinuity.[18]

If this brief passage is taken at face value it implies a progressive increase in the level of organization over geological time. The sequence of development would be predetermined by the germs, but the result would be a gradual ascent towards the human form.

Bonnet's temporalization of the chain drew on the suggestion he had already made that there might be another type of germ destined to provide the soul with a more perfect body at the resurrection. In his *Palingénésie philosophique* of 1769 he expands this process to cover the whole of creation. The first step in the argument was to extend resurrection to the whole of nature. At the same time as humans are resurrected in a higher form, so all the creatures lower in the scale will move a corresponding step upwards in the scale.

The same progression that we discover today between the different orders of organized beings will be observed, no doubt, in the future state of our globe; but it will follow other proportions determined by the degree of perfectibility of each species. Man, now transported to a new abode more suited to the eminence of his faculties, will leave to the ape and the elephant the first place which he occupied among the animals of our planet.

---

[17] See Lovejoy, *The Great Chain of Being*, pp. 277–81, and Roger, *Les sciences de la vie*, pp. 650–1.
[18] Robinet, *De la nature*, vol. 4, p. 6 and quotation from p. 128 (my translation).

In this universal restitution of the animals, there will thus be found a Newton or a Leibniz among the apes, a Perrault of a Vauban among the beavers.[19]

The same mechanism is then extended back in time to create a vision of the history of life as an ongoing sequence of resurrections, allowing all to ascend the scale of perfection.

Bonnet accepted that the earth had experienced a series of great catastrophes in the course of time, of which the Mosaic flood was but the last. He assumed that all life would be wiped out in each upheaval, but when the violence had died down, all souls would be resurrected in new bodies supplied by their germs. His chapter on the topic is devoted mostly to arguing that in each epoch the bodily forms would be pre-adapted to the new conditions, the Creator having anticipated what was needed.[20] He gives only a hint that as conditions improved, so the level of organization would increase. The clearest indication that the steps would always be progressive is articulated later when the idea is extended to the whole universe: 'If His plan required that the sentient beings inhabiting a certain planet should pass successively through various subordinate degrees of perfection, He has pre-established from the beginning the means destined to increase the form of their perfection and has given them the whole extension that their nature can bear.'[21]

Modern commentators have tended to assume that the main message of the *Palingénésie* focused on the historical progression of life. Stephen Gould presents it as an anticipation of the belief that the evolutionary progress of life is recapitulated in the development of the modern embryo (while conceding that for Bonnet himself there was no such development because the embryo simply expanded from a pre-existing miniature).[22] The historians who insist that there is no anticipation of modern evolutionism here are correct in the sense that resurrection after death is not evolution, and the process is nothing more than the unpacking of potentials created by God. Yet the very fact that Bonnet was concerned to show that life had to be adapted to conditions that changed over geological time suggests that we should not dismiss his influence. He was widely read at the time, and was perceived as contributing to the expanding vision of earth history. If his contemporaries focused on the element of progress in his writing, his contribution to the process of temporalization cannot be ignored.

---

[19] Bonnet, *La palingénésie philosophique*, part 3, vol. 1, p. 203 (*Oeuvres*, vol. 15, p. 219; my translation). On these developments see the studies cited in the previous section and my 'Bonnet and Buffon'.

[20] *La palingénésie philosophique*, part 1, chap. 6, vol. 1, pp. 236–62 (*Oeuvres*, vol. 15, pp. 254–84).

[21] *La palingénésie philosophique*, part 14, vol. 2, p. 74 (*Oeuvres*, vol. 16, p. 74; my translation).

[22] Gould, *Ontogeny and Phylogeny*, pp. 17–28.

Evolution in an Age of Revolution

Bonnet's world-view was swept aside when Europe was traumatized by the French Revolution and the convulsions of the Napoleonic era. There was an increasing willingness to accept that nature was a dynamic rather than a static system, and the element of progress hinted at in the temporalized versions of the chain of being began to seem more prescient. The transformist theories proposed by Erasmus Darwin and Jean-Baptiste Lamarck are typical of this movement. Both were convinced that living structures can be modified over many generations to produce new forms of life, and both believed that in the long run the new forms achieved higher levels of organization. Both were also willing to accept that natural forces could produce at least primitive living structures from inorganic matter: the process known as spontaneous generation. The theory of pre-existing germs was swept aside, allowing the development of the individual organism to be seen as a genuinely creative process. What we now call evolution – transformism in the terminology of the time – depended on the gradual modification of the process of individual development over many generations.

All this seems a prelude to the more organic, developmental world-view of the early nineteenth-century Romantics and transcendentalists. Indeed Charles Gillispie's account of Lamarck's transformism identifies him with the Romantic movement.[23] Few historians now accept this interpretation, seeing Lamarck instead as a thinker who expanded the mechanical philosophy of the Enlightenment by accepting that natural forces can play an active, creative role. This is not quite the same as the Romantic image of a dynamic, self-organizing nature, but it allowed the emergence of similar views on progressive development through time. The same approach underpinned the views of Erasmus Darwin, but his vision of progress was less structured than Lamarck's and was thus closer to the modern vision of branching development proposed by his grandson Charles.

Lamarck is remembered as one of the first evolutionists, although few realize just how different his transformism was from the modern Darwinian theory. He is identified with a hypothetical mechanism of adaptive evolution that is seen as the most obvious alternative to natural selection, superficially plausible although later challenged by modern genetics. This is the inheritance of acquired characteristics, often called 'Lamarckism', according to which purposeful modifications developed by the organism in the course of its life can be transmitted to its offspring and hence accumulate over the generations to modify the species. If this alternative were Lamarck's primary contribution,

---

[23] Gillispie, 'Lamarck and Darwin in the History of Science'.

then he could be included among those who promoted the open-ended, unstructured view of the evolutionary process. A diagram included in his main evolutionary text, the *Philosophie zoologique* of 1809, is often seem as an early example of the branching-tree model of the development of life on earth. Yet most modern studies now present Lamarck in a very different light. He did see adaptation to the environment as an important transformative agent, but this was supplementary to a more fundamental force that drove living things steadily to mount a scale of organization that was, in principle, linear. The necessity of adapting to an ever-changing environment merely distorted the regularity of the fundamental hierarchy of organization.

Lamarck's theoretical views extended far beyond the transformism he proposed in the later part of his career and were highly controversial. But he was a working naturalist who made improvements of permanent value in the field of biological classification (taxonomy). Originally a botanist, he published a major survey of the French flora. In 1794 he was appointed a professor at the new Muséum d'Histoire Naturelle established by the revolutionary government and was assigned to the study of invertebrate animals. He improved invertebrate taxonomy in ways that are still considered fundamental today. One modern account of Lamarck's work (by Richard Burkhardt, Jr) is entitled *The Spirit of System*, indicating the importance of the drive to organize in his vision of nature. His *Hydrogéologie* contained an early account of earth history committed to the vision of vast changes over immeasurable periods of time. Here was a framework within which he could, in the middle part of his career, convert his originally static view of the plant and animal kingdoms into a developmental or evolutionary theory. This was expounded in his *Philosophie zoologique* of 1809 and in the introduction to his *Histoire naturelle des animaux sans vertèbres* of 1815–22.

Chapter 5 of part 1 of the *Philosophie zoologique* is devoted to the classification of animals and proclaims Lamarck's endorsement of a basically linear hierarchy of organization. The book's preliminary discourse contains a reference to Bonnet, suggesting that the latter had been unable to confirm the reality of the chain owing to the limited state on knowledge at the time.[24] Lamarck did not accept the chain in its original form, because he saw the hierarchies of the plant and animal kingdoms as entirely separate, not one above the other. But within each kingdom there was a scale of organization, produced gradually

---

[24] Lamarck, *Zoological Philosophy*, p. 12; in the edition cited the famous 'branching-tree' diagram is on p. 179. On the influence of the chain of being on Lamarck's thinking see the second volume of Daudin's *Études d'histoire des sciences naturelles*, esp. part 2, pp. 110–25, and Schiller, 'L'échelle des êtres et la série chez Lamarck'. More generally see Burkhardt, *The Spirit of System*; also Hodge, 'Lamarck's Science of Living Bodies'; Corsi, *The Age of Lamarck*; Jordanova, *Lamarck*; and Laurent, ed., *Jean-Baptiste Lamarck*.

over time, and visible in the major groups if not in the individual genera and species where adaptation had produced additional modifications.

I shall show that nature, by giving existence in the course of long periods of time to all the animals and plants, has really formed a true scale in each of these kingdoms as regards the increasing complexity of organization; but that the gradations in this scale, which we are bound to recognize when we deal with objects according to their natural affinities, are only perceptible in the main groups of the general series, and not in the species or even in the genera. This fact arises from the extreme diversity of conditions in which the various races of animals and plants exist; for these conditions have no relation to the increasing complexity of organization . . . ; but they produce anomalies or deviations in the eternal shape and characters which could not have been brought about solely by the growing complexity of organization.[25]

Lamarck thus recognized the diversification produce by adaptation, but saw it as distorting a linear scale of organization. His 'branching-tree' diagram was not meant as an evolutionary phylogeny; it is a representation of the divergences produced in the chain of being by external factors.

Lamarck did not see the array of living species as the end-points of a branching tree, because he did not think that all forms today share a common ancestry. This is apparent from the fact that he thought that the spontaneous generation of the simplest forms of life has gone on throughout the earth's history and is still at work today. He thought there was an active principle in nature which he identified with current electricity. This was then a newly discovered force, widely seen as an active principle and taken by many as something essential to life (Mary Shelley invokes it to explain how Frankenstein vivifies his artificial creation). Lamarck imagined it being able to work on gelatinous matter to produce very simple life forms. Electricity is also the force that operates the nervous system and hence controls the body, and he proposed that its constant activity tends to increase the level of the body's complexity. In the course of generations, the creatures thus steadily mount the scale of organization.

Lamarck's vision was one of parallel developments up a scale with some limited divergence. As soon as one line of advance sets off up the scale, another is founded by a further act of spontaneous generation, and so on throughout geological time. The oldest lines of development have risen furthest to give the highest animals of today. Middle-ranking creatures originated from lines founded later in the earth's history, while the lowest modern organisms have only recently been generated. Continuity is maintained by the constant activity of spontaneous generation (Lamarck did not accept the reality of species, nor the possibility of extinction). Evolution is an escalator on

[25] *Zoological Philosophy*, p. 58.

which forms rise in parallel, one behind the other, up the same scale of organization. There is some divergence caused by environmental factors, but nothing corresponding to the splitting of a lineage into two or more branches that occurs throughout the Darwinian tree of life. The lines that diverge have always been separate since their first, independent origins.

Lamarck accepted that, as the currently highest form, the human race had evolved from the oldest line of development, speculating that an ape-like creature had eventually developed an upright posture.[26] He cited the chimpanzee as an illustration of the ancestral form, but this was not to indicate that we had evolved from the modern apes, because they would belong to a separate line advancing behind us up the scale. Note the similarity to Bonnet's vision of the higher animals taking our place as we move one step further up at the resurrection. Lamarck would no doubt accept that the precise form of the human species was not predetermined, since some adaptive features were involved, but his system implied that something very like us, with the same advanced mental faculties, was an inevitable product of nature's activity.

Lamarck was also keen to stress that the progressive development of life was relevant to our understanding of human nature. Much of the *Philosphie zoologique* was devoted to his views on the physiology of the nervous system and their implications for mental activity. He insists that since our mental and moral faculties have a material basis, it is the zoologist who can throw most light on them.[27] The human mind is an extension of the mental faculties developed through time in the animal kingdom, and an understanding of its origins will throw light on its functions and hence on the fields we now call psychology and sociology. Lamarck's views on these topics were in line with the group known as the *idéologues*, led by the abbé Condillac and P. G. J. Cabanis. They in turn were linked to Condorcet, a major contributor to the emerging idea of social progress. Lamarck certainly thought that a better understanding of our psychological nature should help us interact socially in a more effective fashion, but he does not seem to have shared Condorcet's optimism, perhaps because his own ideas were dismissed by Napoleon and the leader of French zoology, Georges Cuvier. The *Philosophie zoologique* ends with a complaint about the power of conservatives to block the emergence of innovative ideas.[28]

Cuvier rejected any notion of a linear arrangement, as did the more radical naturalists of the post-revolutionary era. The new approaches increasingly

---

[26] Ibid., pp. 171–2; see Bowler, *Theories of Human Evolution*, pp. 61–2.
[27] *Zoological Philosophy*, p. 287; part 2 of the book is on physiology, including Lamarck's views on spontaneous generation (chap. 6); part 3 is on the nervous system and psychology.
[28] Ibid., pp. 404–5; Jordanova, *Lamarck*, chap. 9, gives a succinct account of his views on the latter topic.

presented development as a divergent rather than a linear process. Yet there were still some naturalists who continued to look for a linear scale hidden within the complexity of living forms. Historians have uncovered a substantial body of French naturalists opposed to Cuvier and have noted their influence across the Channel in Britain. At least some of these opponents took up parts of Lamarck's thinking, including his hope of constructing a linear hierarchy of organization. Prominent among them was Henri de Blainville, who both championed the serial arrangement and used it constructively: he was the first to see the monotremes and marsupials as the most primitive mammals. Edinburgh anatomists including Robert Jameson and Robert Knox used the linear model, and Everard Home interpreted the duck-billed platypus as an intermediate in the chain.[29]

The absolutely linear chain of being was gone, but its influence lingered in these efforts to see a hierarchy linking the most basic categories of the animal kingdom. This was not necessarily evolutionism – de Blainville rejected Lamarck's transmutationism – but the fascination with the idea of a main line of progress in the history of life lingered on and would continue to influence evolutionary theories into the late nineteenth century.

The Law of Parallelism

In the early decades of the nineteenth century the repercussions of the age of revolution were expressed in the sciences as much as in wider cultural values. This was the age of Romanticism and the rise of idealist philosophies, but also a time in which radical materialism challenged the old social order. In the life sciences there were revolutionary developments in comparative anatomy and embryology, while studies of the fossil record provided the first hard evidence of a sequence of development in the history of life on earth. Nature itself was now seen as a creative system, a world-view in which the idea of progress could flourish. Bonnet's germ theory in which the embryo simply unfolded from a pre-existing miniature was replaced by an embryology which took epigenesis – the purposeful development of new structures – for granted.[30]

The chain of being too seemed under threat from the new 'biology' (a term now introduced to denote the systematic study of life). Comparative anatomists such as Cuvier focused on the internal structure of organisms, undermin-

---

[29] See Appel, 'Henri de Blainville and the Animal Series' and Jenkins, 'The Platypus in Edinburgh'. More generally see Corsi, *The Age of Lamarck*, esp. chap. 7.

[30] On these developments see Appel, *The Cuvier–Geoffroy Debate*; Lenoir, *The Strategy of Life*; Nyhart, *Biology Takes Form*; and on the British radicals Desmond, *The Politics of Evolution*.

ing the credibility of the superficial resemblances used to create an illusion of continuity in the chain. He divided the animal kingdom into four basic types and declared that they could not be ranked into a series leading up to the human form. The linear model survived, however, in a modified form among those embryologists who thought they could see a parallel between the sequence of development in the embryo of the highest form in any group and the linear hierarchy of classes they recognized within the group. In the vertebrates this sequence ran from the fish through the reptiles up to the mammals and the human form.

At the same time, studies of the earth's crust created the outlines of our modern view of the planet's history, confirming its vast antiquity and the immense changes it has undergone. Palaeontologists also began to reveal the history of life as recorded in the fossil record. By the 1840s the basic steps in the advance of animal life had been identified, and they corresponded to the same linear sequence seen in embryological development. The apparent parallel encouraged an assumption that there was a link between how the embryo develops towards maturity and how the animal kingdom has advanced towards its highest manifestation in the human form. To conservative thinkers within the Romantic movement the development of life was an expression of the creative activity of the deity or some equivalently purposeful force.

The possibility of retaining a linear framework within the variety of natural forms arose because everyone admitted that some organisms are more advanced than others in the sense that they are further removed from the most basic form of their type. As Karl Ernst von Baer insisted, a human embryo never looks like a mature fish, but fish and human embryos resemble each other up to a certain point in the process of development. In order to mature the fish embryo adds on fewer extra structures that the human does not have, which is why it was possible for some embryologists to think in terms of the human embryo passing through a fish-like stage at an early point in its development. From this assumption arises what was known as the 'law of parallelism', according to which phases in the development of the human embryo correspond to the hierarchy of classes in the vertebrate type. The human embryo is (in a generalized sense) first a fish, then a reptile and finally a mammal. When this is coupled with the revelation that this was the sequence in which the classes appear in the fossil record we get the 'recapitulation theory', in which the development of the human embryo recapitulates the progressive development of life on earth.[31]

The law of parallelism allowed the basis for a linear hierarchy of organization to be preserved even within a system that accepted the branching nature of

---

[31] Gould's *Ontogeny and Phylogeny* provides a detailed history of the theory.

development. All species represent the end-points of a diverging tree of relationships, but because we define one species (our own) as the highest, the line leading towards it becomes the main line of development, the trunk of the tree. All other lines are side-branches leading away from that main line. They no longer participate in the progressive trend and are by definition of lesser importance. Key elements of the chain of being are preserved: its linearity and its end-point in the human form.

The law could, in fact, be applied to any other branch in the tree of relationships, and much work in taxonomy was done on this basis. The Edinburgh anatomist Robert Edmond Grant – a political radical who interacted with Charles Darwin during the latter's student days – applied it in his work with the sponges. This involves the creation of multiple linear series. But as Grant would have been aware from his interest in Lamarck's evolutionism, most attention focused on the hierarchy of the vertebrate classes and humankind's position at the head of that series. Lorenz Oken and his fellow enthusiasts for the idealist view of the world known as *Naturphilosophie* began to argue that the development of the human embryo corresponds to the series to be seen in the animal kingdom as a whole.[32] Naturalists knew there were branches in the tree of life, but most were dead-ends: only the main trunk led upwards to the goal represented by the human form.

J. W. von Goethe was the most influential source of what Robert Richards calls the 'Romantic conception of life', but his morphological and evolutionary theories placed most stress on the diversity of the developments that emerged from the archetypical form. It was Oken''s *Naturphilosophie* that served as the foundation for the law of parallelism. Goethe and Oken fell out over priority in their anatomical discoveries, and at the same time Oken adopted a more mystical style of exposition often based on cryptic declarations intended to startle those of a more mundane disposition. In his *Elements of Physiophilosophy* he stated that: 'The animal kingdom is only a dismemberment of the highest animal, i.e. *Man*.' He insisted on the law of parallelism in its most brutal form: 'Animals are only the persistent foetal stages or conditions of man.' The implication seems to be that the human embryo passes through stages that represent the mature forms of the various grades of the lower animals. Oken also declared that despite the diversity of animal forms and organs 'Each animal ranks therefore above the other; two of them never stand upon an equal plane or level.' This implies that the scale of nature is unilinear; otherwise one could imagine two branches independently reaching equal levels of organization. Although these ideas were developed in the early decades of the century, these quotations are taken from an

---

[32] On Grant and the other British transcendental anatomists see Desmond, *The Politics of Evolution* and Rehbock, *The Philosophical Naturalists*.

English translation published as late as 1847, suggesting that Oken's more provocative views were still considered worthy of attention.[33]

Stephen Gould shows how the serial relationship was endorsed in a far more sober form by J. F. Meckel in 1811. Like Oken, Meckel accepted that nature was governed by a unified set of laws and that these laws directed the course of its development through time. His paper's title is, in Gould's translation: 'Sketch of a Portrayal of the Parallels That Exist between the Embryonic Stages of Higher Animals and Adults of Lower Animals'. The parallels are illustrated with examples drawn from comparison between the human embryo and the lower animals, including invertebrates. The heart in particular passes though fish-like and reptilian stages of development. The same points were made independently by Étienne Serres in a series of papers published from 1824 onwards. Serres was no idealist, being a follower of the more materialistic version of transcendental anatomy active in France. He declared that his work 'has proven that lower animals are, for certain of their parts, permanent embryos of higher classes'.[34] Both wings of the transcendentalist movement in anatomy, German idealism and French materialism, thus converged on the law of parallelism or what E. S. Russell called the 'Meckel–Serres law', although he also added the names of Friedrich Tiedemann in 1808 and K. G. Carus in 1834 to the list of supporters. This law would serve as the basis for the theories of recapitulation that would flourish in the age of evolutionism.

The law of parallelism allowed the Lamarckians' appeal to a linear progressive sequence to be seen as evidence for the existence of an inherent developmental trend in nature with the human form is its high point or goal. The new approach was exploited in early nineteenth-century Britain by radical thinkers seeking to challenge conservative religious and social views. Eventually, though, the new transcendental or philosophical anatomy entered the mainstream of scientific thought. De Blainville's work was translated by Robert Knox, and linear sequences were recognized within many classes by Peter Mark Roget, the young William Benjamin Carpenter and others.[35] However great the diversity of life, it was still possible to see a progressive sequence within the vertebrate type leading up to the human form.

Fossils and Progress

Through the eighteenth century there had been a growing awareness that the complexity of the earth's crust revealed huge changes that were hard to explain

---

[33] Oken, *Elements of Physiophilosophy*, quotations from pp. 492–4; see Richards, *The Romantic Conception of Life*, esp. chap. 11.

[34] E. R. A. Serres, 'Explication du système nerveaux des animaux invertébrés' (1824), p. 378, trans. in Gould, *Ontogeny and Phylogeny*, p. 48; see also Russell, *Form and Function*, chaps. 6 and 7.

[35] See Rehbock, *The Philosophical Naturalists*, esp. pp. 49–52, 59 and 63.

except on the assumption that the planet is much older than the traditional account based on Genesis had assumed. There was a sequence of rock strata superimposed in an order that allowed one to argue that the lowest was the oldest. Massive disruptions after deposition implied major earth movements and volcanic events. By the 1830s the main outlines of the geological periods we still accept today had been established. It was still widely assumed that the periods were defined by catastrophic events that served as the 'punctuation marks' in the sequence, but the sequence itself was becoming clearer.

There was also an expanding knowledge of the fossils contained in the strata, now no longer dismissed as curiosities but accepted as the remains of the earth's former inhabitants. The populations of the successive periods were very different and often included forms unlike anything alive today. This made the possibility of a erecting a genuinely unilinear arrangement of organic forms even more implausible and also undermined the principle of plenitude by confirming the reality of extinction. At the same time, though, the fossil record did seem to reveal evidence of a basic sequence in the introduction of the main vertebrate classes, a sequence which corresponded to the series implied when the law of parallelism was applied to the human embryo. The implications of this model of development aimed towards the mature human form were extended to include the actual progress of life on earth. The result was what Loren Eiseley described as a transcendental, man-centred progressionism very reminiscent of the temporalized chain of being.[36]

Martin Rudwick has provided the most extensive histories of the work done by these pioneering earth scientists.[37] Dramatic changes in the conditions under which rocks were laid down were assumed to be the product of catastrophic events separating long periods of relative stability. William Whewell coined the term 'catastophism' to identify this approach, contrasting it with the 'uniformitarianism' proposed by Charles Lyell in the 1830s. Lyell argued that it was more scientific to assume that all changes in the past were brought about by agents identical to those we see in operation today. He insisted that the earth must be immensely old in order that such slow changes could produce the transformations observed in the crust. The catastrophists wanted a more limited time-span, but they were by no means exponents of a 'young earth', being quite willing to acknowledge that millions of years were involved.

Systematic changes in the populations of invertebrates were used to define the geological periods, but it was the ancient vertebrate forms that were most exciting to the general public. Cuvier studied mammals such as the mammoth and the mastodon – huge and impressive beasts, their fossil skeletons widely

---

[36] Eiseley, *Darwin's Century*, pp. 94–7.

[37] Rudwick, *Earth's Deep History*; for more details see the same author's *Bursting the Limits of Time* and *Worlds before Adam*.

displayed in museums. They were so large that Cuvier was able to convince everyone of the reality of extinction. Now that the earth was better explored it was impossible to believe that such creatures still survived in some remote location. Soon giant reptiles were being discovered, culminating in the first known dinosaurs. By the 1840s the notion of an 'Age of Reptiles' preceding that of the mammals had become commonplace. Below this was a series of formations from which only fish were known, while the oldest fossil-bearing rocks contained only strange invertebrates such as the trilobites. There were occasional anomalies, and Lyell was able to argue that the sequence was not quite so clear-cut, a few primitive mammals appearing in the middle of the Age of Reptiles. On the whole, though, the sequence of creation represented what most commenters regarded as an obvious progression from the simplest forms through the vertebrate classes to the mammals and finally humans. It was taken for granted in the early nineteenth century that there are no human fossils and that we are the most recent creation.[38]

Conservative naturalists, especially in Britain, still assumed that the progression worked through a series of miraculous creations. Extinction was assumed to be sudden, the result of catastrophic upheavals in the earth's crust. If these revolutions were global the planet would have to be repopulated from scratch, so miracles were the only explanation. Continental scientists were less inclined to invoke the supernatural, but their theories of underlying patterns and trends were equally linked to discontinuity. Some even though that the powers embedded in nature itself could generate higher forms of life spontaneously. More often the actual process by which new forms of life emerged was left discretely unexplained. There was still a wide reluctance to accept the possibility that new species could be produced by the transmutation of old ones. The conviction that the history of life displayed a progressive trend leading towards humankind did not immediately lead to the emergence of a consensus in favour of what we call evolutionism.

There were some attempts to explain the progress of life in terms of adaptations to the improving physical environment. But it was also possible to exploit the similarity between the sequences observed in the fossil record and in the development of the human embryo. The parallel became a symbol of our position at the head of creation: this was the 'transcendental, man-centered' vision of the progress of life on earth. Its leading proponent was the Swiss naturalist Louis Agassiz, later one of the founding fathers of American science. As a young man he had heard lectures from Oken, but he had also studied with Cuvier and learnt a more practical approach to the study of fossils. Agassiz was deeply religious, reinterpreting the transcendentalists'

---

[38] See Rudwick, *The Meaning of Fossils* and my own *Fossils and Progress*.

philosophy of an underlying unity of nature in terms of a divine plan of creation. He also took the term 'creation' literally, remaining a staunch opponent of evolutionism throughout his life. He provides a classic example of a naturalist who is well aware that the overall pattern of relationships within the animal kingdom is too complex to be reconciled with a linear chain of being, yet is still anxious to see a main line of development running through the whole towards humankind as the pinnacle of creation. He linked the hierarchical sequence of classes with the development of the higher embryos and the order in which the forms appear in the fossil record, all interlocking aspects of a coherent divine plan which unfolds through the earth's history and is recapitulated in the development of every modern embryo.

Agassiz made his reputation in the 1830s with a comprehensive study of fossil fish. Within this class he was well aware of the diversity of forms. He recognized that the law of parallelism could be applied within each fish family – the highest modern representatives of the family having embryos that recapitulated earlier forms in the series. At the same time, however, he insisted that when we step back to take an overview of the whole sequence of vertebrate creation, we see a linear development pointing towards the human form as the goal of creation. As he declared in an inaugural lecture at the Academy of Neuchâtel in 1842:

The history of the earth proclaims its Creator. It tells us that the object and the term of creation is man. He is announced in nature from the first appearance of organized beings; and each important modification in the whole series of these beings is a step towards the definitive term of the development of organic life. It only remains for us to hope for a complete manifestation in our epoch of the intellectual development which is allowed to human nature.[39]

Note here the link between the idea of progress in earth history and the hope of progress in human history today: it wasn't only the evolutionists who saw the one process as evidence for the other.

Agassiz continued to expound these views in books and articles through the rest of his career. In his *Outlines of Comparative Physiology* (written in collaboration with A. A. Gould), for instance, he stresses that the connections between successive forms is ideal, not material, so we are witnessing a divine plan that announces its end-point from the beginning:

The link by which they are connected is of a higher and immaterial nature; and their connection is to be sought in the view of the Creator himself, whose aim, in forming the earth, in allowing it to undergo the successive changes which geology has pointed out, and in creating successively all the different types of animals which have passed away,

---

[39] Agassiz, 'On the Succession and Development of Organized Beings at the Surface of the Terrestrial Globe', p. 399. On his career see Lurie, *Louis Agassiz*.

was to introduce Man upon its surface. Man is the end toward which all the animal creation has tended, from the first appearance of the first Palaeozoic fishes.[40]

His vision of the plan of creation became more complex in time, and the idea of a linear progress pointing to humanity less clear-cut, but later works such as the *Essay on Classification*, written after he moved to America, still made it clear that the pattern of natural relationships must be seem as a coherent divine plan in which the eventual appearance of humankind is prophesied from the beginning.[41] Progress along the scale of organization has a predictable end-point.

Agassiz was by no means the only exponent of transcendental anatomy to see a link through to the appearance of humanity within the complex development of the animal kingdom. Richard Owen became Britain's best-known anatomist, hailed for his work introducing the new methods and using them to throw light on the relationships between animal forms. He emerged as a staunch opponent of the radical transformism of Lamarckians such as Grant. At first sight, he seems an unlikely candidate to be identified with a linear model of progress. He was best known for his theory of the vertebrate archetype, the underlying basic form of the vertebrate skeleton from which all the actual species can be derived, the modifications branching out in many different directions. Yet in the conclusion to his *On the Nature of Limbs* of 1849 he too falls back on the belief that within the diversity we can see a definite line of development leading towards the human form. He argues that 'recognition of an ideal Exemplar of the Vertebrated animals proves that the knowledge of such a being as Man must have existed before Man appeared. For the Divine mind which planned the Archetype also foreknew all its modifications.' He then concludes:

To what laws or secondary causes the orderly succession and progression of such organic phenomena may have been committed we are as yet ignorant. But if, without derogation of the Divine power, we may conceive the existence of such ministers, and personify them by the term 'Nature', we learn from the past history of our globe that she has advanced with slow and stately steps, guided by the archetypical light, amidst the wreck of worlds, from the first embodiment of the Vertebrate idea, under its old Ichthyic vestment, until it became arrayed in the glorious garb of the human form.[42]

Owen seems here to come perilously close to advocating a kind of divinely planned transmutation, a position he backed away from in the 1850s only to return to after Darwin published.

---

[40] Agassiz and Gould, *Outlines of Comparative Physiology*, pp. 417–18.
[41] Agassiz, *Essay on Classification*, p. 167.
[42] Owen, *On the Nature of Limbs*, pp. 85–6. The edition cited has introductory material by Ron Amundson and Brian K. Hall on the significance of Owen's ideas. See also Rupke, *Richard Owen*.

Owen and Agassiz were quoted side by side on this topic in Hugh Miller's *Testimony of the Rocks*, a popular exposition of the implications of the latest fossil discoveries. Miller was an enthusiastic follower of Agassiz and an equally prominent opponent of all forms of transmutation. He would later extend the law of parallelism to the whole life-cycle of the organism, including senility and death, arguing that within each class we see not progress but decline, so there can be no continuous progression as the transmutationist requires.[43] There were discrete episodes in the history of life marked by the sudden appearance of new types at a time when the previously dominant type had begun to decline. Agassiz could also point to the fact that in many forms of life individual development itself is discontinuous, as in the metamorphosis of insects.

A less idealized element of progression was included in the work of the German palaeontologist Heinrich Georg Bronn. His *Untersuchungen über die Entwicklungsgesetze der organischen Welt* of 1858 was submitted for a prize offered by the Paris Academy of Sciences for the best survey of the trends that could be observed in the fossil record. Like Owen, Bronn was keen to display the many branches that spread out from the primitive origins of each group, leading to a variety of specialized adaptive forms. But he did not believe that the overall pattern could be explained completely in terms of adaptation to the earth's changing conditions. There was, in addition, a distinct progressive trend pushing each group on the whole towards higher levels of organization. Bronn even seems to have believed that, left to itself, this trend would have produced a linear scale – but it was distorted over and over again by the need for organisms to adapt to their environment. There was nothing here to suggest a mystical trend towards the human form, but the diagram Bronn used to illustrate his position resembled a tree with a main trunk leading vertically upwards towards humanity (see Figure 2.2). Lines lead off on every side, but they are side-branches of the main line of development, leaving the impression that there is a linear hierarchy buried within the diversity of animal forms.[44]

Bronn later produced a German translation of the *Origin of Species*, but at this point he was not an evolutionist. He was, however, contributing to a general trend by which palaeontologists were becoming aware that as more fossil discoveries were made, they tended to fill in gaps in the record. In the early decades of the century it had been easy to dismiss transformism by simply pointing to the huge discontinuities in the sequence of fossil forms. Gaps still remained, but most of them were getting smaller as new discoveries

---

[43] Miller, *The Testimony of the Rocks*, pp. 191–2. On Miller's views see Bowler, *Fossils and Progress*, chap. 4.

[44] See Bowler, *Fossils and Progress*, pp. 108–10. I have used the French translation of Bronn's work, *Essai d'une réponse a la question de prix proposée en 1850 par l'Académie des sciences*. See also Gliboff, *H. G. Bronn, Ernst Haeckel and the Origins of German Darwinism*, although Gliboff takes a very different view of the issues from my own.

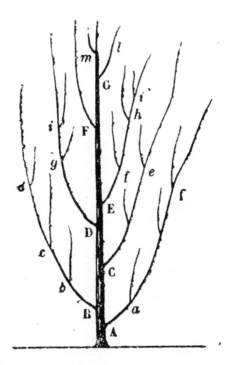

Figure 2.2 Tree-like representation of the development of the animal kingdom with a central trunk defining the main line of advance.
From H. G. Bronn, *Essai d'une réponse a la question de prix proposée en 1850 par l'Académie des sciences* (1861), p. 524

fitted into the sequence and the case for transformism no longer seemed quite so absurd. In the 1840s the argument that the progress shown by the fossil record was evidence for a continuous progressive trend in nature was placed on a new footing and had become the basis for an extended and often acrimonious debate. No one could believe that the history of life on earth was a simple ascent of the chain of being, but few could doubt that there was progressive trend involved. The question of whether that trend might unfold continuously via the gradual transmutation of species now became crucial, although the expectation that the trend had a preferred direction and goal remained.

Problems with Continuity

The original chain of being was seen as absolutely continuous, but for practical reasons most naturalists wanted to believe that species are distinct entities, links in a slightly discontinuous chain. This element of limited discontinuity remained a key feature of most attempts to explain not only the diversity of life but also its progress through geological time. The conservative thinkers who thought that species were created by God had obvious reasons for accepting the reality of their separate forms. But the radicals who argued for a natural process of what would later be called evolution faced a problem. If species are clearly separated from one another, surely they must be formed by discontinuous steps – the theory of evolution by jumps or 'saltations'. In fact this is a false assumption: if evolution is divergent, the process of change can be continuous but will give rise to gaps between the species produced when a line of development splits into two tracks that move gradually further apart.

The catastrophist vision of earth history introduced discontinuity at a more drastic level. The individual populations found in each geological period were certainly distinctive and could thus be seen as the products of discrete creative acts. But studies of the fossil record also suggested that there were a small number of much more dramatic transformations in the kinds of animals inhabiting the earth, episodes in which entirely new classes of creatures with new levels of complexity were introduced. These were the major discontinuities defining the ages dominated by fish, reptiles and mammals preceding the final creation of humanity. This latter step too could be seen as a step towards an entirely new level or organization, in this case defined by the emergence of mind and spirit. The catastrophist model of progress was thus radically discontinuous. It was a step-by-step ascent, each step upwards being represented as the rise to dominance of a new level of organization. Some catastrophists drew an analogy with the days of creation recorded in Genesis.

On such a model history is certainly a development towards a preordained goal, but it is also a process that proceeds by discrete upward leaps. It could even be argued that after each upward leap the higher class that had been introduced faced a gradual decline during the period of its dominance. Progress occurred in sudden advances, each gradually subsiding before the next upward step. This synthesis of the progressionist and cyclical world-view pictured the advance as analogous to the waves surging forwards as the tide comes in. There was an obvious analogy with the rise and fall of empires in human history that did not go unremarked.[45]

The emergence of Darwinism has been presented as the triumph of a more continuous model of development, but the triumph was limited because the discovery that there were apparently sudden breakthroughs in the ascent of life did not go away and eventually had to be incorporated into the modern view of

---

[45] On the cyclic model in palaeontology see my *The Invention of Progress*, chap. 6.

evolution. The 'uniformitarian' view of change as always continuous, proposed in Charles Lyell's *Principles of Geology* (1830–3) and taken up by
Darwin, was unable to displace this one element of the catastrophist position.
Darwin followed Lyell in arguing that the apparently dramatic appearance of
new types was an artefact of the 'imperfection of the geological record': the
slow intermediate phases had simply not left any records in the rocks (or they
had been obliterated by later geological events). In fact the continued expansion of palaeontological evidence did not erode the validity of the initial
perception that there were indeed discrete episodes in the development of life
on earth. In this respect the element of discontinuity promoted by catastrophism represented a valid insight that would be integrated into the Darwinian
world-view only by the palaeontologists of the early twentieth century.

Development and Evolution

In the decades during which Darwin developed his theory in secret there were
occasional efforts to promote the Lamarckian suggestion that new species are
produced not by miracle but by the transmutation of existing forms. These
ideas often incorporated the assumption that the changes would take the form
of abrupt leaps to new levels of organization. The most influential initiative
came in a book published anonymously in 1844 under the title *Vestiges of the
Natural History of Creation*. The author, whose name would not be revealed
until much later, was the Edinburgh publisher and amateur naturalist Robert
Chambers. His publishing house issued books and magazines aimed at the
rising middle classes, and he was committed to the expectation of social
progress. His aim in *Vestiges* was to bolster the case for reform by showing
that progress was a universal trend: politics would merely facilitate the
advance towards better things that was inherent in the laws of nature.
Chambers believed that the experts who insisted that the fossil record was
too discontinuous to support the case for transmutation were short-sighted. If
one stepped back and looked at the overall trend it was obvious that there was a
law of development running through all of nature, a law that should serve as
the basis for our hopes for the future. This was what the historian of the
Vestiges controversy James Secord called the 'popular science of progress'.[46]

There was, however, a curious mismatch between Chambers's view of
biological evolution and his hopes for society. When Herbert Spencer moved
in a similar direction a decade or so later, he ensured that the mechanisms he
invoked to explain biological progress were exactly the same as those he
expected to work in society (in his case the Lamarckian theory, later

---

[46] Secord, *Victorian Sensation*; see also Hodge, 'The Universal Gestation of Nature'.

supplemented by natural selection). Whatever Chambers hoped to achieve in society, his vision of the progress of life on earth provided nothing more than a loose analogy. Species were driven up the hierarchy of organization by an inbuilt trend that was somehow predetermined within the forces governing embryological development. There was no naturalistic explanation of how new characters were shaped in the sense that Darwin was looking for. Progress to a new level was purely automatic, driven by a plan somehow embedded in the fabric of nature by its Creator. Chambers took the idealist vision promoted by Agassiz and Owen and argued that the pattern unfolded by law rather than by a series of miracles. This was exactly what Owen seemed to hint at in his *On the Nature of Limbs*. The claim was controversial not just because it made the Creator's involvement seem more remote, but also because it entailed seeing humankind as a modified form of the highest animals.

Vestiges showed how the earth's formation and development could be accounted for by natural law. The origin of life was also supposed to be a purely natural process; like Lamarck, Chambers assumed that it was brought about by electrical activity and that the formation of very simple living things was still going on today. The fossil record showed how the very first organisms gradually developed over vast periods of time towards higher levels of organization. Every effort was made to minimize the significance of any gaps or apparent leaps so that the whole process could be seen as more or less continuous and hence due to the transformation of forms over successive generations. The ultimate source of new characters was derived from the forces governing embryological development, producing a series of small but still discrete upward steps:

the simplest and most primitive type, under a law to which that of like production is subordinate, gave birth to the next type above it ... that again produced the next higher, and so on to the very highest, the stages of advance being in all cases very small – namely, from one species to another, so that the phenomenon has always been of a simple and modest character.[47]

Chambers envisaged the law of progress operating through extensions of the normal process of embryological development. Perhaps when stimulated by a change in the environment, development could advance to the next stage in the hierarchy, resulting in a small jump to the next highest species.

As to what actually determined the additional structure produced by this effect, Chambers had no answer other than to suggest that it was predetermined by the law itself: in effect the law was nothing more than the unfolding of a plan of development built into nature. Two elements within the first edition of *Vestiges*, soon abandoned in later printings, illustrate the source of this way of

---

[47] [Chambers], *Vestiges*, p. 222.

thinking. A whole chapter was devoted to the circular or quinary system of classification devised by William Sharpe MacLeay. This imagined superimposed circles each of five elements: the animal kingdom was composed of five types, each type of five classes, each class of five orders and so on down to each genus composed of five species. It was a highly artificial system, a perfect example of the obsession with the hope of finding order and regularity hidden in the diversity of nature. Chambers appealed to it because it fitted his belief that there was a predetermined pattern built into the plan of development – even though a system based on circles didn't really correspond with his notion of a developmental scale (which is presumably why it disappeared from later editions of his book).

The other strange feature of the first edition was its use of an idea proposed in Charles Babbage's *Ninth Bridgewater Treatise*, an unofficial addition to a series of books commissioned to display how the wisdom and benevolence of God could be seen in nature. Babbage is remembered today as a pioneer of the computer – he actually built a mechanical computer that could be programmed to undertake a series of tasks one after the other. In his *Bridgewater Treatise* he used an analogy based on his invention to argue that God could have arranged for miracles to occur not through His immediate intervention but by programming (to use the modern term) the laws of nature to jump to a new function at a predetermined point in time. Chambers was able to argue that the short extensions to development required by his theory were built into the laws in the same way. This suggests that he was thinking of progress as the unfolding of a divine plan built into nature at the creation, not the product of laws operating in the normal mechanistic fashion. In principle, anyone who could work out the pattern would be able to predict its eventual products, just as Agassiz and Owen believed.

This leaves the crucial question of just what the shape of the plan or pattern was supposed to be, and here there has been some disagreement among historians. Chambers was aware of von Baer's embryology which made it impossible to construct a linear hierarchy of forms: each class develops in its own way and never passes through adult stages of other classes. But I suggest that Chambers offers us another example of how to imagine a linear hierarchy buried within a system of branching relationships. He gave a simple diagram, modified from one published earlier by the physiologist William Benjamin Carpenter, which shows the classes branching off separately from a vertical stem (see Figure 2.3). But the vertical stem still defines the 'main line' of development leading up to the mammals and then ultimately to humankind. A fish, for instance, develops along the main line of development until it reaches the point defining the fish level in the scale, after which it branches off to form the more detailed aspects of its mature structure. A reptile will carry on further up the scale before branching off, and so on. There is certainly

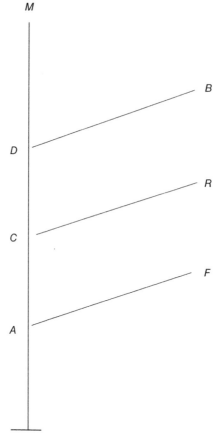

Figure 2.3 Simplified tree of development in the animal kingdom, with branches representing the classes branching from a main line leading towards the mammals. F denotes the fish, R the reptiles, B the birds and M the mammals.
From Robert Chambers, *Vestiges of the Natural History of Creation* (1844), p. 212

branching in the plan, but there is also a linear scale defining the main line of development which leads up to humanity. We are still meant to regard ourselves as the intended goal of creation; the lower forms are just side-branches leading to dead-ends.

In later editions of *Vestiges* Chambers struggled to work out how his system could be applied to explain the complexity of the animal kingdom. He suggested that evolution (as we would call it today) proceeds through a number of parallel lines all advancing through the same hierarchy of complexity within its class. He even hinted that a line can pass from one class to another, in effect making each class polytypic, having a number of independent ancestors. He also suggested that each line must begin with an aquatic form and then evolve terrestrial equivalents. Most of these strange phylogenies are constructed from living forms with little reference to what would make sense in terms of the fossil record.

Chambers's book made no effort to conceal his main point: that the human race was the final product of the most active line of development and that we had originated by a small change from some highly developed animal ancestor (curiously he made no effort to identify this with the apes). Where the traditional view saw the human mind as existing on a higher plane than anything available to the animals, he argued that an increase in the size and complexity of the brain could explain how our mental and moral faculties had emerged from those of the animals. He appealed to phrenology, a popular amalgam of science and social philosophy which held that the structure of the brain determined an individual's mental capacities. If the brain is the organ of the mind, then expanding the brain via an evolutionary jump would explain how we had advanced mentally beyond our animal ancestors. Phrenology was tied in with the ideology of social progress: since its practitioners claimed to be able to determine one's brain structure from the shape of the skull they could, for instance, advise people on the careers they were best suited to.[48] Chambers also argued that there were laws governing how society operates, so that it was possible to identify regularities and thus predict future changes in crime rates and other social activities.

Compared with those of later evolutionists such as Spencer, Chambers's efforts to see social progress as a continuation of biological development were unconvincing. But his overall message was clear enough, and despite his efforts to present development as the unfolding of divine plan his linkage of humans to the animal kingdom was enough to offend many religious thinkers. There was a huge outcry and numerous rebuttals, often based on the discontinuity of the fossil evidence rather than any reasoned response to the theory's deeper implications. The book was a long-running sensation, the last edition (the eleventh) appearing in 1860, after the publication of the *Origin of Species*. Its impact on the public was considerable, and historians generally agree that *Vestiges* paved the way for the reception of Darwinism, if only by absorbing some of the shock that would otherwise have made the response to the *Origin* all the more acrimonious. We also need to be aware that in attracting so much public attention it may have shaped the way Darwin's book was read. People

---

[48] There is a huge literature on phrenology; see for instance Cooter, *The Cultural Meaning of Popular Science*; Young, *Mind, Brain and Adaptation in the Nineteenth Century*; and Van Wyhe, *Phrenology and the Origins of Victorian Scientific Naturalism*. The debate over *Vestiges* is covered in detail in Secord's *Victorian Sensation*.

had come to think of evolution as a purposefully designed progression leading to humankind. That wasn't what Darwin had in mind, but not all of those who read his book (or more often accounts of it published by others) would be able to appreciate that fact.

Developmentalism in the Age of Darwin

The theory of evolution proposed by Charles Darwin differed fundamentally from the vision of progress expounded in the pages of *Vestiges*. For Chambers, as for Agassiz and Owen, the pattern that shaped the history of life was linked to and (for Chambers at least) actually driven by the forces of embryological development. The resulting trends were predetermined and aimed towards the emergence of predictable goals. They had no naturalistic explanation because they were embedded in nature by its Creator, left to unfold before observers who could not penetrate to the underlying processes. Development was an intrinsically teleological process. It was conceded that there were many branches to the tree of life, but the human race was still seen as the high point of creation, defining a main line of development that formed the central trunk of the tree of evolution.

It was this developmental viewpoint that Darwin set out to challenge. He didn't deny that progress occurred, but he did not see it as an inherent force of nature, nor would he accept that it had a privileged axis running towards humankind. We may be higher than the other animals, but there was nothing inevitable about our emergence into a new world of mental and social evolution. Many of the other lines of evolution have progressed too – and many have degenerated – all at the mercy of ever-changing conditions to which they must adapt. Progress is a by-product of evolution, not its central purpose or driving force. Leaving aside the unsettling idea of natural selection acting on 'chance' variations, it was hard for Darwin's contemporaries to accept or even appreciate the full extent of the revolution he sought to bring about. Naturalists concerned with classification might accept the need to see evolution as a divergent tree with no central trunk, but for many ordinary people and most religious thinkers it was hard to throw off the sense that somehow the human race was *meant* to be here as the high point of creation.

There were numerous efforts to retain the impression that evolution was somehow intended to produce the human race, including evolutionary trees with obvious central trunks implying a main line of development. These were, admittedly, mostly in publications aimed at a wider public; trees constructed by working biologists were less obviously teleological.[49] Theories of human origins, including surveys of the increasing number of hominid fossils, were

---

[49] For a good selection see Pietsche, *Trees of Life*.

more likely to present the material as evidence of a trend towards the fully developed human form. Evidence of the tendency to think in developmental terms can be seen in the theories of non-adaptive evolution and parallelism proposed by non-Darwinian biologists.

The extent to which developmentalism survived is bound up with the more general question of how scientists and the wider world tried to cope with the radical implications of Darwin's theory. I have argued elsewhere that our perception of how successful the 'Darwinian revolution' was at the time is skewed by a tendency to assume that everyone who pledged support immediately became a Darwinian in the modern sense of the term. In fact many who figure prominently in the defence of the theory retained doubts about the adequacy of the mechanism of natural selection and were less inclined than Darwin to face up to the possibility that evolution was completely open-ended and unpredictable. Even Thomas Henry Huxley, known as 'Darwin's bulldog', thought that selection would need to be supplemented by some process of predetermined variation that pushed groups to evolve in a particular direction. Some time ago I coined the term 'pseudo-Darwinism' to denote this position, and although it has not been widely accepted I still maintain that it is a valid way of drawing attention to the fact that not everyone was able to accept the implications of a theory in which adaptation to the environment is the only mechanism of change.[50]

The most contentious candidate for inclusion among the pseudo-Darwinians is the German biologist and social campaigner Ernst Haeckel. He proclaimed himself a supporter of the new theory and used it to promote his opposition to traditional religion based on a philosophy of 'monism' – the view that mind and matter are just different manifestations of a single substance whose activities are the sole determinant of how nature operates. Haeckel liked Darwinism because it seemed to support his claim that there was no divine purpose to be seen at work in the universe. He used it in a number of books that were popular both in Germany and in English translation. Yet historians disagree fundamentally over just how close Haeckel's thinking was to Darwin's. Robert Richards and Sander Gliboff argue that the links are substantial (although Richards blurs the issue by suggesting that Darwin himself owed much to the Romantic way of thinking). Others, including Michael Ruse and myself, think that the links are superficial and conceal Haeckel's commitment to a developmental way of thinking based on the inevitability of progress.[51]

---

[50] The term 'pseudo-Darwnism' was coined in my *The Non-Darwinian Revolution*. For more detailed accounts of the evolutionary debates discussed in this section see also my *The Eclipse of Darwinism* and *Evolution: The History of an Idea*.

[51] In addition to Gould's *Ontogeny and Phylogeny* and Gliboff's *H. G. Bronn, Ernst Haeckel and the Origins of German Darwinism* see Richards, *The Tragic Sense of Lifet*; di Gregorio, *From Here to Eternity*; and Richards and Ruse, *Debating Darwin*.

The problem of interpretation is compounded by the very different areas of research that Darwin and Haeckel engaged in. Haeckel was a morphologist: he studied the structure of animals, especially their embryos, and his evolutionism was based on using this work to reconstruct the history of the animal kingdom, supplementing and to some extent bypassing the evidence from the fossil record. Darwin did morphological research too, notably in his study of the barnacles, but his real concern was how animals and plants interact with and are shaped by their environment. He had little interest in trying to reconstruct the details of how the various forms had evolved over geological time ('phylogenies', to use the term coined by Haeckel).

Haeckel also coined the term 'ecology', but he had limited interest in the topic. He paid little attention to other issues that were crucial to Darwin's way of thinking, including animal breeding and island biogeography (Gliboff notes that in translating the *Origin*, Bronn found the material on artificial selection particularly troubling). It is also obvious that whatever his endorsement of natural selection, Haeckel was far more interested in the Lamarckian theory of the inheritance of acquired characteristics. He preferred to believe that most new characters are produced by the organism's purposeful activity in response to the environment, not 'random' variation. This allowed him to argue that variation tends to be progressive; he knew that some evolution is degenerative but seldom focused on the topic.

It is partly this tendency to assume that most new characters represent an advance in the level of organization that leads the historians who question Haeckel's Darwinian credentials to see him as having a more developmental viewpoint. Lamarckism was also linked to the recapitulation theory, Haeckel's main contribution to phylogenetic research. For newly acquired characters to be inherited, it was thought that they had to be added on at the end-point of embryological development, thus preserving the old adult stage as a clue to ancestry.

Here we enter an area which has been fraught with misunderstanding. Stephen Gould's monumental study of the recapitulation theory focused on this Lamarckian component, especially visible in the work of a group known as the 'American neo-Lamarckians'. The leading figures here were Edward Drinker Cope and Alpheus Hyatt, and their efforts to reconstruct phylogenies were based on the claim that embryology preserves adult ancestral forms. They also saw the development of each animal group not as a branching tree but as a series of lineages advancing in parallel through the same developmental hierarchy, as though aimed at a predetermined goal. Since it was applied separately to each group, this project was not intended to restore a single linear chain of being, but it did retain the focus on linear patterns and it did see evolution as based on predetermined trends. Parallelism became a key feature of many non-Darwinian theories, and for Cope, at least, the Lamarckian theory allowed one to believe that evolution was the expression of God's purpose.[52]

Gould thought that Haeckel's recapitulationism was cast in the same mould, but I now believe this to be a mistake that has confused the situation for decades. Richards and Gliboff are correct to argue that Haeckel's use of the principle was much less rigid, being compatible with von Baer's non-linear embryology because it acknowledged only the retention of ancestral *embryonic* stages. Nor was there any suggestion that evolution was driven along predetermined lines by inbuilt developmental patterns. In the end, though, I am still not convinced by the claim that Haeckel was a good Darwinian. One could argue that for him the Lamarckian process represented a progressive force that could be applied in many divergent branches in the tree of life. It would be Darwinian in the sense that it was compatible with an open-ended, unpredictable vision of evolution, although the progressive factor would certainly be stressed far more than Darwin himself would have allowed. But on some occasions at least, Haeckel does seem to encourage his readers to see the line of descent leading though to the human form as the main line of development. Here, at least, there is the retention of something like the temporalized chain of being.[53]

It may be that this impression is an unintended by-product of Haeckel's approach to his subject. In parts of his popular works, including *The History of Creation* and more especially *The Evolution of Man*, he focused on showing his readers how their own species has emerged, which inevitably involved picking out the line that led towards the human form for special attention. This would create the impression of a predetermined line of advance, although in fact the 'main line' would be established by hindsight, identifying the transitions that became crucial for our own origins without implying that the steps were predetermined. Nevertheless, Haeckel's works and the other popular books which followed his lead helped to create a whole genre focused on the process of evolution leading towards humankind. Such books give the impression of a linear chain of developments leading towards a goal, even if there is no intention to imply that the human form is the predetermined end of creation. Haeckel also used extensive evidence from embryology rather than the fossil record, again creating an impression that development towards maturity was somehow analogous to the process of evolution.

---

[52] Cope's theory is outlined in his *The Origin of the Fittest* and *The Primary Factors of Organic Evolution*. For his religious views see his *The Theology of Evolution*. On Cope and the other American neo-Lamarckians see Gould's *Ontogeny and Phylogeny* and my *The Eclipse of Darwinism*, chap. 6.

[53] Pietsche, *Trees of Life* includes nineteen of Haeckel's trees, of which three have a main stem leading to humankind (although several are for the plant kingdom only). The 'knarled tree' is from *The Evolution of Man*, vol. 1, plate XV, facing p. 188.

Haeckel provided several diagrams that show the tree of life with a central trunk running up to the human form. The most famous resembles a gnarled tree with braches leading off to the side at various levels towards the other animal groups (see Figure 2.4). The impression that there is a main line of progress aimed at a particular goal, and that all the other animals are just modified descendants of key stages in the ascent towards humankind, is hard to miss. There are many other phylogenetic diagrams in Haeckel's works that are perfectly compatible with an open-ended branching tree model. But in his *History of Creation* he listed twenty-two stages in 'The Chain of the Animal Ancestors, or the Series of the Progenitors of Man'. Again an apparently linear progress towards a predetermined goal emerges from the diversity of life.[54] Darwin himself would not have felt comfortable with any means of presentation that gave the impression that the human species is the goal of evolutionary progress.

The same approach is visible in the work of other Darwinians. Anton Dohrn antagonized Haeckel by proposing an alternative to the latter's views that the vertebrates had evolved from the tunicates (sea-squirts). He proposed instead that they had evolved from arthropods and that the tunicates were a degenerate offshoot of the vertebrate stock. He then implied that more or less all the lower forms of life had degenerated at one point or another from a single progressive line of development: 'we gain the picture of a single lineage, which concealed within itself the germ of all remaining high, highest, and also lowest productions, whose descendants here span the universe with sense and thought ...'.[55]

Dohrn's friend, the British zoologist E. Ray Lankester (who also wrote on degeneration), invoked the model of a single hierarchy in the animal kingdom from which all the lower types had branched off one after the other. As a good Darwinian, Lankester warned against the simple-minded assumption that some of the lower species today are 'living fossils' preserving earlier stages of development. Humans had not evolved from the chimpanzee or the gorilla but from some less specialized ancestral primate. Yet it was still possible to see the main line of development in outline:

the general doctrine of evolution justifies us in assuming, at one period or another, a progression from the simplest to the most complicated grades of structure; that we are warranted in assuming at least one progressive series leading from the monoplast to man; and that until we have special reason to take a different view of any particular case we are bound to make the smallest amount of assumption by assigning to the various

[54] Haeckel, *The History of Creation*, vol. 2, pp. 278–94.
[55] From Michael Ghiselin's translation of Dohrn's *Der Ursprung der Wirbelthiere* (1875), 'The Origin of Vertebrates and the Principle of Succession of Functions', p. 77. On the debate over vertebrate origins see Bowler, *Life's Splendid Drama*, chap. 4.

PEDIGREE OF MAN.

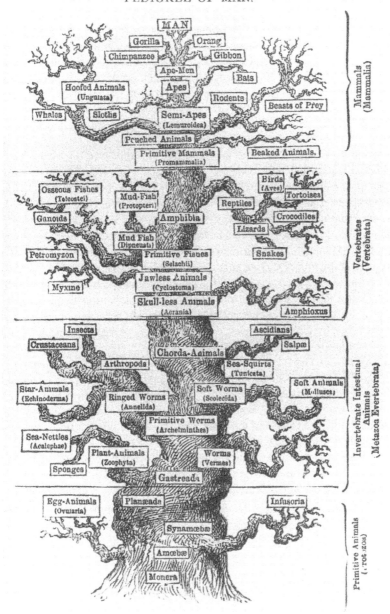

Figure 2.4 Tree-like representation of the evolution of the animal kingdom. Although gnarled like a real tree, the central trunk still implies a main line of development towards humans.
From Ernst Haeckel, *The History of Creation* (1876), vol. 2, facing p. 188

groups of organisms the place will fit on the supposition that they do represent in reality the original progressive series.[56]

If there were still efforts to see a linear hierarchy in nature, there were few explicit attempts to revive Agassiz's claim that the human form is the intended goal of creation. For all that progress was the mantra of the age, so explicit a return to the traditional way of thinking would seem inappropriate to all except the most conservative thinker. Even Darwin argued that progress was inevitable in the long run, although he saw it as an irregular by-product of the relentless action of natural selection. In the course of his effort to present Darwin as a developmentalist, Richards notes a reference to the idea that 'Man is the one great object, for which the world was brought into the present state', but this is in one of his early notebooks and there is little to suggest he ever took it seriously.[57]

Outspoken opponents of Darwinism including Richard Owen and St George Mivart appealed to the orderliness of nature to defend the view that evolution was the expression of a divine power, but they seldom imagined linear patterns leading directly towards the human form. Their 'theistic evolutionism' was meant to retain the old vision of nature as a divine artefact but update it to accept a non-materialistic version of evolutionism. This was the position staked out in Chambers's *Vestiges*, and we have seen that in the late 1840s Owen seems to have shared the view that the unfolding of the divine plan was aimed at the production of humanity. As evolutionism became more popular in the decades after Darwin, he backed away from this position to concentrate on demonstrating regular patterns in the evolution of a diverse range of organisms. Mivart also took this approach in his highly anti-Darwinian *Genesis of Species* of 1871. Significantly, both were now anxious to argue for a separation between humans and apes in order to preserve our unique spiritual character. Arguing for a main line of evolution pointing towards humanity would have undermined their efforts – in this case worrying about continuity of development clashed with the idea of linear progress.[58]

Yet there were still some efforts to defend the idea that humanity is the goal. As late as 1884 the Glasgow professor of anatomy John Cleland claimed that 'man is a terminus, and not only *a* terminus, but *the* terminus of the advance of vertebrate life'.[59] Those who wrote for a wider readership were more likely to stress the idea of progress and to present the human race as – if not the predetermined goal – then certainly as the highest product of evolution,

---

[56] Lankester, 'Notes on the Embryology and Classification of the Animal Kingdom', p. 440.

[57] Richards and Ruse, *Debating Darwin*, p. 164, see *Charles Darwin's Notebooks*, p. 409.

[58] On theistic evolutionism see Bowler, *The Eclipse of Darwinism*, chap. 3, and more generally on these debates Desmond, *Archetypes and Ancestors* and Moore, *The Post-Darwinian Controversies*.

[59] Cleland, 'Terminal Forms of Life', p. 359.

thereby defining the most significant line of development. In America John Fiske converted Herbert Spencer's system of progressive evolutionism into a 'cosmic philosophy' in which the process was underpinned by divine power and led inevitably to the appearance of humanity and towards a civilization based on a liberalized Christianity. Fiske suggested that the extension of childhood in the human race promoted socialization and the emergence of the family and saw this as an example of several 'critical points' in evolution.[60] Joseph LeConte also wrote of 'critical periods' in the advance of life, giving a diagram showing waves representing the rise and fall of successive classes in the history of life. He insisted that although evolution threw off many branches, all except one led off to specialization and stagnation. Humanity was the divinely planned outcome of this main line of development.[61]

A curious offshoot of the tendency to see evolution as the unfolding of predictable trends was the introduction of the concept of 'racial senility'. Here the analogy between evolution and the individual life-cycle was extended to its obvious conclusion: each line of development led inevitably to an ultimate decline and to death. Palaeontologists such as Alpheus Hyatt in America thought they could distinguish developmental sequences within the history of each class, each rising to a maturity and then declining towards degenerate forms and eventual extinction. In this profoundly non-Darwinian version of evolutionism, all the animal classes exhibited similar overall patterns of rise and fall. This was linked to the claim that within each class there were many parallel lines, each following the same predetermined path but perhaps at different rates. Hyatt's favourite examples were in the Ammonites, but the cyclic model of evolution was promoted by experts on a number of other classes including the dinosaurs.

The theory of parallel evolution driven by inherently developmental trends was anything but an endorsement of evolutionary progress, and indeed its implications were profoundly pessimistic. If applied to the human race it would imply that we too are predestined to decline to racial senility and extinction. Here we see the downside of the parallel drawn between evolution and the individual life-cycle. Those with a more optimistic view focused on the first phase of development up to maturity, conveniently ignoring the later phases. Faith in progress was retained by focusing on the upward steps in the development of the vertebrate classes.[62]

---

[60] Fiske, *Outlines of Cosmic Philosophy*; see part 3 on the wider implications of the system. Spencer's views actually come closer to the more open-ended view of evolution and are discussed in Chapter 7 below.

[61] LeConte, *Evolution: Its Nature, Its Evidences and Its Relation to Religious Thought*, p. 15 and p. 19. See Stephens, *Joseph LeConte*.

[62] For details of Hyatt's work and other examples of the concept of racial senility see my *The Eclipse of Darwinism*, chaps. 6 and 7, and *Life's Splendid Drama*, chaps. 7 and 9.

There were numerous popular contributions to the kind of evolutionary survey focused on the line of development leading to humankind. This was the age of what Bernard Lightman has called the 'evolutionary epic', and while some of these narratives were written from the naturalistic perspectives others were intended to convey a message more in tune with traditional beliefs. It was the struggle to develop the mental and ultimately the moral faculties that constituted the main driving force of evolution, and humanity was obviously the most successful outcome of this trend. Arabella Buckley, an enthusiast for spiritualism, published her *Winners in Life's Race* in 1882 to promote this optimistic vision of evolution. While she recognized that the birds deserved an honourable mention for their colonization of the air, there was no doubt that it was the mammals, the primates and ultimately humans which represented the triumph of vertebrate evolution.[63] Religious thinkers shared these sentiments and inevitably gravitated towards the non-Darwinian theories. In 1894 Henry Drummond's *Ascent of Man* enjoyed huge success promoting the view that the struggle for existence was not the driving force of evolution; instead it was the drive to cooperate that led animals to progress, leading inevitably to the appearance of humanity's moral character.[64]

Buckley's enthusiasm for spiritualism was shared by the co-discoverer of natural selection, Alfred Russel Wallace. Although he saw selection as the only valid mechanism of evolution (even Darwin had allowed a minor role for Lamarckism), Wallace made an exception for the human mind, which he came to regard as being of supernatural origin. While well aware of the diversity of life, he was forced to accept that the main purpose of evolution was to produce a being capable of being endowed with a spiritual dimension. For those who accept the existence of a spiritual world 'the whole purpose, the only *raison d'être* of the world – with all its complexities of physical structure, with its grand geological progress, the slow evolution of the vegetable and animal kingdoms, and the ultimate appearance of man – was the development of the human spirit in association with the human body'.[65] This view came to the fore in his last book, *The World of Life: A Manifestation of Creative Power, Directive Mind and Ultimate Purpose*, published in 1911. There was almost no mention of natural selection in this account of the progressive development of life towards a spiritual goal. The diversity of life was presented as a necessary background for the appearance of humankind and our future progress. The whole universe has been formed 'firstly, for the development of life culminating in man; secondly, as a vast school-house for the higher education

---

[63] On Buckley and evolutionary epics see Lightman, *Victorian Popularizers of Science*, chap. 5.
[64] On the response of religious thinkers including Drummond see Moore, *The Post-Darwinian Controversies*.
[65] Wallace, *The World of Life*, p. 391. On Wallace's beliefs see Fichman, *An Elusive Victorian*,

of the human race in preparation for the enduring spiritual life to which it is destined'.[66]

By the time Wallace published *The World of Life* its presentation would have begun to seem dated, its focus on our spiritual destiny all too obviously a product of Christianity's traditional message that humankind is the primary focus of the Creator's attention. From the earliest efforts to imagine a structure for the world of life and to investigate how that structure might have developed though earth history, the assumption that humanity was somehow the both the high point of creation and the intended goal of history had shaped naturalists' thinking. As the plausibility of the simple chain of being collapsed, this vision was preserved by insisting that the tree of life had a central trunk and that the human race lay at the head of this main line of development. This vision was now threatened by Darwinism, which made it much harder to think that the diversity of life concealed a man line of advance with a predetermined goal. Whatever Wallace's hopes in his declining years, the twentieth century would witness the triumph of this alternative way of visualizing evolution. For some, this meant the elimination of any hope of seeing evolution as progressive, but for others it would necessitate a reconfiguration of the idea of progress along less rigidly structured lines.

As we shall see in Chapter 6, this wasn't quite the end of the story. There have been occasional efforts to defend the claim that humanity is the predetermined outcome of evolution in the twentieth century. Some of these, indeed, are by eminent biologists, but they all reflect the conceptual driving force of what began as the linear model of progress represented by the chain of being. In effect this driving force is a conviction that somehow we humans are intended to be here by some higher power. Whether that power is divine or merely embodied in material nature, its intentions cannot be denied without depriving us of our crucial status in the world. A world that did not end up inhabited by something like ourselves is unthinkable.

Wallace's concern to argue that humanity's spiritual powers must have originated though supernatural action points to another problem that had been implicit in the linear model from the beginning. The original chain of being had been conceived as absolutely continuous, which would imply that there could be no break in the sequence between apes and humans: the apparent gap must be bridged by intermediates. This implication was also incorporated into any theory in which evolution was conceived of as a continuous process. Concern that this element of continuity would undermine our distinctive status had driven efforts to insist that species are discrete entities and that evolution would have to work in steps or jumps, the last of which would be the introduction of our mental and moral powers.

---

[66] Wallace, *Darwinism*, p. 477.

Darwin and the evolutionists challenged this sense of human uniqueness in the name of continuity. As the nineteenth century drew to a close, the crucial nature of the evolutionary step up to humanity was highlighted by the discovery of fossils that threw light on our immediate origins. Developments in evolutionary biology began to dovetail with new insights in areas such as anthropology and archaeology which challenged traditional assumptions about the original state of humanity. The resulting debates revisited issues that had emerged in the first efforts to explore the implications of continuity in the chain of being, issues that reveal a much darker side to its influence.

3   The Hierarchy of Humanity

The linear view of nature assumed that humanity represented the high point of the chain of being and – when the chain was temporalized – the goal of progressive creation. But those who were committed to this vision of the world had to think about the intersection where the highest level of the animal kingdom was supposed to grade seamlessly into the human form. Recognizing the apes as the highest animals raised the uncomfortable prospect of them looking very much like degraded humans. But if humans have immortal souls and animals do not, that physical resemblance must be purely superficial as far as our spiritual destiny is concerned. The claim that there is indeed a sharp distinction in the intellectual, moral and spiritual realms was one way of dealing with the issue, but this position seemed increasingly suspect as better knowledge of the great apes became available. It was all too easy to invoke another aspect of the chain, the notion of a complete continuity between the highest animals and the lowest humans, looking for examples of the latter among the non-white races that Europeans were now encountering – and subjecting – around the globe. Here the chain of being exposed its darker side, creating the framework for an ideology that would underpin Europeans' sense of their own superiority for centuries to come.

A sense of cultural superiority did not necessarily translate into the assumption that the non-white races were mentally or biologically inferior. It was quite possible to erect a hierarchy of cultures while supposing that those less developed had merely been retarded by exposure to poorer environments. But as naturalists and anatomists studied and measured the physical characteristics of races it seemed obvious to some that there must be a link between low levels of cultural development and inferior mental powers, the latter being inferred from allegedly smaller cranial capacities and superficially ape-like features. Here was the basis for the creation of a hierarchy of races with Europeans at the top – a small-scale application of the principles inherent in the chain of being.

The supposedly continuous hierarchy stretching up from the apes to the highest form of humanity via a series of lower human or semi-human types emerged in the eighteenth century before there was any sense that the pattern represented a temporal sequence in which the successively higher forms emerged. The image lay waiting to be absorbed into the early theories of biological evolution that were still based on the linear model of progress and which still took it for granted that humans, especially white humans, were the goal of creation. In principle, the Darwinian theory based on the model of a branching tree of relationships undermined the whole idea that humanity – let alone the white race – was the destined goal of evolution. The different human races, just like the various ape species, were the end-points of divergent branches of evolution and could not be arranged into a simple hierarchy. One branch might have developed further along some abstract scale of mental development, but there was no sense in which the lower forms were somehow trying to ascend a path towards the highest.

The full implications of the Darwinian position would not be recognized until the mid-twentieth century. Instead the belief that there had been a single pattern of human development remained dominant, along with the assumption that the non-white races could be arranged in a corresponding hierarchy. There were two ways of explaining why the various grades of development were still visible in the world today. The simplest depended on the possibility that at various points in the ascent populations had become separated, and those exposed to poorer environments had ceased to participate in the subsequent ascent. The allegedly inferior races were perceived as 'living fossils', preserving earlier stages intact through to the present in less favoured parts of the world.

The concept of evolutionary parallelism offered a somewhat more complex way of preserving the image of the white race as the head of creation by translating the hierarchy into an abstract scale of development which independent populations or species could ascend – but at different rates so that some will have reached higher levels than others. This model provided the evolutionists of the late nineteenth century with a means to evade the full implications of Darwinism, allowing them to accept the idea of a 'tree of life' at a superficial level while pretending that there was a sequence of developmental stages that all branches were trying to ascend. Multiple branches of the human family were all striving to reach the same goal, but with varying degrees of success. Some of the differences, especially those in facial appearance, were interpreted as evidence of different levels of development from an ape-like origin.

As fossil hominids were discovered they were fitted into the pattern of development along with the various living races of humanity. A whole range of disciplines from archaeology to anthropology and psychology sought to explain the emergence of humanity from the apes in a way that preserved at least some element of the old goal-directed model of development. All too often, the non-white races were still depicted as the less successful efforts in the race to progress up the scale of creation.

The Hierarchy of Races

Charles Bonnet was perhaps the leading proponent of the chain of being in the eighteenth century, but his belief in the existence of a soul distinct from the body distanced him from the ideological applications of the chain that became prevalent towards the end of the century. William Bynum's retrospect on A. O. Lovejoy's book pointed out that the latter's focus on the history of ideas deflected attention from the ways in which the chain was exploited by naturalists and anthropologists seeking to objectify differences within the human species. Europeans were convinced of their own superiority, and the scale of nature leading down from humans to apes became a convenient way of portraying the 'lower' races as intermediates. The use of anatomical features such as skull size and shape to define levels of intellectual ability became commonplace. The chain of being now acquired a darker significance: it might be the Creator's rational plan of nature, but those humans alleged to have smaller brains and ape-like features were condemned by their place in the scale to an inferior social position. The hierarchy of races became a justification for slavery and would later gain even more influence when adopted as the model for the evolution of humans from an ape ancestor. Its origins, however, lay in the originally static version of the chain. The first chapter of Nancy Stepan's account of the origins of race science is entitled 'Race and the Return of the Great Chain of Being'.[1]

The chain was taken seriously as an organizing principle in the first volume (1790) of William Smellie's *Philosophy of Natural History*: 'There is a graduated scale or chain of existence, not a link of which, however seemingly insignificant, could be broken without affecting the whole.' Humans were, of course, the high point: 'In the chain of animals, man is unquestionable the chief or capital link, and from him all the other links descend by almost imperceptible gradations.' The focus on continuity more or less required there to be a descending range of characters in the human races, with the Hottentot being identified as the lowest. The gap between the lowest race and the orang-outang, 'it is humiliating to remark', was very small. In his second volume Smellie again commented on the low intellect of some races, including the natives of Australia, although he made no effort to establish a systematic ranking.[2]

Charles White's *An Account of the Regular Gradation of Nature in Man* of 1799 included a version of the complete chain of being derived from Bonnet but used anatomical evidence to construct a hierarchy of races based on brain

---

[1] Stepan, *The Idea of Race in Science*, chap. 1, and Bynum, 'The Great Chain of Being after Forty Years'.
[2] Smellie, *The Philosophy of Natural History*, vol. 1, quotations from pp. 520–3.

size and degree of similarity to the ape. White was one of the first to exploit what we now know to be a misrepresentation of the work of the Dutch anatomist Petrus Camper, a misreading that was to become a staple foundation of nineteenth-century views on race. Camper introduced a measuring technique based on the facial angle of a skull, the angle between the horizontal and a line from the lower jaw passing over the nose. A low facial angle indicated a receding forehead and hence a smaller brain. The figures given by White are: monkey 42°, orang 58°, negro 70°, Chinese 75°, European 80° (his figures for Classical Greek and Roman facial angles are even higher). Plate 2 of his book also depicts an increasingly ape-like physiognomy for the lower races.[3]

Thanks to the work of Miriam Claude Meijer and others we now know that Camper himself did not use his facial angle to measure the capacity of the cranium: he was more interested in the different jaw shapes of the races. He also had no intention of erecting a hierarchy of races and argued strongly that all should be seen as belonging to a single species. White insisted that he was opposed to slavery, but nevertheless saw the races as distinct species that could be arranged in a sequence stretching down from the European to something close to the ape. A similar misinterpretation of Camper's facial angle also found its way into Georges Cuvier's survey of comparative anatomy, from which it spread into the literature to become virtually universal within the anthropological debates of the later nineteenth century.[4]

The fascination of a linear sequence linked to the apes seems to have affected thinkers who had no intention of seeing the 'lower' races as relics of an ape ancestry. White thought the races were separately created, yet still saw the 'lowest' as having ape-like features as well as a lower cranial capacity. Whether it was explicitly derived from the chain of being or not, there was something about our similarity to the apes that encouraged the imposition of a linear scale on the diversity of racial types. The scale was readily available for exploitation as soon as the idea of progress through time became popular.

The Origin of Races

Europeans had a long history of prejudice against the non-white races they exploited around the globe. Seeing them as not fully human was an obvious excuse for treating them as inherently destined for servitude or extinction. Black Africans were the initial choice for the lowest form of humanity, eventually replaced by the aborigines of Australia. To place this ranking on

---

[3] White, *An Account of the Regular Gradation of Mans*; see pp. 16–17 for Bonnet's chain and pp. 50–1 on the facial angle.
[4] On Camper's work and the misinterpretations of his position see Meijer, *Race and Aesthetics in the Anthropology of Petrus Camper.*

an apparently scientific footing, it was justified by an appeal to various characters of which colour, hair character and the size of the brain were the most important. The chain of being re-emerged in this limited area on the basis of the alleged fact that whites had bigger brains than blacks and were thus more intelligent. The link back to the chain metaphor is evident in the fact that the black races were also depicted as more ape-like. In this area, at least, theories based on progressive evolution merely reinforced a linear model already established on more traditional conceptual foundations.

The pervasiveness of this remnant of linear thinking is evident from the fact that it also straddled the divide between those who favoured a single origin for the human species, the monogenists, and the supporters of polygenism who claimed the races were distinct species. Even those who thought the black races had degenerated from the originally created white form imagined the process as a slide down to a more ape-like character.

Linnaeus had defined four distinct human races, and the eminent German anatomist J. F. Blumenbach five – but Blumenbach thought the white or 'Caucasian' race (his term) was the original creation, with the others having degenerated in different ways. The leading defender of monogenism, James Cowles Pritchard, thought that the original human form was negroid and explicitly criticized anatomists' efforts to depict blacks as more ape-like. Even Arthur de Gobineau – for all his conviction that the Aryans were the only race capable of developing a civilization – was critical of the anatomical criteria. Those anatomists who argued for a wide gap between the ape and human forms were also predictably suspicious of the claim that the back races were intermediates. Richard Owen, for instance, conceded that the brains of black Africans and Australians were slightly smaller than those of whites but insisted that the variation within the human species was trivial compared with the gulf between human and ape.[5]

As the nineteenth century progressed polygenism became more popular and the effort to create a racial hierarchy defined by anatomical characters more active. The two positions could be combined by supposing that separate species ascended the scale in parallel but at different rates. The polygenists became increasingly obsessed with the hope of ranking the races according to their cranial capacities, which they thought could be measured more easily than the corresponding levels of mental ability. In the early decades of the nineteenth century the new science of phrenology promoted the claim that mental powers were dependent on brain structure. Although it was dismissed

---

[5] Pritchard, *Researches into the Physical History of Man*, p. 67 and p. 233. The facsimile of the first edition cited in the Bibliography includes an essay on Pritchard's anthropology by George W. Stocking, Jr. Gobinau's *The Inequality of Human Races*, chap. 10, offers a critical account of physical anthropology. On Owen's position see Rupke, *Richard Owen*, chap. 6.

as a pseudo-science by many conservative thinkers, its claim that the brain was the organ of the mind was taken increasingly seriously. Basing a hierarchy on craniometry, the study of cerebral capacity and skull shape, became attractive because it was a surrogate for the assumed ranking based on intellectual and moral faculties. Anatomists rushed to measure skulls and their brain cases, using methods that are now seen as classic illustrations of how an apparently scientific methodology can be distorted by the preconceptions of those who apply it.[6]

Camper's facial angle was widely misused as a way of depicting apes and the 'lower' races as having smaller brains and ape-like features. A low facial angle indicated a receding forehead and hence the smaller capacity of the front part of the brain. For the 'better' type of European the angle was almost ninety degrees, for apes it was much lower, and the figure for negroid skulls was intermediate. Images based on this technique became commonplace through the nineteenth century; one was used, for instance, by the Scottish anatomist Robert Knox, depicting a black African as intermediate between an ape and a European. His *Races of Man* of 1850 deals mostly with psychological characters, however, and did not attempt a systematic ranking. Knox argued that the races are distinct species, and his views inspired later polygenists who exploited alleged differences in cranial capacity to argue that the races had evolved to different levels in the hierarchy of development.

In Britain a group of anatomists inspired by Knox and led by James Hunt founded the Anthropological Society of London in 1863 to promote polygenist views. Their journal published articles by Hunt and others insisting that the human races were separate biological species but also ranking them in a series based on cranial capacity and structure, and on the degree of resemblance to apes. Hunt openly argued for a version of transformism based on parallel lines of development. There were close links with the Anthropological Society of Paris, where Paul Broca promoted similar views, again based on non-Darwinian evolutionary trends. The London society published a translation of Broca's work critical of the popular assumption that the different races could interbreed freely. It also translated a book by Georges Pouchet which not only argued the polygenist position but insisted that the races could be arranged in a linear sequence. Pouchet referred back to Bonnet's views on the chain of being and argued that if we want to look down the 'steps of the human ladder' we should look to those 'races placed so low, that they have quite naturally appeared to resemble the ape tribe'. There was also a translation of the work of the German anatomist Karl Vogt, which claimed that the

---

[6] Gould, *The Mismeasure of Man*, esp. chap. 3, is the classic critique of craniometry. For broader studies of race science see, in addition to Stepan's *The Idea of Race in Science*, Stanton, *The Leopard's Spots* and Haller, *Outcasts from Evolution*.

different human races had each independently evolved from a different ape species. Apparently there were developmental forces at work which aimed at the production of the human form.[7]

Evolution and Race

The rising popularity of polygenism muddied the waters considerably during the debates sparked by the publication of Darwin's *Origin of Species* in 1859. It now became widely accepted that humans had evolved from apes, but the exact nature of the transition was disputed, as was the position of the various living races. Darwin's own views, articulated in his *Descent of Man* in 1871, included the radical suggestion that the separation of the human and ape stocks was not a matter of one surging ahead in the race to gain greater mental powers. He argued that the earliest distinct human ancestors had stood upright in order to move onto the open plains. The apes had stayed in the trees and stagnated. On this model the expansion of the brain was not the driving force of our development. It was a by-product of an adaptive shift, and if that transition had not been made, we would not have evolved. Most of Darwin's contemporaries preferred to retain the assumption that the expansion of our brain and mental powers had been the driving force of the ascent towards the white race as the pinnacle of evolution.

With the exception of Alfred Russel Wallace the Darwinians accepted the common view that the black races were of inferior intelligence, but they were committed to the belief that all the races belong to the same species. Darwin himself campaigned actively against the polygenists because their theory was often used to justify slavery.[8] As E. Ray Lankester warned, the popular notion that we had evolved from chimpanzees was erroneous because on a theory of divergent evolution the apes too had been modified from the common ancestor. But that ancestor would have been ape-like, and it was still possible to believe that as the human stock added its new faculties there were groups who had been left behind and retained more of the ape character. In effect the 'lower' races were living fossils, relics of earlier stages in the process. This retained a more or less linear view of the sequence, although there was a substantial gap between the highest ape and the lowest surviving human type. It was hoped that fossil discoveries would eventually fill in this gap.

A model of human origins based on parallel lines of development along a sequence from ape to European humanity was part of the evolutionary

---

[7] Broca, *On the Phenomena of Hybridity in the Genus Homo*; Pouchet, *The Plurality of the Human Race*, quotation from pp. 15–16; Vogt, *Lectures on Man*. See Gould, *The Mismeasure of Man*, chaps. 2 and 3, and my *Theories of Human Evolution*, pp. 55–6 and 130–3.

[8] See Desmond and Moore, *Darwin's Sacred Cause*.

ideology promoted by Ernst Haeckel, the stages in the ape-human transition
continuing the ascent he traced out through the whole animal kingdom. The
original German editions of his *Natürliche Schöpfungsgeschichte* appeared
with images showing a more or less continuous sequence of forms from ape
to European. The English translation under the title *The History of Creation*
does not contain these images, but the text does make it clear that of the twelve
living 'species' of humanity, the eight comprising the 'wooly-haired' races are
at a lower stage of development and are more like the apes. Haeckel claimed
they were incapable of developing a civilization and were probably doomed to
eventual extermination. He conceded that the term 'species' was ambiguous in
this case – he wasn't claiming that the races had a completely separate origin –
and he did see a gap lower down in the record where the original ape-man, for
which he coined the name *Pithecanthropus*, had become extinct. Separate
forms of *Pithecanthropus* had independently evolved into the modern human
races, founding the main language types in the process. Haeckel's efforts to
downgrade the black races and imagine their extermination in the ongoing
struggle for existence have been seen as a significant contribution to the race-
centred social Darwinism that would have horrific consequences in the next
century. Blaming Haeckel for Nazism is far too simplistic, but his promotion
of a linear hierarchy of races cannot be denied.[9]

The polygenists who argued for the races having completely separate origins
also invoked strongly directed trends governing evolution in order to explain
how several different lines emerging from an ape ancestry had independently
acquired human characteristics. When they moved to accept the general idea of
evolution, it was in a completely non-Darwinian form. They denied that close
similarity between forms implied recent divergence from a common ancestor
and opted instead for what became known as parallelism, the existence of
inbuilt variation-trends forcing multiple lines of ascent along the same hierarchy of development. On this model, the fact that some human races or
species retained more ape-like characters than others could be explained
by supposing that some lines had advanced further along the scale than
others: the 'lower' forms *resembled* ancestors but were not actually ancestral
to the higher.

In America craniometry was applied to construct a hierarchy of races by
Samuel George Morton and his followers Josiah C. Nott and George
R. Gliddon. Studies of black soldiers recruited to fight in the Civil War were
used to argue for their inferior mental faculties and resemblance to the apes.

---

[9] Haeckel, *History of Creation*, vol. 2, chap. 23, esp. pp. 307–8; for the original illustrations see
Richards, *The Tragic Sense of Life*, pp. 224–8 (they are widely reprinted elsewhere). Richards
argues strongly against those historians who see Haeckel as a major source of Nazi ideology; see
his appendix 2, pp. 489–512.

Louis Agassiz also endorsed the polygenist position. One prominent American polygenist was the palaeontologist Edward Drinker Cope, a leading figure in the neo-Lamarckian school which drew inspiration from Agassiz to create an anti-Darwinian approach to evolution. Cope rejected Darwin's view that individual variation is essentially random and insisted that it worked by directed additions to the growth process, thus allowing embryological development to recapitulate past evolutionary stages. From this starting point he erected a theory which focused on inbuilt trends driving species along linear patterns of development. Parallelism explained the resemblances that the Darwinians attributed to divergence from a common ancestor. Such a model made it possible to argue that the human races had independently evolved from the apes, some retaining more embryonic (and therefor more ape-like) characters than others.[10]

Darwinism is frequently blamed for encouraging Europeans to adopt a sense of their own racial superiority, but it was the non-Darwinian evolutionists who promoted the claim that the races have separate origins. Their theories of parallel development allowed them to imagine distinct lineages independently acquiring human characters, something that would be most unlikely on the basis of truly divergent evolution. Parallelism allowed the polygenists to endorse the traditional hierarchy in which the races were arranged in a sequence leading from the apes up to the Europeans. Their approach also implied that the white race was in a sense the goal towards which all the other lines were tending – but not so rapidly.

In many respects the developmentalist model of evolution represented the continuation of a conservative way of thinking about nature. Agassiz remained convinced that species (and hence the separate human races) are divinely created. Chambers's *Vestiges of the Natural History of Creation* explained the trends he imagined as divinely preordained, and Cope began from the same position. Both were theistic evolutionists, seeing the ascent of life as the unfolding of a divine plan towards predetermined goals. It was easy enough to imagine that the human form was somehow the intended outcome of the process. This way of thinking survived into the early twentieth century and represents the last vestige of the great chain of being's influence.[11]

The Neanderthal Phase

The initial phase of the Darwinian debate took place in the absence of any hard evidence for the transition from an ape ancestry to the human type. Although

---

[10] On the Americans see Haller, *Outcasts from Evolution* and Gould, *Ontogeny and Phylogeny*, chap. 4. Cope's papers are collected in his *The Origin of the Fittest*; see my *The Eclipse of Darwinism*, pp. 121–7.

[11] For a full development of these claims see my *The Eclipse of Darwinism* and more especially *The Non-Darwinian Revolution*.

Darwin himself avoided the topic until he published his *Descent of Man* in 1871 it was widely assumed that his theory implied an ape ancestry for humankind. Critics asked why there was no fossil evidence for the transition – hence the emergence of the term 'missing link'. In the early nineteenth century it was assumed that there were no human fossils because our species had appeared only at the very last stage in the earth's history. Darwinism implied that there must have been an extended process of evolution towards the human form, but in the absence of fossils the debate had to be conducted with indirect evidence. In the 1860s discoveries of very ancient stone tools began to the taken seriously and it was accepted that the human species had been in existence for a long period before recorded history began.

Fossils of what became known as the Neanderthal race (named after the location of the first widely discussed specimen) were at first greeted with suspicion. As T. H. Huxley noted in his *Man's Place in Nature* of 1863, there were good reasons not to see the Neanderthal type as a link in the chain leading to modern humans. The skull might have some ape-like features but the capacity of the brain-case was fully as large as a modern human's. Only after a series of discoveries at Spy in Belgium in 1886 was it accepted that the Neanderthals were a well-established type of early human. In the 1890s Eugène Dubois discovered remains in Java of an even more primitive form, for which he borrowed Haeckel's name for the hypothetical ape-human intermediate, *Pithecanthropus*. But this too seemed anomalous because although the cranium had a capacity intermediate between ape and human, the associated thighbone suggested that the creature had walked fully upright. Today we call this species *Homo erectus* and it is regarded as ancestral to ourselves, but in the late nineteenth century no one expected the 'missing link' to be fully upright in its posture: surely it should be a shambling ape-man.

Curiously, in the last decade of the century fossils at first dismissed as anomalies began to be taken seriously as steps in a sequence linking the great apes to modern humanity. The linear sequence of development widely applied to the living races was now extended back in time via the fossil types to give a complete chain of being extending upwards from the apes to the highest form of humanity – assumed to be the Europeans. This was what was later called the 'Neanderthal phase of man' theory in which *Pithecanthropus* and the Neanderthals were successive steps on the path towards modern humanity. Both had skulls with heavy brow-ridges, giving them an ape-like appearance, with the Neanderthal brain being much bigger. Given that they were a tool-making species, the Neanderthals could be seen as the last step before the emergence of the lowest modern human races. This position was endorsed by Haeckel – who had coined the term *Pithecanthropus* – in his *The Last Link* of 1898. Some went even further: in his *Ancient Hunters and Their Modern*

Representatives of 1911 William Johnson Sollas suggested that the Australian aborigines were little more than living Neanderthals.[12]

By erecting a more or less linear hierarchy of stages from the ape to the 'highest' modern race, the Neanderthal-phase theory identified the expansion of the brain as the key defining factor in the process of transformation. The evidence that *Pithecanthropus* already walked upright was sidelined to focus attention onto the emergence of humanity's most important characteristic as the driving force of progressive evolution. The claim that brain-expansion led the way in the ascent from ape to human thus preserved a linear model of development analogous to the old chain of being. There was an almost teleological implication embedded in the theory, an impression that somehow nature was striving to produce the human form and its higher mental powers. As the palaeoanthropologist Arthur Keith (who did not accept the theory) later wrote of Haeckel, his approach depicted the apes and the earlier hominids as 'abortive attempts at man production'. Gabriel de Mortillet, architect of the sequence of stone-age cultures now being accepted by archaeologists, also saw the Neanderthals as embodiments of the earliest human form he had predicted. They were 'precursors of man on the chain of being'. It was as though evolution had a goal and was constantly striving to achieve it.[13]

The advance was not completely smooth, and some populations had often been left behind. As Gustav Schwalbe insisted, the Neanderthal remains actually unearthed by archaeologists were probably not our direct ancestors: they were a relic of an earlier phase in the process that had survived in some isolated region.[14] They were thus analogous to the 'lower' races still living today. There was room in the model for the extinction of races by more progressive types, a form of social Darwinism promoted by anthropologists who often had little interest in Darwin's original version of individualistic natural selection. Sollas's *Ancient Hunters* was a pioneering work in this vein. He insisted that natural selection would ensure the elimination of backward races while elsewhere dismissing Darwin's theory as an 'idol of the Victorian era' that was incapable of explaining how new characters emerged.[15] The extinction of the late-surviving Neanderthals was an illustration of what was happening to the 'lower' races in the world today.

---

[12] Haeckel, *The Last Link*; Sollas, *Ancient Hunters*, p. 170, although he gave up this interpretation of the Neanderthals in later editions of this book. The phrase 'Neanderthal phase of man' was coined later by Ales Hrdlička; for details of these theories see Bowler, *Theories of Human Evolution*, chaps. 3, 4 and 7.

[13] Keith, *The Construction of Man's Family Tree*, p. 10; de Mortilllet, *Le préhistorique*, p. 104.

[14] Schwalbe, *Die Vorgeschichte des Menschen*; for an English summary of his views see his 'The Descent of Man'.

[15] For the appeal to natural selection at the racial level see Sollas's *Ancient Hunters*, p. 383, and on the 'idol of the Victorian era' p. 405. On this version of social Darwinism see my *Theories of Human Evolution*, chap. 9, and *The Non-Darwinian Revolution*, chap. 7.

When Sollas published his book in 1911 a new interpretation of the Neanderthals was emerging that would dominate palaeoanthropologists' thinking through the first half of the twentieth century. Perhaps the Neanderthals were not remnants of our own ancestors, but an independent line of hominid evolution that had also developed mental powers sufficient to acquire cultural traits including toolmaking. In 1909 Marcellin Boule used a newly discovered Neanderthal skeleton from La Chapelle-aux-Saints to create what became a widely accepted image of the race as shambling ape-men far too primitive to serve as ancestors of the modern races of humankind. His reinterpretation recognized that the origin of humanity had to be seen as a process in which evolution had diverged into multiple branches, a position which at first sight seems to undermine the plausibility of a linear scale of development. Indeed Boule and his followers dismissed the Neanderthal-phase theory precisely because it represented an outdated linear model.

Yet by acknowledging that the Neanderthals had independently acquired some human characteristics, the new interpretation implied a significant level of parallel evolution in the two branches. There was still a built-in trend, now seen to be driving multiple lines of development in the same direction – towards modern humanity. Because one line had advanced more rapidly than the others, it alone had achieved the ultimate goal. Boule wrote of the human stock having rivals, ape-like forms that would seek 'to evolve towards types of greater perfection'. In the end, however, 'the direct descendants of our primitive ancestors alone would seem to have reached the end of this race towards the goal of progress'.[16] This interpretation of the Neanderthals became the basis for what has been called the 'presapiens theory', the claim that there must be another, as yet undiscovered, line of human ancestry leading towards *Homo sapiens*. The theory has been criticized by later commentators as an evasion of the whole evolutionary paradigm, but it is better seen as a manifestation of the prevailing faith in parallelism and the belief that there was a directional trend at work.[17] It was perhaps the most sophisticated product of the attempt to retain the logic of the linear chain of being within a world that had been reluctantly forced to admit the branching nature of evolution. The hierarchy of development became not a literal sequence of fossil and living forms, but a scale of hominization along which several lines had ascended independently and to different levels.

The Neanderthals were widely supposed to have been a distinct species wiped out by the modern humans who had overtaken them – an interpretation

[16] Boule, *Fossil Men*, p. 110; see Hammond, 'The Expulsion of the Neanderthals from Human Ancestry'.

[17] The argument that the presapiens theory was in denial of evolutionism was made by Brace, 'The Fate of the "Classic" Neanderthals'; for a discussion see my *Theories of Human Evolution*, pp. 87–105.

well in line with the imperialist ideology of the age. This modification of Sollas's view was promoted by Sir Arthur Keith in his *Antiquity of Man* of 1915. His claim that the Neanderthals were a separate branch of human evolution was still being endorsed in the 1930s by palaeoanthropologists including Henry Fairfield Osborn and, at the start of their careers, Wilfrid LeGros Clark and the young Louis Leakey (see Figure 3.1).[18] Osborn derived the human family not from the apes but from a separate 'dawn man' inhabiting central Asia. As the descendants of the dawn man had spread around the world, they independently evolved bigger brains – although the white race had advanced further than the others.[19]

The popularity of the presapiens theory helps to explain one of the more embarrassing episodes in the history of palaeoanthropology, the Piltdown fraud of 1912. The remains found at Piltdown consisted of part of a cranium that could be interpreted as similar to the modern human form and a jaw that was ape-like. No doubt national pride encouraged many British experts to take the combination seriously as an early human type, but it also fitted the expectations of those like Keith who believed that there had been a line of human evolution separate from the Neanderthals in which the development of the brain had advanced more rapidly. A whole generation of anatomists was misled by this fraud, giving an indication of how committed they were to the belief that modern humanity stands at the head of a hierarchical pattern of development along which many lines of evolution had been driven to ascend.[20]

The Scale of Mental Development

Given the scarcity of hominid fossils in the Victorian era, the first generation of evolutionists had been forced to look elsewhere for evidence on which to base their thinking about how the human mind had developed. The craniometric techniques used by the anatomists who studied racial types and fossil hominids were a surrogate for their real concern, which was to estimate the groups' levels of mental development. However controversial phrenology may have been, its claim that the brain was the organ of the mind was now generally accepted. Measuring the capacity and shape of a skull was supposed to reveal the level at which various mental faculties could operate. Obviously it would be better to gain direct information about those faculties, and there was a wide range of anecdotal and usually highly biased information about the intellectual

---

[18] Osborn, *Man Rises to Parnassus*; LeGros Clark, *Early Forerunners of Man*; Leakey, *Adam's Ancestors*; again for details see my *Theories of Human Evolution*, chaps. 4 and 6.

[19] See Rainger, *An Agenda for Antiquity* and Regal, *Henry Fairfield Osborn*.

[20] On the eventual exposure see Weiner, *The Piltdown Hoax*; among the many later commentaries are Blinderman, *The Piltdown Inquest*; Millar, *The Piltdown Men*; and Spencer, *Piltdown: A Scientific Forgery*.

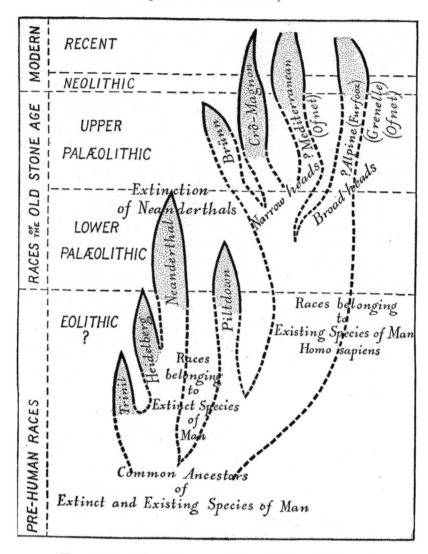

Figure 3.1 Parallelism in human evolution. The Neanderthal type is represented as a line of development that separated from true humans in the distant past and independently acquired a tool-making culture. The Piltdown type appears as a more recent offshoot which also developed alongside the hypothetical presapiens race leading to the modern human races.
From Henry Fairfield Osborn, *Men of the Old Stone Age*, 3rd ed. (1927), p. 491

capacities of non-European races. Extinct hominid species couldn't be observed in action, but archaeological evidence could be used to throw some light on their skills and social lives.

Evolutionism required a progressionist account of the origins of the higher faculties. In his *Descent of Man* Darwin made the case for there being no difference in kind between the mental powers of the higher animals and those of humans, only an increase in the level of performance. Belief in an immortal soul as the true source of our mental and moral faculties was coming under increased challenge. To throw light on the issue Darwin studied the development of his own children, showing how the mental functions emerged in the course of time. The child was not just a miniature adult, because its faculties appeared in a sequence as its growing brain interacted with the world around it.

Darwin's followers began to explore the possibility that this development followed the sequence by which our ancestors had acquired the higher faculties in the course of their evolution. The recapitulation of phylogeny by ontogeny could now be extended to the realm of the mind, establishing a hierarchy of developmental stages that encouraged evolutionists to see the emergence of the human mind as a goal-directed process analogous to individual growth. As it grew in popularity in the late nineteenth century this model would impact on not only on psychology and education theory, but also on the understanding of criminal behaviour and the prevailing ideology of race. Via its influence on Freud it would also play a role in undermining the optimistic implications of the whole developmental project.[21]

The first steps were taken by Darwin's disciple George John Romanes, who set out to confirm that rudiments of human mental faculties can be observed in the higher animals. He then turned to the study of infant behaviour, using personal observation and anecdotal reports by parents of their children's development. In an age when experimental psychology was in its infancy, it was still possible to build a substantial theoretical structure on what would eventually be shown to be extremely shaky foundations. In his *Mental Evolution in Animals* of 1883 Romanes combined the two projects to create a hierarchy of fifty developmental stages that could be observed in the animal kingdom and the growing child. Although the animal exemplars did not form a purely linear scale, the impression of a main line of development was hard to miss. Stage 28 was reached by the human infant at thirteen months and corresponded to the apes (and dogs). The infant reached stage 27 a month earlier, and this paralleled the mental powers of monkeys (and elephants). Working down the scale one passes through the lower mammals, the birds, the reptiles and amphibians and then through the invertebrates, reaching down to

---

[21] For details see Gould, *Ontogeny and Phylogeny*, chap. 5; Shuttleworth, *The Mind of the Child*; and Morss, *The Biologizing of Childhood*, chaps. 2–4.

the echinoderms (sea urchins etc.) with mental powers equivalent to those of the newborn baby.

Romanes's exposition was remarkably free from any implication that the progressive development of mind in the animal kingdom is driven by a goal-directed progressive force. But as the implications of the parallel between child development and evolution began to be explored, the link with the ideology of progressionism became more apparent. Sally Shuttleworth's survey of Victorian beliefs about child developments notes a number of efforts to use the recapitulation theory as a link between psychology and evolutionary progressionism. James Crichton Brown warned parents not to assume that their children were little men and women: they were passing through stages recapitulating evolution, including a phase corresponding to our savage ancestors. James Sully's *Studies of Childhood* of 1895 made the same point but interpreted the process in a positive light: 'It gives a new meaning to human progress to suppose that the dawn of infant intelligence, instead of being a return to primitive darkness, contains from the first a faint light reflected on it from the lamp of racial intelligence which has preceded.'[22]

Sully endorsed the Lamarckian theory in which the experiences of ancestors could accumulate to shape the development of the modern individual. In America the leader of the neo-Lamarckian school, Edward Drinker Cope, extended the link between ontogeny and phylogeny to the mental powers of the human races. If evolution works by extending the process of individual growth, this must apply to the mind as well as the body. The 'lower' human races had failed to benefit from the more progressive trends, so that they were left at the level of mentality that a European child passes through in its infancy. All this was proposed in the context of a world-view in which Cope stressed the role of consciousness in allowing animals to choose the direction of their future evolution, the whole process being presented as the unfolding of the Creator's purpose. That the emergence of the modern European race was the culmination of this purpose was taken for granted.[23]

In later decades G. Stanley Hall and James Mark Baldwin became the most prominent American exponents of the link between mental development at the individual and racial levels. Hall's *Adolescence* of 1904 interpreted a whole list of instinctive behaviour patterns shown by the child as relics of episodes in the evolution of the species. Fear of water, for instance, looked back to the time when the amphibians first moved onto dry land. Adolescence itself, with all its traumas, reflected a dramatic episode in the emergence of modern humanity, but it also expressed an element of flexibility that held the hope of further

---

[22] Sully, *Studies of Childhood* (1903 ed.), p. 9.

[23] On Cope's views on human evolution see Haller, *Outcasts from Evolution*, pp. 187–202, and Bowler, *The Eclipse of Darwinism*, chap. 6.

progress: our tentative knowledge 'is sufficient to generate a deep optimism in the hope that man is yet in the making, and that perhaps his present stage is at the same time a point of departure of a yet higher one related to all that adolescence now gives, as it is to the stages that have preceded'. Drawing on Cope, he argued that creative responses to the environment would allow the race to move 'toward a more perfect development of its higher and later acquired powers'. This might take place in sudden bursts corresponding to the foundation of new levels of organization.[24]

Baldwin's applications of the link between evolution and mental development were more sophisticated, and unlike most of the recapitulationists he accepted that advances in the study of heredity were throwing increasing doubts on the plausibility of the Lamarckian position. Yet he too was drawn to Cope's claim that intelligence is the crucial driving force in progressive evolution. He proposed the mechanism of 'organic selection', sometimes known as the 'Baldwin effect', to show how natural selection could be led along beneficial channels by the organisms' purposeful behaviour. Acquired characters are not inherited directly but provide a breathing space during which selection can determine which characters will be selected.[25]

Cope had made his views on the teleological implications explicit in a book entitled *The Theology of Evolution*. Romanes struggled with religious doubts and at the end of his life came back to belief. It would be wrong to suggest that the detailed parallels drawn by the recapitulationists were all driven by similar sentiments, but it is suggestive that their theory was routinely linked with the idea that mental activities play a key role in evolution and a belief that the human species is the high point that progress has reached to date. Haeckel's rhetorical technique of listing a sequence of ancestral forms leading up to the human encouraged his readers to see that line of evolution as the main trunk of the tree of life. In the same way, emphasizing the parallel between the development of the child and the past evolution of the human species promoted the assumption that both processes were directed towards the goal of personal or racial maturity. The heyday of developmental psychology in the decades around 1900 represents the last episode in which the idea of progress as an ascent towards a predetermined goal held sway over the collective imagination.

Descending the Scale

The application of recapitulation theory to mental development had obvious implications for child studies and educational policies. Jean Piaget seems to

---

[24] Hall, *Adolescence*, p. 50. On Hall's idea see Gould, *Ontogeny and Phylogeny*, pp. 139–48.
[25] Baldwin's evolutionary psychology is described in Richards, *Darwin and the Emergence of Evolutionary Theories of Mind and Behavior*, chap. 10.

have accepted the link (along with the inheritance of acquired characteristics), although Gould concedes that he did not see any direct causal link between the evolutionary past and the modern child's development.[26] But the darker side of the theory's implications became apparent as Hall and others drew connections between the child's mental states and the thought processes of 'savages'. If non-white races were seen as frozen at an early stage in human mental evolution, their mentality would represent a stage through which the white child must pass.[27] Rudyard Kipling's poem 'The White Man's Burden' contained the much-quoted depiction of colonized savages as 'Your new-caught, sullen peoples/Half-devil and half child'. Such sentiments capture the widespread influence of the association now being drawn between primitive savagery and the immature behaviour of the child.

It wasn't just the savage who was now caught in the net cast by recapitulationism. In 1900 Hall's colleague Alexander Chamberlain published a massive survey of the literature in the field under the title *The Child: A Study in Evolution*. Despite ending with an identification of the child as the 'evolutionary being of our species, he in whom the useless past tends to be suppressed and the beneficial future to be foretold', he included chapters on 'The Child and the Savage', 'The Child and the Criminal' and 'The Child and the Woman'.[28]

The feminist implications of the latter topic would take us on a detour too wide for convenience, but Chamberlain's account of what became known as 'criminal anthropology' points towards an important extension of the developmental model. His main source was the work of Cesare Lombroso, who published a series of studies in which criminals from the European population were identified as throwbacks to a primitive state. Their criminal behaviour was a natural product of their arrested mental development, leaving them stuck at a level that resembled that of the savage or the naughty child. Nor was the retardation purely mental: Lombroso argued that one of the benefits of his system was that delinquents could be identified by physical features which were reminiscent of the apes.[29]

The last years of the nineteenth century witnessed an outburst of concern that Western society was threatened with degeneration. Some of these worries arose from a fear that modern society has become so complex that it is imposing too much stress on people. Lombroso's theory highlights a very different fear more in line with the linear, developmental model of progress:

---

[26] Gould, *Ontogeny and Phylogeny*, pp. 144–7; see also Morss, *The Biologizing of Childhood*, chap. 4.

[27] On this aspect of Hall's theory see Muschinske, 'The Nonwhite as Child'.

[28] Chamberlain, *The Child*, chaps. 8–10; the quotation is from p. 464.

[29] On Lombroso and recapitulation see Gould, *Ontogeny and Phylogeny*, pp. 120–5. More generally see Chamberlin and Gilman, eds., *Degeneration*, and Pick, *Faces of Degeneration*.

the possibility that even the white races might begin to slide back down the scale of development that they had struggled so hard to mount. In 1895 H. G. Wells's *The Time Machine* exploited the recapitulationist theory's view that primitive humans – often identified with modern 'savages' – were childlike in their thinking. He imagined the human race degenerating in two directions, towards the ape-like Morlocks and the childish Eloi. In his *Civilization and Its Discontents* of 1930 Sigmund Freud added to the fear of decline by suggesting that the pressures of modern life might result in our whole culture becoming neurotic.

Freud's psychoanalytic theory was actually based in part on the recapitulationist view of mental development. He suggested that the higher levels of mental activity added in the last stages of evolution might be unable to control the more primitive instincts still buried in our psyches. His theories had an enormous influence on twentieth-century thought and culture, but were presented in a way that concealed their origins in the developmental approach. As Frank Sulloway, Harriet Ritvo and others have now shown, Freud's disturbing vision of the conscious mind being unable to cope with the darker imaginings of the subconscious was based on the belief that the higher functions of consciousness were added onto deeper and more primitive levels of thought that have been preserved in the subconscious. In effect, the subconscious is the savage and the child buried in all our minds as a relic of our evolutionary ancestry. The leading exponent of the recapitulation theory, G. Stanley Hall, invited Freud to lecture at Clark University, and the subsequently published text was dedicated to Hall. In such publications the references to recapitulation are subdued, but a long-unpublished 'Phylogenetic Fantasy' that Freud wrote in 1915 has survived and makes clear his dependence on the belief that evolution has worked through the adding-on of successive stages of mental development.[30]

What Freud did was to warn that the nineteenth century's complacent assumption that our higher functions can control the buried relics of the past is misguided: neurosis is the sign of a mind in which the ancestral forms have broken through the barriers that the later additions seek to impose. His explanation of mental illness assumed that we all carry the ancestral potential buried within our minds waiting to challenge the dominance of the conscious mind. It was all very well to imagine a hierarchy of progressive stages leading to maturity, but the developmental model contained a hidden threat. Freud's psychoanalytic theory recognized that primitive stages of development could

---

[30] On Freud and recapitulation see Sulloway, *Freud, Biologist of the Mind*; Ritvo, *Darwin's Infleunce on Freud*, chap. 5; Morss, *The Biologizing of Childhood*, pp. 43–8; and Gould, *Ontogeny and Phylogeny*, pp. 155–64. Freud's 'Five Lectures on Psycho-analysis' is dedicated to Hall, but typically says little about the recapitulationist element in his thinking.

still break out to threaten the higher intellects of the white race. The chain of being could be read downwards as well as upwards, and basing one's theory of evolution on the existence of predetermined trends aimed at a desired goal left open the possibility that the high point reached at maturity might not be maintained, let alone built on further. The linear model thus contributed to the growing feeling that the whole idea of progress rested on shaky foundations.

During its heyday, however, the idea that modern humans represent the goal towards which life on earth has been developing helped to shape a much wider vision of historical change. Applying evolutionism to our immediate origins necessarily forced a reconsideration of human history, because it defined the initial state from which the development of civilization began. That the process by which we ascended to that state constituted a similar ascent of a hierarchy towards a predetermined goal seemed only natural. Application of the developmental model to human origins would have direct effects on the emergence of prehistoric archaeology, and this in turn would influence the efforts of philosophers and social thinkers who were trying to make sense of the pattern of human history.

4   Progress to Paradise
Christianity, Idealism and History

The original form of the idea of progress in the biological realm was defined by the ascent of the divinely planned chain of being towards humankind. There was an obvious parallel in the study of human history, covering a variety of interpretations based on the assumption that there has been a progressive trend towards a well-defined future goal – the differences arising from disagreements over the nature of that goal and the mechanism by which it would be achieved. Despite these differences, this approach always entailed the attempt to identify the stages by which society or culture advanced towards its culmination. It also included the identification of our present position in the sequence. The stages of development created an equivalent of the temporalized chain of being, an equivalence noted by several modern historians of the idea of progress. As Frank E. Manuel says of Condorcet's vision, 'The *Esquise* sets forth the historical great chain of being', a parallel also noted in Keith M. Baker's detailed study of the same thinker.[1] Since the change is seen as a progress towards a predetermined goal the analogy with individual development also becomes appropriate: we can talk of the youth and the maturity of a civilization or of the race (leaving the pessimists to worry about old age and death).

Condorcet's approach was typical of the utilitarian way of thought which emerged in what became known as the Age of Enlightenment and which focused on the achievement of happiness in this world. But there were rival schools of thought which sought to preserve something of the traditional Christian view that spiritual perfection should be the goal. Theologians and philosophers sought to define and redefine what it means to achieve that goal. Yet despite their fundamental differences, these approaches all retained a developmental viewpoint; they simply differed over the nature of the goal. The metaphor of ascending a ladder step by step towards the goal lay at the heart of most efforts to understand the development of society and culture through into the nineteenth century.

---

[1] Manuel, *The Prophets of Paris*, p. 62, and Baker, *Condorcet*, pp. 356 and 379.

The search for a progressive trend in human affairs focused on developments during the period of recorded history from the great empires of the ancient world onwards. The Bible showed that humanity became spiritually flawed almost immediately after it was created. But if we had begun in paradise, perhaps we could regain that state by following Christ's message and seeking to improve our moral character. The rival utilitarian approach extended the scale of social evolution by imagining how the first societies emerged from an original 'state of nature'. Although there were idealized visions of savage society, it became increasingly common to see this as a period when life was, in Hobbes's terms, 'nasty, brutish and short', thus making progress from this state an important component of the whole upward trend. By the nineteenth century anthropologists and archaeologists were hoping to reconstruct these early stages of the scale using evidence derived both from the remains of stone tools and from the study of modern savages. Here elements of the scale of biological development outlined in the previous chapters dovetailed with the study of social and cultural progress. The linear hierarchy seen in the chain of being was extended continuously into the origins and progress of societies.

Goals and Trends

On the question of defining the end-point of progress, the most fundamental division was between those who thought mainly in terms of happiness in this world – defined by material comfort and social harmony – and those who prefered a more spiritual vision of what humanity can hope to achieve. Condorcet is usually seen as a pioneer of the former position, founding a tradition that leads through Comte and the utilitarian social philosophers to Herbert Spencer and the anthropologists who saw the ascent from savagery to commercial civilization as a sequence through which all societies could hope to evolve. The opposite pole is represented by those thinkers who derived their inspiration from the legacy of the Christian vision of history in which God has ordained a means by which humanity can rise again to a state of spiritual perfection. When the Fall is taken into account this is a cyclic model of history: we are simply trying to recover that state of grace enjoyed by Adam and Eve in paradise before the fatal apple was eaten. But from the seventeenth century onwards attention increasingly focused on the upward trend towards spiritual perfection. From the early millenarians to liberal Christians in the modern world this vision creates a spiritually based idea of progress. As Carl Becker and other historians remind us, Condorcet and his fellow enthusiasts for material progress were in some respects simply redefining the nature of the Heavenly City and assuming that we may be able to build it ourselves.[2]

---

[2] Becker, *The Heavenly City of the Eighteenth-Century Philosophers*; Baillie's *The Belief in Progress* stresses the role of religious thought more generally.

Some of the materialistic or utilitarian exponents of the idea of progress shared the hope of the perfectibility of humankind, a hope originally defined in terms of spiritual perfection but easily transformed into one based on adjusting human nature to encourage more harmonious social interactions. But as R. V. Sampson notes in his survey of the idea of progress, anyone who believes in the perfectibility of humankind or of human societies must have some idea of what will constitute the perfect state.[3] Even if one does not believe that the perfect condition is actually achievable, the concept will at least define the way forward. It will also tend to place constraints on how one can make sense of what has been achieved so far. This is why the image of an ascending scale of developmental stages leading to a predetermined goal, an inevitable utopia, is so powerful.

Using the term 'utopia' raises the issue of how speculation about some hypothetical perfect state in the future relates to the various ideas of progress. The linear model requires the postulation of a goal that could be regarded as a utopia in the accepted sense of the term. But to imagine a utopia is not necessarily to endorse the idea of progress. The perfect state may be a device to highlight the imperfections of the present, with no sense that it is attainable, let alone the inevitable goal towards which history is progressing. There is thus a significant gulf between certain kinds of utopian literature and theories of progress. Our study does not concern itself with speculations which imagine a perfect future state but make no effort to suggest how it might be achieved or to connect any hypothetical transition with past events. Progress is utopianism in which the ideal state is not only predicted to emerge in the future but also supposes that history is moving inexorably towards that end-point via the present situation.

The assumption that there is a well-defined goal to be achieved leads to the question of whether there can be an 'end of history' after which nothing else of significance can occur. The Christian vision of a future state equivalent to that initially enjoyed by Adam and Eve in paradise implies that no further progress would be necessary or even possible. Theories that assume we are being guided – directly or indirectly – towards some ideal spiritual utopia retain this concept of a goal beyond which there can be no further change, defining an end of meaningful history. Such a closed vision of the future was widely attributed to Hegel, although modern scholarship is inclined to dismiss this interpretation as an over-simplification imposed on his teaching by his followers. The same point is made in the case of Marx's materialist reinterpretation of the Hegelian scheme. The expectation of an end to history is a product of these goal-directed visions, even when the goal is seen as a theoretical one towards which we should aim but which we may never achieve.

---

[3] Sampson, *Progress in the Age of Reason*, chap. 3.

A utopia may allow significant creativity in artistic and other fields that have little capacity to disturb the smooth operations of society. Increasingly, however, the spiritual vision of progress was replaced by the more utilitarian approach favoured by Condorcet. Marxism also shows how materialists could still operate with the expectation that a clearly defined utopia can be attained. They could still imagine that social progress advanced inexorably towards the goal, but their focus on happiness in this world created a presumption that technological innovation played a role in achieving the goal. This focus on technology would undermine the credibility of plans to achieve a utopia based purely on spiritual development, but by raising the prospect of endless innovation it would also make it impossible to believe that history could come to a static end-point.

Biological models of development were often applied to human history. The most general application of the analogy presents the transition from childhood to youth and maturity as a paradigm for the overall history of the human race. If savages are equivalent to children, the ancient civilizations marked the period of youth, and the modern world has at last achieved maturity. Here the imagery is optimistic: the focus is on the rise to maturity as a process that is inevitable and predetermined, and which by definition leads to a state that is more significant that the germ from which it evolved. The only problem with this image is that it ignores the obvious fact that for the individual maturity is inevitably followed by old age, senility and death. Applied on the large scale of world history, this would allow pessimists such as Spengler to argue that whatever progress our civilization has achieved, it is doomed to decline and perhaps to disappear.

Optimistic thinkers could use a more restricted form of the analogy by applying the cyclic model to the civilizations that make up the main stages in the progress of the race. Just as some evolutionary biologists saw the rise and fall of the major groups in the fossil record as evidence of successive pulses of developmental energy, historians frequently recognized that cultures or civilizations could be seen as having a built-in life-cycle in which development to maturity was followed by decline towards degeneration and destruction. The hope of overall progress rested on the possibility that the achievements gained during the mature phase of one culture could be transmitted to the originators of the next phase of progress, to be used as the foundation for even greater triumphs during their own maturity. Progress was both episodic and cumulative.

The belief that development would occur in the form of a series of discrete pulses, each episode of advance being followed by a temporary eclipse, was a manifestation of the wider view that historical change is best seen as discontinuous rather than gradual. A classic expression of this approach is Hegel's dialectic, in which the tensions inevitably generated in one phase form the essential foundation for the synthesis that introduces the next. The 'liberal Anglican' view of history promoted by Thomas Arnold also saw the rise and fall of civilizations as key steps in God's providential scheme for human spiritual advance. The historians who focused on the trend towards spiritual awareness were well aware of the tensions that could be generated by the more negative aspects of human nature and could appreciate how they could interrupt the flow of progress to limit the scope of each step forward. It was the assumption that there was an overarching providential scheme at work that allowed them to hope that the episodes would be cumulative, not an endless cycle.

The Marxists adapted this model to the materialist perspective by focusing on class-based conflict. On the whole, however, the more utilitarian view of progress was less likely to see cyclic patterns as an inbuilt component of the driving force. The cumulative effects of human effort and ingenuity might be subverted by external pressures, as when the barbarian invasions led to the 'Dark Ages' for instance, but the expectation that cultures simply 'ran out of steam' seemed less plausible when progress was seen as the result of each generation's efforts to improve itself. One empire might replace another, perhaps, but only when a new and more innovative culture arose to challenge its predecessor. This resembles the evolutionists' view of how the major biological groups succeeded one another in the fossil record.

The utilitarian vision of progress will be the subject of the next chapter. We begin with the efforts to create a vision of progress based on the search for moral and spiritual perfection. But underlying both of these fairly conventional histories lies a deeper theme defining ideas about the shape of historical change: the legacy of the linear model of development analogous to the chain of being.

Millennium and Utopia

Scholars who study the idea of progress agree that it began to take shape only when European culture gained a new sense of self-confidence in the seventeenth century. To a significant extent this was a consequence of what is conventionally known as the Scientific Revolution and the technological innovations that were, directly or indirectly, linked to the new methods of studying the material world. Here at least there seemed to be no question that modern Europe had surpassed the achievements of the ancient world. But the growing expectation that there would be social improvements derived from our increased control over nature was not initially conceived as the continuation of a historical trend of progress from the savage 'state of nature' in which the first humans had lived. On the contrary, most of the natural philosophers who contributed to these innovations saw future progress as the restoration of the state of dominance over the material world originally granted to Adam – but also the state of spiritual grace which Adam and Eve had lived in before the Fall. Their vision of a future utopia was not one of mere technical mastery and social cohesion, but a spiritually motivated return to paradise.

Predictions of future progress emerged in the context of the seventeenth-century debate over the relative merits of ancient and modern learning.[4] Few were prepared to argue that the ancients could be surpassed in areas such as poetry and the arts, but in philosophy – especially as it related to the study of nature – and the technical arts there seemed clear evidence of modern advances. It was no longer a case of merely trying to equal the ancients, because their achievements had been surpassed by advances including the introduction of gunpowder, the magnetic compass and printing. Given the new scientific methodology it seemed reasonable to assume that we will continue to advance and apply our knowledge of nature for the betterment of the human condition. At first the hope of future progress was seldom seen as the continuation of a long-standing historical trend. The ancients were still respected in some areas, and the collapse of learning in the Dark Ages made it hard to see anything like a cumulative process.

Recognition that there might be a cumulative progressive trend (with temporary reversals) emerged in the eighteenth century, often derived from the utilitarian approach characteristic of the Enlightenment. Freed by their rejection of the Christian message from the myth of paradise, the philosophers of France and Scotland could more easily imagine society evolving by stages from a primitive state of nature. The end-product of their reflections was the linear sequence of technological and social developments codified most effectively by Condorcet. A similar model was also accepted by religious thinkers such as Joseph Priestley, who certainly realized the transforming effects of new technologies but still saw social progress as a spiritual transformation that would re-create paradise on earth. Priestley did see progress as something with roots in history, although his vision was less systematic than that of Condorcet and the more radical optimists.

The quarrel over the relative merits of the ancients and modern was particularly acute in Britain, and here Priestley's views were anticipated by a number of writers who endorsed the impact of the new developments in science and technology within a religious context linked to millenarianism – the traditional Christian prediction of a period of human perfection that would come before the end of the world. This would, in effect, re-create the paradise from which we had been expelled at the Fall, making it a cyclic rather than a linear process. But since the Fall was a dramatic event at the very beginning, the rest of

---

[4] The classic account is Jones, *Ancient and Moderns*.

history could still be seen as having progressive elements in both ancient and modern times. Francis Bacon not only proclaimed the superiority of his empiricist methodology over ancient speculations; he also saw our efforts to control nature as the regaining of the original dominion over the world granted to Adam at the creation. The Fall was not complete, and its effects could be partially reversed, both spiritually and materially: 'For Man, through the Fall, lost both his state of innocence and his lordship over the created world. Both of these losses can, even in this life, be partially repaired; the former by Religion and Faith, the latter by Arts and Sciences.'[5]

Historians such as Peter Harrison and William Eamon have shown how a host of seventeenth-century writers, many from a Puritan background, echoed Bacon's optimistic world-view. In a chapter entitled 'Eden Restored' Harrison notes Henry Reynolds's use of the metaphor of a chain to describe the progress by which we 'by links of that golden chain of Homer, that reaches from the foote of Iupiter's throne to the Earthe, more knowingly and consequently more humbly climb up to him, who ought to be indeed the only end and period of all our knowledge and understanding'.[6] Later in the century natural philosophers including Robert Hooke and Robert Boyle took up the call to promote the new science as a guide to both material and spiritual betterment. Although most of their hopes focused on the future, we can see in the naturalist John Ray's *Wisdom of God Manifested in the Works of the Creation* of 1691 how the movement could be depicted as part of a more general trend: 'the bountiful and gracious Author of Mans Being and Faculties, and all thing else, delights in the beauty of his Creation, and is well pleased with the Industry of Man in adorning the Earth with beautiful Cities and Castles, with pleasant Villages and Country Houses, with regular Gardens and Orchards and Plantations' – all of which Ray contrasts with the primitive conditions surviving in parts of the world such as Scythia and America.[7]

In Ernest Tuveson's account of this movement Thomas Burnet emerges as a key player. Burnet is known to historians of the earth sciences as the author of one of the first attempts to argue that Noah's flood – while clearly providential – was the result of natural causes. His *Sacred Theory of the Earth* of 1691 (the Latin original was published ten years earlier) suggested that the original paradisiacal state of the earth was broken up in the flood, leaving a ruined planet for fallen mankind to inhabit. He also predicted that the millennium would follow the eventual dissolution of the planet by fire. There are hints in

---

[5] Bacon, *The Novum Organon*, book 2, conclusion, p. 306 in the edition cited.

[6] Henry Reynolds, *Mythomyses* (1632), quoted by Harrison, *The Bible, Protestantism and the Rise of Natural Science*, p. 230. See also Eamon, 'From the Secrets of Nature to Public Knowledge'.

[7] Ray, *The Wisdom of God*; I am quoting the second edition of 1692, pp. 156–7. Harrison quotes the slightly different wording of the first edition, pp. 117–18; see *The Bible, Protestantism and the Rise of Natural Science*, p. 243.

the *Sacred Theory* that our understanding of these events has improved over time, and Tuveson sees more substantial indications of a progressionist view in Burnet's next book, his *Archaeologiae philosophicae*. Here he suggested that both the Fall and the subsequent recovery were extended processes, the latter implying a sequence of cultural developments which Tuveson sees as an anticipation of the social evolutionism of the following century. These are but brief hints, however, and the book was so heterodox that Burnet was forced into retirement; he is remembered far more for the image of our present state as fitted to an ugly, ruined world, with the millennium coming only after its future destruction.[8]

In eighteenth-century France speculations moved increasingly towards the more materialistic view of progress, although there was one significant effort to link the idea to a future regeneration of religion. In the early part of the century the abbé de Saint-Pierre became notorious for his schemes to improve French society and ensure perpetual peace among nations (the latter project being taken seriously by Rousseau). In 1737 Saint-Pierre published a short piece entitled 'Observations sur le progrèz continuel de la raison universelle'.[9] He exploited the analogy between social history and the life-cycle of an individual, arguing that we are only just emerging into the maturity of the race. He warned, however, that the analogy was unreliable because it implied that there could not be indefinite progress in the future. In the past progress had all too often been blocked by ignorance and superstition, which had prevented those who could reason from influencing events. We are now emerging from the era of barbarism, and Saint-Pierre called for more efforts to promote the study of politics and morals. In the end, though, his recipe for progress retained the traditional Christian perspective: the best use of reason was to convince everyone to take seriously what their fate would be in the next world rather than this.

In Britain, there were more widely publicized efforts to promote the idea of progress towards a spiritually improved future, although the millenarian concerns of the previous century had begun to fade. A problem of interpretation similar to that for Burnet arises in the case of a figure regularly held up as a

---

[8] Tuveson, *Millennium and Utopia*, chap. 5. On the progresss of knowledge see e.g. Burnet, *The Sacred Theory of the Earth*, book 2, chap. 9, p. 203 in the reprint of the 2nd ed. cited, and *Archaeologiae philosophicae*, preface, p. 3 in the original and p. iii in the 1729 English translation. Tuveson does not indicate the edition from which his other quotations are taken, but it is neither of the two I have seen. There are numerous recent accounts of Burnet's geological speculations: see for instance Davies, *The Earth in Decay*, pp. 68–74.

[9] Saint-Pierre's 'Observations sur le progrèz continuel de la raison universelle' can be found in his *Ouvrages de morale et de politique*, vol. 11, pp. 257–316; on the life-cycle analogy see pp. 274–5. The somewhat bizarre spelling in the title is characteristic of the piece. The accounts of Saint-Pierre's schemes in Druet *L'abbé de Saint-Pierre*, Suriano, *L'abbé de Saint-Pierre ou les infortunes de la raison*, and Bois, *L'abbé de Saint-Pierre* do not mention this work.

pioneer of the idea of progress, David Hartley. His *Observations on Man* of 1749 is seen as the source of the psychological doctrine of the 'association of ideas' subsequently used by the utilitarians to explain how we learn about the world. This is coupled with the claim that the application of the principle in education can lead to a cumulative improvement in the human race's intellectual and moral nature. Thus in his chapter on the 'Rule of Life' Hartley suggests that if properly managed our behaviour 'corrects and improves itself perpetually'.[10] David Spadaforda, who follows the conventional view of Hartley as a progressionist, nevertheless concedes that there is a difficulty with this interpretation, as pointed out by Margaret Leslie.[11] Her reading of the *Observations on Man* sees Hartley arguing for rational education as a means of improving the individual personality only, the extension to a wider progress for the race being a misinterpretation promoted by Joseph Priestley's later exposition of his thinking. Spadaforda accepts that much of the language in the original is open to alternative readings but still thinks there is enough emphasis on continual improvement to suggest the idea of progress. All accept that there is an element of millenarianism in Hartley's vision of the future, but the question is whether the future paradise will come as a result of our efforts over the generations, or only after a providential reconfiguration of the earth itself.

A short article by Ronald B. Hatch clarifies the problem created by Priestley's *Hartley's Theory of the Human Mind* of 1775. Although this reprints much of Hartley's original text, there are amendments, substantial cuts and one significant (and unacknowledged) addition, all of which enhance Priestley's own hopes for progress. In the process Hartley's explicit suggestions that education cannot produce a permanent improvement of the race and that the millennium will come only after the destruction of the earth by fire are ignored. The addition is the short fifth section of the fifth chapter, 'On the Practical Applications of the Doctrine of Necessity', in which Priestley argues that the cumulative effects of education 'must greatly accelerate our progress to humility and self-annihilation. And when men are far advanced in this state, they may enjoy quiet and comfort notwithstanding their past sins and frailties; for they approach the paradisiacal state, in which our first parents, though naked, were not ashamed.'[12] There is nothing to show that these are not Hartley's own words, and in the absence of the material on a post-apocalyptic

---

[10] Hartley, *Observations on Man*, vol. 2, chap. 3, proposition 49, p. 207. I have used the second edition of 1791, which appears to be a reprinting of the first.

[11] Leslie, 'Mysticism Misunderstood'; see Spadaforda, *The Idea of Progress in Eighteenth-Century Britain*, pp. 149–63. Leslie traces the assumption that Harley believed in progress back to Halévie's *The Growth of Philosophic Radicalism*, but it is already there in Stephen's *History of English Thought in the Eighteenth Century*: see vol. 2, pp. 55–9.

[12] Priestley, *Hartley's Theory of the Human Mind*, pp. 366–7; see Hatch, 'Joseph Priestley'.

millennium the reader is led to assume that he too believed in gradual progress towards a re-created paradise. In fact, he had lamented the decline in moral standards and insisted that there could be no complete happiness before the destruction of the world.[13]

If Priestley's enthusiasm led him to exaggerate Hartley's commitment to progress in this world, that enthusiasm was obvious throughout his own work. He occupied an uneasy position straddling the utilitarians' hope that technological progress would improve the human condition and the Christian vision of an ascent towards spiritual perfection. His work in chemistry and his links to the Lunar Society's industrial entrepreneurs confirmed the former allegiance, his position as a radical Nonconformist minister the latter.[14] He thought that Christianity had been corrupted by the early Church and rejected supernaturalism apart from the belief in miracles. Like the Baconians of the previous century, he assumed that material and spiritual progress would go hand in hand, leading towards paradise, as displayed in the quotation above. History reveals the hand of providence, but shows that God preferred to act through human actions. As Priestley wrote in his *Essay on the First Principles of Government* of 1768, the intention of divine providence is 'that mankind should be, as far as possible, self-taught; that we should attain to everything excellent and useful, as the result of our own experience and observation'.[15]

Twenty years later Priestley used his *Lectures on History and General Policy* to bolster his hope of future progress by invoking a vision of past advances that, in effect, guaranteed continuation of the trend: 'There can hardly be a more entertaining object to a speculative mind than to mark the *progress of refinement* in the ideas of a people emerging from a state of barbarism, and advancing by degrees to a regular form of government.'[16] Much of his survey was focused on the period from the Middle Ages, charting the emergence of more liberal social arrangements and the rise of manufacturing. But as the last quotation above indicates, he was also concerned to imply a wider dimension by which societies advanced from savagery or barbarism to higher levels of organization. He was not always consistent on the cause of progress, at one point insisting that humans are naturally sociable while later stressing the need for our naturally selfish and sensuous natures to be

---

[13] Hartley, *Observations on Man*, vol. 2, chap. 4, pp. 366–81, and the conclusion, pp. 438–55.

[14] On the links to industrialization see Uglow, *The Lunar Men*. Stephen's *History of English Thought in the Eighteenth Century* discusses some of Priestley's theological controversies; see vol. 1, pp. 364–9.

[15] Priestley, *An Essay on the First Principles of Government*, pp. 142–3. Schofield's biographical studies of Priestley do not address the issues discussed here in any detail, but on history see his *The Enlightenment of Joseph Priestley*, p. 187 and pp. 209–11, and *The Enlightened Joseph Priestley*, pp. 253–9.

[16] Priestley, *Lectures on History and General Policy*, p. 325 (italics in original).

overcome in the rise to 'politeness'.[17] But however it was worked out, the trend was obvious – and obviously providential:

Let the person, then, who would trace the conduct of Divine Providence, attend to every advantage which the present age enjoys above ancient times, and see whether he cannot perceive marks of the things being in progress towards a state of greater perfection. . . . That the state of the world at present, and particularly the state of Europe, is vastly preferable to what it was in any former period, is evident from the very first view of things. A thousand circumstances shew how inferior the ancients were to the moderns in religious knowledge, in science in general, in government, in laws . . . in arts, in commerce, in the conveniences of life, in manners, and in consequence of these, in happiness.[18]

There was, however a sting in the tail: Priestley was also convinced that history would unfold in a manner that would fulfil the biblical prophecies, so he believed that the achievement of an earthly paradise would be the prelude to the end of all things, Armageddon. Most of his readers, however, may have missed this aspect of his vision.[19]

Priestley shows how the idea of progress as a historical trend towards a perfect state emerged independently in England. In contrast to with his Scottish and French contemporaries, his vision was still motivated by the assumption that all was guided indirectly by providence – although his emphasis on happiness as well as spiritual perfection indicates his utilitarian leanings. The historical dimension is certainly present, but the stages of development from the state of nature to modern society are not as clearly defined as in the writings of the Scottish and French scholars. They, however, had abandoned the religious dimension still active in England, and their efforts are dealt with in the next chapter. Convinced that liberty was the source of happiness, Priestley welcomed the revolution in France. He was soon identified as a dangerous radical; his house was burned by the Birmingham mob, and he subsequently left for America. The idea of progress would take on a new lease of life in the following century.

Idealism and History

It was in Germany that the most innovative response to the crisis within the Enlightenment project emerged. Immanuel Kant's intellectual revolution challenged the psychological assumptions on which Hartley, Priestley and their French contemporaries had built their hopes of reforming human nature. The

---

[17] Ibid., pp. 272 and 425.   [18] Ibid., p. 531.
[19] The apocalyptic aspects of Priestley's vision are not clearly expressed in the texts studied here. I am indebted to John Christie of the University of Oxford for alerting me to this additional dimension of his religious beliefs.

image of the mind as a passive recipient of sense-impressions implicit in Hartley's 'association of ideas' was rejected. For Kant the mind itself played a role in creating our perception of the world. From this insight emerged an idealist philosophy which, when linked to Romanticism, would have major implications for historical studies and the doctrine of progress. As the Napoleonic invasions traumatized and ultimately reformed the German states, a new way of imagining the forces at work in history took shape.

This intellectual revolution affected all areas of German thought, but in the area of history it is associated most closely with the work of J. G. Fichte and G. W. F. Hegel. I approach this area with some trepidation, having always found the esoteric language used by these thinkers rather impenetrable. I hope that what follows will not do too much injustice to them and the scholars who work in the field. The most recent scholarship challenges the traditional interpretation of the idealists' views on history and progress. The theme of the 'end of history' is particularly relevant here: many of Hegel's followers took it for granted that his vision of history assumed that the ascent towards a world of perfect freedom would have a definite end-point in the emergence of a state in which everyone could express themselves fully without oppressing others. It was sometimes claimed that Hegel thought this state had already been achieved in the reformed Prussia of his own day. The goal had been achieved, and no meaningful change could occur in the future.

Recent scholarship by Eric Michael Dale and others suggests that this represents a misreading of Hegel's intentions. The progress to be seen in history was not something driven by an external spiritual power towards a predetermined goal: it was dependent solely on human activity. For that reason the goal would probably never be achieved, so the 'end of history' was an idealized state defining the direction of travel. It was not something that could ever be fully achieved, and certainly not something already in existence. If anyone actually promoted the idea of an end to history it was Fichte, whose thought is widely assumed to have darker implications arising from his appeal to German nationalism.

Kant himself pioneered not only the new interpretation of the mind's activity but also a new approach to history. Already in 1755 his *Universal Natural History and Theory of the Heavens* had proposed what later became known as the 'nebular hypothesis' to explain the origin of the solar system. He suggested that a rotating cloud of dust and gas would gradually condense under the force of its own gravity, the majority of the material eventually forming the sun while smaller accretions would end up in orbit as the planets, including the earth. Such a process would create order out of chaos solely through the operations of natural laws. It became the inspiration for many later ideas of universal evolution, including that proposed in Robert Chambers's *Vestiges of the Natural History Creation*.

In 1784 Kant turned to social evolution with a short essay proposing his *Idea for a Universal History with a Cosmopolitan Aim*. He introduced the approach that would inspire the idealist historians: the goal of history was a state in which rationally applied human freedom would be extended to the whole world. History is goal-directed, its course being shaped by divine providence: 'One can regard the history of the human species in the large as the completion of a hidden plan of nature to bring about an inwardly and *to this end*, also an externally perfect state constitution, as the only condition in which it can fully develop all its predispositions in humanity.'[20] There is thus a 'guiding thread' in history which Kant explicitly compares to the goal-directed development of the embryo. The species needs to develop over an extended period because an individual cannot achieve everything in a single lifetime. He uses the term 'germ' – originally used to denote the fertilized ovum – to make this point: society develops 'toward the condition to which all germs nature has placed in it can be fully developed and its vocation here on earth can be fulfilled'.[21] Elsewhere he refers to his theory as one of 'epigenesis', again using an embryological term denoting the sequential development of a germ towards maturity (as opposed to Bonnet's theory of pre-existing miniatures). The process is not direct, of course, because humans are wilful and a certain amount of struggle and suffering is needed to drive them onward towards the goal, an insight that has been compared with Hegel's 'cunning of reason'.[22]

Johann Gottlieb Fichte took Kant's new insight into how the mind works and brought out its most radical implications. He made the conscious self the foundation of his philosophy, leading to the charge that he celebrated the power of the individual will, in effect anticipating the position later developed by Nietzsche. Fichte was also accused of atheism, a charge that seemed plausible for his early writings. Later works talked of a supreme intelligence, allowing supporters to argue that he was merely reformulating the traditional idea of God.[23] For all his focus on the self, Fichte recognized that to achieve anything of significance the individual had to acknowledge the existence of other minds and their right to self-determination. Freedom became central, but only freedom within a cultural and social framework that made it available to all. The claim that it was German culture alone that had to ability to impose this vision on the world has led to accusations that Fichte's thought fuelled the growth of German nationalism. His claim that the highest form of society

---

[20] Kant's text is translated in Rorty and Schmidt, eds., *Kant's Idea for a Universal History with a Cosmopolitan Aim*, pp. 9–23, quotation from proposition 8, p. 19.

[21] Ibid., proposition 9, p. 22.

[22] See Ameriks, 'The Purposive Development of Human Capacities', and on the cunning of reason Allinson, 'Teleology and History in Kant'.

[23] These various charged are rebutted in the prefaces to the earliest English translations, *The Popular Works of Johann Gottleib Fichte* and Fichte, *The Destination of Man*.

allowed freedom only in the sense that all accepted the framework of the state has similarly led to the charge that he anticipated fascism. A recent study by David James argues that these interpretations of Fichte's political doctrines are misreadings based on a failure to appreciate the wider context of his idealist philosophy.[24]

This reinterpretation builds on an analysis of Fichte's main contribution to the philosophy of history, his popular lectures *The Characteristics of the Present Age* delivered in 1804–5. These propose a predetermined sequence of cultural developments which has obvious relevance for the idea of progress, although the link is complex and the work certainly does not imply that the modern age is to be seen as the summit of perfection. To understand the present state we need to define its fundamental idea, which in turn requires that we identify its position in the sequence of ages that represent universal history. This presupposes the existence of a World-Plan from which we can deduce the nature of all epochs: to this extent history is to be understood not from known facts about the past but through a framework imposed on it *a priori*. The facts merely reveal that there are disturbing elements preventing the smooth working-out of the plan. We are looking for the key steps in the progressive life of the race, not the individual, and 'the End of the Life of Mankind on Earth is this – that in this Life they may order all their relations with FREEDOM according to REASON'.[25]

Fichte's World-Plan evolves in five stages, each age centred on an idea that may have originated with an individual but comes to be accepted by all. The first epoch is a state of innocence dominated by Reason, but operating though instinct rather than conscious choice. Somehow this breaks down (no details of this equivalent to the Fall are given), and in the second epoch instinct has to be replaced by a ruling authority, leading to a state of 'progressive sin'. This gives way to the third epoch, in which authority breaks down, leading to 'unrestrained licentiousness' – a state of 'completed sinfulness'. Things improve in the next stage, the epoch of Reason and Science, when 'Truth is looked upon as the *highest*, and loved before all other things'. Finally we achieve that epoch of Reason as Art, based on a 'fitting image of Reason – the state of complete justification and sanctification'. In one sense the sequence is a cycle: we return to a remade state of Reason but have given it new meaning because now the human race has self-consciously recognized its benefits rather than simply acting by instinct. The religious connotations of this are quite explicit: progress is a return to the original state in which we were created. The human race

---

[24] James, *Fichte's Republic*.
[25] Fichte, 'The Characteristics of the Present Age', in *The Popular Works of Johann Gottlieb Fichte*, pp. 1–271, quotation from lecture 1, p. 5.

'builds a paradise for itself, after the image of that which it has lost; – the tree of life arises; it stretches forth its hand to the fruit, and eats, and lives in immortality'.[26]

Although Fichte recognizes that the transitions can be affected by external factors, he insists that the basic sequence can be worked out *a priori* and is thus predetermined: 'all the Phenomena of Time, without exception, are regarded as a necessary and progressive development of the One, Ever-Blessed, Original Divine Life'. It is this aspect of his philosophy of history that leads Dale to argue that it is Fichte who actually proposes the idea of a definite 'end of history' usually attributed to Hegel.[27] Because we can identify the place of our own age in the sequence, we can predict the future:

we have delineated the Present Age as a necessary part of the great World-Plan on which the Earthly Life of our Race is arranged, and have endeavoured to disclose its secret significance; we have sought to understand the phenomena of the Present by means of this Idea, to bring them forth as the necessary results of the Past, and to predict their immediate consequences in the Future ...[28]

It is clear that our own age is *not* the final paradise; we are in the third epoch of 'unrestrained licentiousness', and Fichte goes into some detail on the evils we experience. Christianity will inspire the next step forward, and we are already beginning to see the benefits of its influence. The world is organized into states that provide personal freedom and recognize each other's right to exist. No Christian can be made a slave, although non-Christian heathens can justifiably be enslaved as a means of spreading the message of Germanic civilization to the world.[29] Here the darker side of Fichte's vision of progress becomes apparent.

For many later commentators Georg Friedrich Wilhelm Hegel became the archetypical proponent of the claim that history can be represented as the ascent of a predetermined hierarchy of development towards a definite goal, often depicted as the 'end of history', beyond which no significant change can take place. This interpretation re-emerged in Fukuyama's late twentieth-century thesis, which may have played a role in prompting the reassessment of Hegel's position by a generation of scholars increasingly convinced that his earlier reputation was based on a misunderstanding of his (admittedly complex) writings. The misrepresentation seems to have begun with Nietzsche, who despised what he thought was Hegel's philosophy, and Engels, who went on to borrow the key element of the dialectic for his materialist version of progressionism. In the early twentieth century the reading by Alexandre

---

[26] Ibid., pp. 9–10.
[27] Ibid., p. 258; see Dale, *Hegel, the End of History, and the Future*, chap. 6.
[28] Fichte, 'The Characteristics of the Present Age', p. 254.   [29] Ibid., pp. 205–6.

Kojève followed this pattern and paved the way for Fukuyama's revival of the 'end of history' thesis.

The rethinking of Hegel's position began in the 1970s with the work of George O'Brien and Burleigh Taylor Wilkins and has been expanded in recent decades by Joseph McCarney, Peter C. Hodgson and Eric Michael Dale.[30] Their arguments rest in part on the claim that the image of Hegel as a simple progressionist reflects a failure to understand his position in the context of his wider idealist philosophy and his residual (if unorthodox) efforts to retain a theological perspective on history. Hegel's writings can be ambiguous, and his use of metaphors derived from biological development sometimes invites misunderstanding. Critics such as Nietzsche may even have been right to argue that for his position to be fully consistent it ought to have included the claim that there was an end-point when the purpose of history would be accomplished. The situation is rendered more complex by the fact that the texts of his most relevant pronouncements on history are derived from sources not published in his own lifetime.[31]

Hegel certainly wanted to see a purpose unfolding in history, a purpose defined by his idealist philosophy as a gradual if irregular ascent towards self-conscious spiritual freedom: 'Universal history ... is the exhibition of spirit in the process of working out what it is *potentially*.' This realization is the expression of a divine purpose, the only true theodicy – a justification of God in history, the achieving of His work.[32] For Hegel there was no original paradise; pre-civilized humanity had lived in a state of stupor. The stages of development from this beginning are clearly spelled out, including a distinct geographical dimension as cultures move from East to West. The state codifies the spirit of each age, and is thus far more than a political entity. Hegel draws on Johann Gottfried Herder's view that the development of each culture must be judged on its own terms: we cannot dismiss another because it does not conform to the values of our own. Herder himself hoped for eventual progress but made no coherent effort to present history as a developmental sequence. His thought had more influence on the emergence of a model in which there is no predetermined pattern of progress, discussed in Chapter 9.

Hegel saw each age as having its own unique culture, often symbolized in the life of a 'world-historic individual' – Napoleon was his contemporary

---

[30] O'Brien, *Hegel on Reason*; Wilkins, *Hegel's Philosophy of History*; McCarney, *Hegel on History*; Hodgson, *Shapes of Freedom*; Dale, *Hegel, the End of History, and the Future*.
[31] I have used only English translations, which are based on originals recorded from lectures that Hegel gave in 1822–5 and 1830–1. The first translation of his lectures, Sibrae's *Lectures on the Philosophy of History*, was published in the nineteenth century; more recent translations are *Lectures on the Philosophy of World History* by Nisbet and *Lectures on the Philosophy of World History*, ed, Brown and Hodgson.
[32] *Lectures on the Philosophy of History*, trans. Sibrae, p. 18 and p. 77.

example – whose personal faults do not prevent them from achieving their destiny. Here we see an example of what Hegel called the 'cunning of reason': events of spiritual or moral significance emerge from the apparently contingent events of human life. Each major state or culture represents an expansion in the level of freedom accorded to the individual; for Hegel the gradual elimination of the master–slave relationship is the key advance. The ancient empires of the Orient knew that only one person was free; the Greeks and Romans knew that some are free; now the Germanic nations under the guidance of Christianity were beginning to recognize the possibility that all are free. Further developments would by implication include the extension of the process to the whole world.

All this sounds very much like the scheme that Fichte worked out *a priori* and sought to impose on the facts of history. Modern readings of Hegel suggest however that his sequence of developments was not derived from first principles but was meant to emerge from the facts themselves. There is indeed a sequence of steps by which the Absolute seeks to create a spiritually developed humanity, but the ascent has not been inevitable or automatic, and any goal that can be dimly perceived may never actually be achieved. The progress that can be seen represents the struggles of real people trying to visualize and actualize their aspirations against the countervailing forces of human greed and other failings. As Burleigh Taylor Wilkins puts it, if there is a law of progress, it ceases to operate if humanity fails to recognize and execute it.[33] Tensions have always arisen and have repeatedly brought down empires – their initial cultural impetus generates counteracting forces that eventually undermine them, hence the dialectical nature of history. Progress occurs only if the moral insights are passed on to serve as the foundation on which another people will build the next onward step.

Most important of all (given some of the later uses of Hegel's thought), there can be no neatly defined goal representing the end of history. Identifying the potential culmination of the process by which freedom emerges defines a direction of travel that we are now beginning to perceive, but any future steps are not guaranteed, and we cannot predict what will actually happen next. Nor is it necessary to assume that nothing of significance would ever happen subsequent to the complete achievement of freedom, were that ever possible. The world might be composed of states all embodying the highest principle, but there still might be meaningful interactions between them.[34]

The complexity of Hegel's thought on these issues can be illustrated through the metaphors he uses in an effort to explain himself. In his early lectures he does speak of a 'ladder of stages the spirit climbs', an image that suggests a

[33] Wilkins, *Hegel's Philosophy of History*, pp. 136–7.
[34] This last point is made by Ahlers, 'The Dialectic in Hegel's Philosophy of History'.

rigidly predetermined ascent that would fit the older interpretation of his position but sits uneasily with the way he is now understood.[35] The suggestion that successive civilizations have moved from East to West carries the impression of movement from dawn to dusk – a viewpoint widely accepted at the time even though the implication might be that sunset marks the end of the cycle in extinction.

Hegel was aware of a similar difficulty with the use of biological metaphors linked to embryological recapitulation and the life-cycle of the individual. He describes the earlier phases in the development of culture as the childhood, youth and manhood of the race, but immediately warns that if this metaphor were continued the present would have to be regarded as the old age of the spirit, so the analogy is no longer acceptable.[36] This cyclic model can be applied to individual civilizations, since these inevitably decay as the contradictions emerging within them undermine their original impetus. But their spiritual gains can be passed on to help build the next step forward, so in this sense the cyclic model breaks down; Hegel notes the contrast with the ancient story of the Phoenix, which regenerates from its ashes, although not in a higher form.[37]

The metaphor comparing human history to the development of the germ or fertilized ovum also has dangers, although of a different kind. Hegel uses the comparison because he does want to suggest that history will in the long run tend to advance in a particular direction:

But the principle of development has further implications, for it contains an inner determination, a potentially present condition which has still to be realized. This formal determination is an essential one; the spirit, whose theatre, province and sphere of realization is the history of the world, is not something which drifts aimlessly amidst the superficial play of contingent happenings, but is in itself the absolute determining factor; in its own peculiar destiny, it is completely proof against contingencies, which it utilizes and controls for its own purposes.

The development is not automatic or rigidly predetermined, hence:

that development which, in the natural world, is a peaceful process of growth – for it retains its identity and remains self-conscious in its own expression – is in the spiritual world at once a hard and unending conflict with itself. The will of the spirit to fulfil its own concept; but at the same time, it obscures its own vision of the concept, and is proud and full of satisfaction in this state of self-alienation.[38]

---

[35] *Lectures on the Philosophy of World History*, ed. Brown and Hodgson, p. 156. See Hodgson, *Shapes of Freedom*, p. 39.
[36] *Lectures on the Philosophy of World History*, trans. Nisbet, pp. 130–1.
[37] *Lectures on the Philosophy of History*, trans. Sibrae, p. 76.
[38] *Lectures on the Philosophy of World History*, trans. Nisbet, p. 128; see also *Lectures on the Philosophy of History*, trans. Sibrae, p. 58.

Spiritual development may have a preferred direction of travel, but the process is far more complex and less predictable than the development of the embryo towards maturity.

It is perhaps easy to see why Hegel's writings were perceived as an expression of a philosophy of history in which progress towards some final goal is inevitable. In a sense, progress is inevitable, but in another it is not. Once the image of him as the exponent of predetermined unilinear progress was created by critics such as Nietzsche, it could easily become embedded in the literature in a way that led generations of historians to assume that he had sought to impose his own idea of what should happen on the recalcitrant facts shown in the records of the past. Modern scholarship has exposed the unfairness of this interpretation, while allowing us to appreciate how his use of metaphors led to confusion and over-simplification. We can now see that the image of Hegel as the archetypical progressionist convinced that he can predict the end of history is a legend or perhaps a myth. Yet from the viewpoint of the history of ideas we need to take the misinterpretation seriously. If this is what Hegel meant to later generations, the over-simplified vision does play a significant role in shaping how the idea of progress developed.

The Pulse of Providence

Despite his insistence that Christianity has played a key role in the most recent development of the human spirit, Hegel's idealism led to him being criticized as a pantheist. His views remained influential among a wide range of thinkers, but historians were increasingly wary of what they saw as his attempt to impose a preconceived scheme on the facts they studied. The more empiricist school associated with the name of Leopold von Ranke was suspicious of grand theories in general and of the idea of inevitable progress in particular (although Ranke himself was by no means devoid of interest in the wider lessons to be learned from history).[39] In the middle decades of the nineteenth century the most active efforts to defend a theory of spiritual progress came from those historians who worked within the traditional Christian world-view. For them, divine providence could still be seen as an active agent directing humanity's spiritual advance. Their approach was much closer to that of the idealists than to the utilitarian version of progress, and the metaphors they used were often the same. Like Hegel, they were inclined to see the rise and fall of nations as analogous to the life-cycle of a biological organism, and to see the succession of cultural developments as a series of waves or pulses advancing

---

[39] See for instance Georg G. Iggers's introduction to Ranke, *The Theory and Practice of History*, and Ranke's lecture 'On Progress in History', which forms chap. 5 of this selection. The same point is made by Gilbert, *History: Politics or Culture?*

steadily in the same direction, each new development building on what the previous one had achieved before its inevitable decline.

In Germany this vision was articulated in lectures given by Friedrich von Schlegel in 1828. Schlegel had been born a Protestant but had converted to Catholicism in 1805, an event which his biographer thought may have ensured that he did not become ensnared in Hegelianism.[40] He rejected the idea of the perfectibility of humankind while insisting on the 'high dignity and divine destination of man'. This would be achieved thanks to an original divine communication which put the earliest civilizations on an upward course and was still evident in the foundation of Christianity, which revived the original message and was 'the sole principle of the subsequent progress of mankind'.[41] The early civilizations had great energy, but this was not well employed, so history could be seen as a series of creative outbursts. As in Hegel, the comparison with the life-cycle of the individual was used to illustrate the rise and fall of each culture: 'the historical observer can accurately distinguish the different periods of national development – the first period an artless, yet marvelous childhood – the next of the first bloom and flush of youth – later, the mature vigour and activity of manhood – and at last the symptoms of approaching age, a state of general decay, and second childishness'.[42] In the modern period the Reformation represented a failed attempt to reform the Church, but there would eventually be a true revival that would allow humanity to achieve its divinely preordained destiny.

The providential vision of progress was retained by Baron Christian Bunsen, who did much to transmit the new German thought to British readers. In his *Christianity and Mankind* he noted 'The belief of mankind in a moral order of the world, and in the progress of the human race' and argued that although nations disappear in the course of history a new one always appears and 'carries on the torch of divine light'.[43] In Britain itself, the most active proponents of this approach were the members of what has been called the liberal Anglican school of history, the most prominent of whom was Thomas Arnold. These historians were products of the 'Germano-Coleridgian' philosophy, which sought an alternative to the prevalent utilitarianism of the time and linked idealism to the revival of religion. As headmaster of Rugby School, Arnold wanted to direct the moral direction of the nation, and for him and his supporters progress meant not industrialization but spiritual development. For historical inspiration Arnold turned to Barthold Niebuhr's history of Rome, which had identified stages of development analogous to the individual life-

---

[40] See James Burton Robertson's 'Memoir of the Literary Life of Frederick von Schlegel' introducing his translation of Schlegel's *The Philosophy of History*, p. 14.
[41] Ibid., pp. 210–11 and p. 276.   [42] Ibid., p. 310.
[43] Bunsen, *Christianity and Mankind*, vol. 3, pp. 3–4 and 35.

cycle. Arnold himself wrote a history of Rome but was equally interested in the Greeks, and his 1835 edition of Thucydides contained an appendix 'The Social Progress of States', which argued that all cultures pass through a cycle from childhood to manhood and old age.[44]

This cyclic view of cultural history was not enough, however, to satisfy the Christian view that there must be some cumulative spiritual progress from one civilization to the next. In an 1840 essay Arthur Penrhyn Stanley asked whether states tended to mature and decay like individuals but argued that it was a mistake to imagine them running out of energy: their eventual decay was always the result of human wickedness. Yet the hand of providence could be seen at work in the fact that Christianity had allowed a new element of vitality to enter into the more recent phases of European history. Like waves in an incoming tide, each culture moved a little further in the direction of spiritual progress.[45] A similar image was explored in Arnold's *Introductory Lectures on Modern History* of 1842. Here he presented Greece as the source of intellect, Rome as the source of law and government, and Christian Europe as the modern champion of spiritual progress. He was quite explicit in declaring 'that modern history appears to be not only *a* step in advance of ancient history, but *the* last step; it appears to bear the marks of the fullness of time, as if there would be no future history beyond it'. When looking around the world it was hard to see another race that had the potential for further development, which suggested that we are living in the last period of the world's history: 'So if our existing nations are the last reserve of the world, its fate may be said to rest in their hands – God's work on earth will be left undone if they do not do it.'[46] This was more than a vision of the end of history: it was a call to the Christian nations to ensure that they would bring the intended end about.

The same pattern of cumulative moral and spiritual progress was popularized by Charles Kingsley in novels such as his *Hypatia* of 1853 and most explicitly in his Cambridge lectures published as *The Roman and the Teuton* in 1864. Here the role of providence was displayed in the complex web of events that allowed the Teutonic barbarians to destroy the Roman Empire just at the point where it could pass on the legacy of Christianity to the invaders and provide them with a spiritual energy they could not have developed by themselves. The Teutons were essentially childish: 'For good or evil they were great boys; very noble boys; very often very naughty boys – as boys with the

---

[44] Arnold in his edition of Thucydides, *History of the Peloponnesian War*, vol. 1, pp. 621–2. See Forbes, *The Liberal Anglican Idea of History* and my own *The Invention of Progress*, pp. 52–9.

[45] Stanley, *Whether States, Like Individuals, after a Certain Period of Maturity, Inevitably Tend to Decay*, pp. 46–7.

[46] Arnold, *Introductory Lectures on Modern History*, pp. 36 and 39.

strength of men might be'. This was a valid metaphor because 'races, like individuals . . . may have their childhood, their youth, their manhood, their old age, and natural death'.[47] In the modern world Christianity had emerged as the only hope of civilizing the world, providing a justification for the expansion of European, and especially British power around the globe, an imperialist ideology that Kingsley celebrated in novels such as *Westward Ho!*

Similar views were expressed in Frederic Farrar's survey of the historical development of languages, *Families of Speech*, based on lectures delivered in 1869. Along with the new techniques of archaeology, philology offered insights into the development of cultures in prehistoric times. There was much debate over the links between language and race, but many early scholars assumed that language identified a racial group. We shall see how philology provided a classic model of divergent evolution, but Farrar's conclusion mirrored Kingsley's vision in which a succession of races added their contribution to the sequence of progressive impulses by which God's purpose had been achieved. The Egyptians created order out of chaos, the Semites had added religious inspiration, and the Aryans contributed the more active virtues. 'It is to the result of their *combined* work – to the science and strength of the Aryan inspired and ennobled by the religious thoughts that were revealed to the Semite – that the immediate future of the world belongs.'[48] Farrar did envision the possibility of another race carrying the work forward in the future, but insisted that modern Europeans had earned their place in history by promoting the spiritual development of humanity.

The assumption that the expansion of European culture would benefit the whole world underpinned the ideology of the age of imperialism, but was increasingly unlikely to be expressed in terms of divine providence. Academic historians were in any case becoming wary of grand theories such as the idea of progress, although Lord Acton, who became Regius Professor of History at Cambridge in 1895, did continue the tradition. He was wary of Hegelianism, with its tendency to see history as a 'manifestation of a single force, whose works are all wise, and whose latest work is the best'.[49] In his view it was those writers approaching the topic from a Christian perspective who now upheld the hope of spiritual progress. In his inaugural lecture he hoped that the study of history

---

[47] Kingsley, *The Roman and the Teuton*, pp. 5–6.

[48] Farrar, *Families of Speech*, p. 186; a branching tree of language evolution is shown facing p. 106. On the links between philology and other areas including natural history see Alter, *Darwinism and the Linguistic Image*. We shall return to this topic in Chapter 9.

[49] Acton, 'German Schools of History', in his *Selected Writings*, vol. 2, pp. 325–64, quotation from p. 338.

will aid you to see that the action of Christ who is risen on mankind whom he redeemed fails not, but increases; that the wisdom of divine rule appears not in the perfection but in the improvement of the world and that achieving liberty is the one ethical result that rests on the converging and combined conditions of advancing civilization. Then you will understand what a famous philosopher said, that history is the true demonstration of religion.[50]

Acton agreed at one level with Hegel's view that freedom was the goal of progress, but for him it would be achieved through acceptance of the guiding hand of providence.

Christianity and Cultural Evolution

Even before the catastrophe of the Great War it had become increasingly hard to see history as a record of cumulative spiritual progress. Religious thinkers had always been wary of treating industrialization as a measure of social development, and it was now even more apparent to them that technology was a threat rather than a benefit to the race. Yet liberal Christians did still want to find evidence of progress, and they now turned to a wider canvas to paint their pictures. In the age of Darwinism, it was widely assumed that evolution was necessarily progressive, and as we have seen in the previous chapters there were many writers who rejected Darwinism in favour of mechanisms that seemed more likely to generate advances towards higher levels of mental and moral capacities. These advances might well come in discontinuous surges, just like those seen in the earlier phases of human history. Evolutionism could thus extend the range of discontinuous history, making up for the uncertainties that turned most historians away from the search for general laws.

As the previous chapters have shown, much of this literature focused on the history of life on earth, a topic that could now be extended by the revelations in human prehistory being made by archaeology. Some religious evolutionists did combine the two areas, an important contribution being made by the American thinker John Fiske, whose work is significant because it was presented as an extension of Herbert Spencer's philosophy, all too often dismissed as the source of 'social Darwinism'. Fiske's *Outlines of Cosmic Philosophy* of 1874 made additions to Spencer's position but insisted that they were conceived firmly within Spencerian principles.[51] He saw humanity as the supreme

---

[50] Acton, 'Inaugural Lecture on the Study of History', in his *Lectures on Modern History*, pp. 1–28, quotation from pp. 12–13; also in his *Essays on Freedom and Power*, pp. 25–52, quotation from p. 36. For details of Acton's work see Gertrude Himmelfarb's introduction to the latter collection.

[51] Fiske, *Outlines of Cosmic Philosophy*, vol. 1, preface.

product of biological evolution, and liberal Christianity as the culmination of the moral and social progress that constitute civilization. His chapters on human prehistory drew on the latest archaeological evidence to depict moral and cultural progress as a constant change for the better. The inhabitants of Europe had advanced faster and further than those of other regions, leaving the non-white races with smaller mental capacities as well as lower levels of culture.[52]

In principle Fiske endorsed the Spencerian view that progress led to a 'definite, coherent heterogeneity of structure and function through successive differentiations and adaptations'.[53] This meant that evolution had to be seen as divergent rather than linear – the Darwinian 'tree of life'. Yet the outcome was not just adaptation to the environment. In the end, evolution passed successively through to higher levels of organization, biological, mental and social, culminating in modern Christian civilization. Most evolution was adaptive and divergent, yet there was a clear main line leading to the white race and Christian culture. Understanding how Fiske squared this circle throws valuable light on the transition from a linear view of evolution to something more open-ended.

The basis of Fiske's insight – which throws light on Spencer's own more materialistic position – was that there have been certain key breakthroughs in evolution which he called 'critical points'. Here organization advanced to a new level, as with the appearance of warm-bloodedness in mammals and birds, the expansion of the human brain to give the higher mental functions, and the emergence of complex societies. Fiske even suggested that the expansion of mental powers had now led to minds so coherent they can transcend the death of the physical body.[54] Evolution within each level of organization might be branching and divergent, but usually only one or a few lines would reach the critical point where they could transcend to the next level. The others would remain trapped at the lower level, while the innovators would serve as the focus for the next wave of divergent evolution. On such a model one could reconstruct the key stages of development without implying that the actual forms that made the breakthroughs were predetermined.

Fiske's concept of critical points allowed him to accept the tree-like model of evolution and still claim that only certain lines of development counted in the upward march of progress, in effect defining the 'main line' by hindsight. The same point may explain how both Ernst Haeckel and Spencer could simultaneously argue for branching evolution and for a linear scale of

---

[52] Ibid., vol. 2, esp. chaps. 18–20.
[53] Ibid., vol. 1, p. 337; see also the statement of Spencer's position on pp. 350–1.
[54] This last point is developed most fully in Fiske's *The Destiny of Man Viewed in the Light of His Origin*.

development – but neither articulated it as clearly as Fiske. His belief in immortality and his Christian faith obviously defined for him what the goal of evolution must be – but his position implied that progress could be defined by abstract levels of organization rather than specific biological or cultural forms. Substitute the ideology of free-enterprise individualism for Christianity and we get Spencer's materialistic alternative.

Fiske's cosmic philosophy was but one version of the attempt to reconcile progressive evolution with a liberalized Christianity. We have seen how a variety of religious thinkers such as Henry Drummond reached wide audiences through the late nineteenth century with their message of spiritual evolution. Historians may have largely abandoned the search for general laws, but biological and social evolutionism allowed popular writers to continue arguing for moral progress. They bypassed recent events to focus on the bigger picture represented by the history of life on earth and the record of human prehistory, in effect insulating their position from ongoing contemporary threats. Curiously, most of them preferred to focus on biological evolution rather than human prehistory. Drummond's *Ascent of Man*, for instance, refers briefly to Fiske on human progress, but the book's material on this area is confined mostly to abstract reconstructions of moral developments. The same can be said for much of the Christian literature on this theme published in the early twentieth century.[55]

This failure to make use of the immense developments in prehistoric archaeology that occurred in the late nineteenth century may reflect the fact that the discoveries were mostly related to technological developments in stone toolmaking. The anthropologists who tried to fill in the associated cultural developments were often materialistic in their approach (for which reason this whole episode is dealt with in the next chapter). As the Duke of Argyll, a noted critic of Darwinism, argued in his *Primeval Man* of 1869, progress in tool-making did not necessarily imply moral or spiritual progress, and for all the archaeologists could show our distant ancestors may have been as morally advanced as we are. Perhaps it was awareness of this point that encouraged most exponents of the spiritual model of evolution to neglect the area of human history: it was easier to concentrate on the huge developments leading up to our appearance.

The Legacy of Hegel

Although losing favour among academic historians, Hegel's thought did retain some influence into the twentieth century. His idealist vision continued to

---

[55] Drummond, *The Ascent of Man*, p. 291. For the later literature see my *Reconciling Science and Religion* and *Monkey Trials and Gorilla Sermons*.

inspire philosophical movements in both Europe and America. Those who focused on history are conventionally divided into the left- and right-wing Hegelians. On the left his philosophy was seen as a challenge to the status quo and as a guide towards meaningful social change, leading eventually to the emergence of Marxism (to which we shall return in the next chapter). The right-wing Hegelians endorsed the existing social order both through political conservatism and through an attempt to show how idealism could provide a new foundation for traditional Christianity. They invoked the role of divine providence much in the same way as the liberal Anglicans, extending the vision of an ascent towards spiritual perfection and forging links with non-Darwinian evolutionism.

Wider philosophical movements inspired by idealism arose in both America and Europe. The groups known as the Ohio Hegelians and the St Louis Hegelians brought the message to the New World, seeing American culture as the vehicle by which the message of progress towards freedom could be inspired by Christianity and spread to the whole world. The leading figure in the Ohio group was John Bernard Stallo, who attacked materialism in the physical sciences to promote the vision of nature as a purposeful evolution of the spirit. Deeply concerned with the social tensions of the time, he seized on the idea of the dialectic to explain why the progress seemed irregular: 'Partial wrongs and obliquities square themselves into universal right, special evils coalesce into general good, particular errors are integrated into collective truth, and the successive generations are nothing but terms in a series before which Eternal Reason places its integral sign.'[56] Elsewhere he insisted that progress was predetermined by the spiritual underpinning of the universe. History is spiritual progress, and 'All the phases of life are prefigured in the origin, and every succeeding phase is immediately contained in and produced by the preceding one; – so in universal life.'[57] Peter Kaufmann's *The Temple of Truth* of 1858 predicted the eventual spiritual perfection of the human race and saw the United States as the 'saviour nation' which would promote universal peace and encourage all nations to join a 'UNITED STATES of the WHOLE EARTH'.[58]

Stallo's writings were an important influence on Ralph Waldo Emerson, inspiring him to draw up a scheme of cultural progress from the Greek tendency to deify nature through the Christian vision of the need to rise above

---

[56] Stallo, *State Creeds and Their Modern Apostles* (1972), quoted in Easton, 'Hegelianism in Nineteenth Century Ohio', p. 362. More generally see Easton's *The Ohio Hegelians* and DeArney and Good, eds., *The St. Louis Hegelians*. Works by Stallo, Conway and Kaufmann are reprinted in Good, ed., *The Ohio Hegelians*.
[57] Stallo, *The Principles of the Philosophy of Nature* (1848), partially reprinted in Goetzmann, ed., *The American Hegelians*, chap. 11, quotation from p. 168.
[58] Kaufmann, *The Temple of Truth*, reprinted in Good, ed., *The Ohio Hegelians*, vol. 1, p. 290.

nature to his own efforts to encourage a return to nature.[59] The St Louis
Hegelians were active later in the century. Their *Journal of Speculative
Philosophy*, founded by William Torrey Harris and Henry Conrad
Brokmeyer, ran from 1867 to 1893, and their teachings were also promoted
by Denton J. Snider. Like the Ohio group they were concerned with the
tensions arising within American society but saw idealism as a justification
for hope. Snider insisted that humans were driven not just by economic
interests but also by the need for recognition. Like Kaufmann he saw the
United States as the basis for a universal society that would fulfil the objectives
of the world spirit, arguing that this was a development that Hegel himself
could not have foreseen.[60]

In Germany Hegel's right-wing followers sought to present his philosophy
as a new foundation for Protestant theology while at the same time helping to
establish the myth that he had regarded the Prussian state as the end-point of
social progress.[61] His views continued to inspire quasi-religious enthusiasm
for the idea of progress throughout the rest of the century. Eduard von
Hartmann's pre-Freudian philosophy of the unconscious saw this aspect of
the mind as a manifestation of the world spirit and celebrated 'the supreme
wisdom of the unconscious and the perfection of the world' by suggesting that
'the world is contrived and guided as wisely and as well as possible'.[62] In
Britain Hegel's philosophy was little known until presented enthusiastically in
James Hutcheson Stirling's *The Secret of Hegel* in 1865. Stirling inspired a
whole generation of philosophers to explore the idealist position. They were
little concerned with history, however, although Stirling himself did propose
Hegelianism as an alternative to Ernst Haeckel's monistic-Darwinian vision of
evolution.[63]

At the end of the century Hegelian language was applied to the history of
British imperialism by J. A. Cramb, professor at Queen's College, London.
During the Boer War in South Africa he published his *Reflections on the
Origin and Destiny of Imperial Britain* (later reissued as the Great War raged).
His overall thesis was reminiscent of that propounded earlier by the liberal
Anglicans, but the underlying philosophy was that of the world spirit's pro-
gress rather than divine guidance. For Cramb, Greece expressed beauty, Rome
power and Egypt mystery. The British Empire was a synthesis of metaphysical

---

[59] See Richardson, *Emerson*, chap. 80.
[60] See Snider's 'The Kings of Capitalism' (1901) and 'A Critique of Hegel's Theory of the State',
reprinted in DeArney and Good, eds., *The St. Louis Hegelians*, vol. 2: *Cultural, National, and
World Unity*, pp. 20–49 and 120–5.
[61] See Toews, *Hegelianism*, chaps. 4 and 6.
[62] Hartmann, *Philosophy of the Unconscious*, pp. 356 and 360.
[63] Stirling, *Hegel not Haeckel*. On later idealism in Britain see Mackintosh, *Hegel and
Hegelianism* and Haldar, *Neo-Hegelianism*.

speculation and practical application, highlighting the 'world-historic signifi-cance of the English reformation'. War was an essential feature of the histor-ical process by which higher powers emerged and spread their message to the world. Each started with a fund of spiritual power, made its contribution and then declined, but not before making its benefits available to the next. Britain was the currently most progressive impulse and should be conscious of its destiny as a manifestation of the world spirit, the 'unseen force from within the race itself'.[64]

Looking Forward

Liberal Christians continued to argue that social evolution would continue the trend exhibited in the ascent of life, both areas being guided by a force that would improve the moral sense. Christ's teachings had enlivened moral debates to ensure that social evolution was a process leading inevitably to a better, more equitable society. In turn this would allow the emergence of a morally superior type of humanity. Sometimes called 'Christian socialism', this reformist movement gained significant ground in both Britain and America during the closing years of the nineteenth century. The most successful expression of this utopian vision was Edward Bellamy's *Looking Backward* of 1888, which outlined the process by which the future utopia would be achieved. Social evolution had a moral direction and would produce a human race worthy of the God who had founded the process.

Christian socialism emerged as an antidote to the Spencerian version of 'social Darwinism' that reached its apogee in the Gilded Age of late nineteenth-century America. As Howard Segal has shown, some utopian visions drew heavily on the hope of technological improvements that would alleviate the problems arising from the headlong rush to industrialize the economy.[65] But technology was not by itself the answer; the new social order would have to come about through political action that would ensure equality of wealth, opportunity and justice for all. As Sarah Churchwell has recently pointed out, the 'American dream' was originally conceived in terms of a society in which all would be able to achieve personal fulfilment, not material prosperity.[66] Some saw the future paradise in terms of a return to nature, as in W. H. Hudson's novel *A Crystal Age* of 1887. For those with religious convictions, the achievement of utopia was the goal of Christ's moral teaching.

---

[64] Cramb, *Reflections on the Origin and Destiny of Imperial Britain*, p. 5; see also *The Origins and Destiny of Imperial Britain*, p. 5.

[65] Segal, *Technological Utopianism in American Culture*. On British novels of prediction see Clarke's introductions to the texts reprinted in his edited collection *British Future Fiction*. Vol. 2 of this collection includes Hudson's *A Crystal Age*.

[66] Churchwell, *Behold, America*.

Putting that message into practice would involve a political campaign aimed at a socialist revolution. This activity could be seen as a rationalization of the mechanism of social evolution that had driven the past history of the human race, pushing it more rapidly towards its intended goal. Here was a parallel to the Hegelian vision of history as a process which mounted step by step towards a utopia in which God's purpose for humankind would be achieved.

Bellamy's *Looking Backward* was an account of life as he imagined it might be in the Boston of the year 2000. (It reads at first as a 'sleeper awakes' novel, a device later used by H. G. Wells to great effect, but turns out in the end to be merely a dream.) Bellamy was both a social activist disgusted by the inequalities of the society he actually lived in and a deeply religious man, but he was also a writer of considerable ability, hence the effectiveness of his predictions. His future certainly involved some technological improvements, for example the system for distributing live music to every household via the telephone network. But most of the transformations in everyday life have been produced by political action which aims to give everyone equal access to the state's resources and equal responsibility to contribute to the production of those resources. The system of distribution involves giant shops where purchasers order goods to be delivered directly to their homes and pay for them with credits recorded on cards (actually paper coupons, so no anticipation of the modern credit card).

The result is a society in which everyone is happy and fulfilled, but which also has the potential to encourage the emergence of a superior form of humanity. In his preface Bellamy notes that he and his fellow enthusiasts have little interest in the past: they focus on the future, although like the Hegelians they see their goal as the culmination of a progression aimed at a predetermined end-point:'The almost universal theme of the writers and orators who have celebrated this bimillennial epoch has been the future rather than the past, not the advance that has been made, but the progress that shall be made, ever onward and upward, till the race shall achieve its ineffable destiny.'[67] In a chapter presented as a sermon delivered in the year 2000 to celebrate a century of progress, the spiritual dimension of the process is made clear. Humanity has 'entered on a new phase of spiritual development, the very existence of which our ancestors scarcely suspected'. There is a sense of the 'unbounded possibilities of human nature' allowing us to 'believe the race for the first time to have entered on the realization of God's ideal of it, and each generation must now be a step upward'. This will be 'the fulfilment of evolution, when the divine secret hidden in the germ shall be perfectly unfolded'.[68] Thus, despite his lack of interest in history, Bellamy makes plain his commitment to a model

[67] Bellamy, *Looking Backward*, pp. 35–6.   [68] Ibid., p. 206.

of progress that is developmental, based on the step-by-step ascent of a hierarchy of stages leading to a predetermined goal.

Bellamy's book was widely read and inspired a number of utopian speculations published during the progressive era around 1900. It marked the high point in the influence of the theologically inspired developmental model of social evolution and linked this view of history to an increasing interest in future progress. The twentieth century would witness a steady decline in that influence, driven partly by the growing evidence that technology was the most powerful force promoting social change. Although Bellamy's future utopians enjoy some technological advances, their lives are improved much more by their spiritual activities, and his book doesn't present invention as a force that might continue to transform society. There were many at the time who thought that science itself had discovered all that could be known about the forces of nature, so discovery had effectively come to an end.[69] If this were the case it might be expected that invention too would similarly reach a limit.

In fact, of course, both scientific research and technological innovation exploded into new levels of activity in the early twentieth century. As Churchwell points out, the 'American dream' was transformed into the worship of the unlimited power to acquire wealth and material goods. The utopian vision became increasingly focused on technological innovation as the solution to all society's problems, encouraging a wide range of speculations about innovations that might have the power to change the world. Instead of a clearly defined spiritual goal to be attained with the help of a developmental historical trend, the future became an open-ended range of possibilities at the mercy of human inventiveness. The expectation of an end of history represented by a morally and spiritually improved humanity declined in influence accordingly, although as we shall see in Chapter 6 it did not disappear altogether.

---

[69] See Badash, 'The Completeness of Nineteenth-Century Science', although Badash argues that there were many scientists who did not accept this view.

5   Ascent to Utopia
The Quest for a Perfect Society

The seventeenth-century supporters of 'modern' learning held that it was superior to ancient thought in that it was based on a rational study of the material world. The natural sciences had begun to advance as never before, with consequent benefits in the form of new technologies that promised well-being for all. It was assumed that the drive towards material prosperity would go hand in hand with a revival of religion so that the future would resemble paradise in every sense of the term. In the eighteenth century the philosophers of what became known as the Age of Enlightenment severed the link between this-worldly and other-worldly concerns. They saw organized religion as an outdated superstition, to be swept away rather than reinforced by the greater appeal to reason. With a few exceptions such as Priestley, those who hoped for material progress divorced their expectations from the search for a spiritual paradise on earth. Utility became crucial: the rise of mechanized industries would produce more goods to improve people's living standards and hence their general well-being. By the nineteenth century those still hoping for spiritual progress had begun to see industrialization as a curse whose effects were all too obvious in the slums of the over-crowded cities. The utilitarians insisted that once the necessary social adaptations had been made the benefits would become available to all. The result might not be paradise on earth, but it would be a utopia in which all could live a pleasurable life.

Materialism and the Ladder of Progress

Enlightenment thinkers demanded freedom from royal and aristocratic con-straints on ordinary people's behaviour. Superficially, their programme resem-bled the Hegelians' call for freedom as the basis for a perfect society. In fact, however, the two ideologies differed on what they understood freedom to mean. Even in France the *philosophes* began to see the British system of free enterprise as the key to industrial progress and as a model for how society should develop. Where Hegel saw freedom as something that had to be exercised within a culturally homogeneous state to provide the benefits of complete self-fulfilment, the Enlightenment thinkers favoured an approach based on laissez-faire, complete freedom from state control. As the critics of this position recognized, however, the freedom to prosper through one's own efforts also meant the risk of falling into poverty without any social support to offer relief.

Recognizing the unconvincing nature of the claim that all would eventually learn how to live successfully in an unregulated society, Marx and Engels formulated a new version of the drive for material progress. They grafted the utilitarian focus on well-being for all onto the Hegelian dialectic, replacing free enterprise with the vision of a rationally ordered classless society that would control the distribution of goods. They too hoped that people would eventually adapt, and that the state in anything like its old form would eventually wither away as cooperation became instinctive. Their followers in the twentieth century focused instead on the centralized control of production as the key to prosperity for all. Only when the Soviet alternative collapsed did it again make sense to see free enterprise as the only way forward to a New World Order.

Even in the early eighteenth century those who hoped to create a more prosperous society still tended to see this as something that was being brought about by an unparalleled application of reason, not as the culmination of a long-standing historical trend. Only towards the end of the century did political and economic thinkers in France and Britain begin to bolster their hopes for the future with the argument that a sequence of previous steps forward suggested a general progressive trend. History and inferences about prehistory became the basis for the idea of material progress. The assumption that in the distant past humanity had lived in a 'state of nature' without organized social structures and with limited technologies created a starting point for a scale of economic and social developments that stretched up to the Industrial Revolution now under way. It seemed natural to imagine a progressive driving force in social evolution derived from the slow but inevitable tendency for human reason to recognize the challenges presented by the natural world and develop technological and social innovations to improve everyday life. Here was a new conception of the ladder of progress.

The utilitarian vision of a society in which all would be happy was, up to a point, the equivalent of the idealists' spiritual paradise, and both approaches depended on a transformation of society that would eliminate past injustices. For the idealists technological improvements might be of benefit, but they were hardly crucial and soon came to be regarded as a hindrance to progress as the evils of the Industrial Revolution became apparent. The materialists always placed more emphasis on technology as a means of improving the human situation, and their vision of social reform was driven in part by a recognition that society would have to adjust to industrialization. In the twentieth century it would become apparent that technological innovation could change society in unpredictable ways, undermining the original hope of a well-defined goal of progress.

The utilitarian focus on technology and associated social developments encouraged the extension of the sequence back into prehistory. Even before archaeological evidence emerged, the introduction of pastoralism, agriculture and finally commercial-industrial activity was seen as crucial. There was less emphasis on comparisons between civilizations and the cycle of individual growth. Materialist thinkers did use developmental language derived from biology, including the depiction of modern 'savages' as having childlike minds and the claim that industrial societies were more mature. But they seldom appealed to the image of a society declining inevitably into senility as it ran out of energy.

To the advocates of material progress it was obvious that European civilization was surging ahead and that others were left behind. 'Savages' were increasingly regarded as primitives, relics of the earliest stages of social development surviving into the present. A distinctive feature of the utilitarian-materialist view of progress was the appeal to parallel evolution as a means of explaining the survival of these earlier stages. Because the course of development was defined in terms of social structures that had adjusted to technological innovations, it was possible to see the hierarchy as an abstract sequence defining steps that any society had to pass through on the way to a fully mature civilization. Multiple lines of development representing groups defined by race or geographical location could independently pass through the same developmental sequence, and any that occupied the same stage would look remarkably similar. If the rate of progress differed – perhaps because of environmental factors – all the various stages of development would be displayed in the modern world, making the whole pattern visible to us today.

This linear model effectively sidelined the question of cultural diversity. Where Hegel acknowledged the distinctive characters of the successive cultures in his scheme, anthropologists such as E. B. Tylor who worked within the utilitarian tradition claimed that one could see identical societies in many parts of the world where the same level of development had been reached independently. Archaeologists equated stone-age cultures with modern 'savages', arguing that some races had never progressed beyond the lowest level. Debates raged over whether a people stuck at an early stage of social evolution could be educated so that they could rise eventually to the mature level of industrial society achieved in the West. Although often identified with Darwinism, this model of cultural evolution was based on non-Darwinian principles, since it depicted the process as the ascent of a ladder rather than a branching tree of increasing diversity.

The success of the linear model was ensured in part by the confidence of its proponents (of either the laissez-faire or Marxist persuasions) that their own society was close to reaching the final goal. The viability of the model also rested on the assumption that innovation is a process constrained by a kind of inherent necessity that made each upward step more or less inevitable. The utilitarians believed that human ingenuity would always conquer the challenges posed by the environment, but they assumed that at each point in the process there was only one solution to the immediate problem, one new technology that would work in practice. Hunter-gatherers naturally moved to pastoralism and then to agriculture, agriculturalists equally naturally tended to develop industry and commerce. The ages of stone, bronze and iron succeeded one another inevitably in the archaeological record. Marxists too assumed that invention was predetermined by the nature of the challenge presented at each point in history.

For the early steps in technological progress this assumption seemed perfectly sound. How else would the early hunter-gatherers have sought a new source of food other than by domesticating the animals they hunted and the plants they gathered? And surely the discovery of metals could only have come after the emergence of technologies based on the most obvious raw material, stone. Only in the twentieth century would it become clear that the logic of this argument breaks down when a certain level of industrial development is achieved. In a complex world with many inventors seeking to improve many industries, there is no way of identifying the most 'obvious' next step because there are multiple competing possibilities.

The *philosophes* of the French Enlightenment saw the rising influence of reason as the driving force of progress and argued that the future structure of society should be determined by a more thorough application of the same principle. Industry merely provided the material goods needed for well-being; the real desideratum was a social order that guaranteed everyone a decent share of the wealth to enjoy. But for Priestley and the more materialist exponents of industrial progress in Britain, innovation was best promoted with an economy that left individual entrepreneurs free from state intervention. They set the scene for the utilitarians of the nineteenth century to develop the philosophy of free-enterprise individualism as the guarantee of progress and to argue that it was the goal of social evolution. In a dramatic ideological reconfiguration, the Marxists co-opted the Hegelian model of coherent developmental stages to preserve the vision of a prosperous future based on a rationally ordered society.

Enlightenment and Progress

As Bury and other historians have argued, the idea of progress began to take shape in the seventeenth century when supporters of the 'modern' learning realized that areas such as science and technology had advanced beyond anything available to the ancients.[1] Bacon and his followers still hoped that progress in our control over material nature would go hand in hand with spiritual renewal, but Continental exponents of the new science such as Bernard de Fontenelle were less willing to see religion as a force for good. They saw it as a residue of ancient superstition that needed to be swept aside by the power of reason. The hope of spiritual progress remained active but became increasingly divorced from the ideology of those who focused on science and industry as forces for the betterment of life. The majority of Enlightenment thinkers moved towards a model of progress based on reason's power to produce both material prosperity and a better-organized society that would spread the benefits to all.

In his *Advancement of Learning* Bacon had coined his much-quoted aphorism *Antiqitas saeculi juventus mundi*, 'Antiquity is the youth of the world', introducing the analogy between the history of civilization and the growth of the individual.[2] Despite the implication that there might be a later phase of decline into senility, the analogy was widely taken to mean that the modern world had achieved a state of maturity that might become permanent, leading to endless future progress. Bacon identified the inventions that confirmed the moderns' superiority over the ancients: printing, gunpowder and the magnetic compass, the two latter becoming symbols of the West's power to dominate the wider world. Now the scientific method that he and his followers had introduced would ensure that our ability to understand and control nature would be applied more systematically to ensure future progress.

Cartesians such as Bernard de Fontenelle also insisted that the scientific method would ensure that there would always be new discoveries. Fontenelle too used the analogy with the individual life-cycle but insisted that once freed from the shackles of superstition the expansion of knowledge would ensure there could be no descent into senility.[3] For many, though, it was the technological progress already flagged by Bacon that promised a brighter future. Early members of the Royal Society frequently speculated about the amazing innovations that would come. Joseph Glanville, for instance, predicted flights around the world and even to the moon. To the people of the future 'it may be as ordinary to buy a pair of wings to fly to remote regions; as now a pair of boots to ride a journey. And to confer at the distance of the Indies by sympathetic conveyances, may be as usual to future times, as to us in a literary

---

[1] Bury, *The Idea of Progress*, chaps. 4–6, and also Jones, *Ancients and Moderns*.
[2] Bacon, *The Advancement of Learning*, book 1, section 5, p. 1.
[3] Fontenelle, *Digression sur les anciens et modernes*, pp. 161–76 in the volume cited, which combines this with the *Entretiens sur la pluralité des mondes*; see pp. 171–2 for the life-cycle analogy.

correspondence.' The human lifespan would be extended through medical discoveries, and agriculture would transform desert regions into a paradise.[4]

Little thought was given to the possibility that these inventions might transform our lives to such an extent that they would actually shape the way society functions. In part this was due to the hope that rational planning would be applied to government to create a new social order controlled by experts on human nature. We have seen how the abbé de Saint-Pierre became notorious for his schemes to transform French society for the better. Sebastian Mercier's optimistic vision of life in the year 2440 was based more on the assumption that society would be rationally ordered than on the expectation of industrial progress.[5] For the French *philosophes*, at least, it seemed obvious that – humanity being what it is and will remain – the majority of ordinary people would need to be governed by their betters. The trick was to ensure that the rulers would now take expert advice to ensure a happy life for all.

To create the idea of progress as an inevitable trend the expectation of future developments had to be connected with an image of the past that represented it as a sequence of episodes in which reason gradually overcame the powers of ignorance and self-interest that had led the majority into superstition and the rulers into despotism. Many of the early *philosophes* found it hard to conceive a clear idea of progress, their uncertainty being exemplified in the historical writing of Voltaire. Although Voltaire is occasionally portrayed as someone who treated other cultures as illustrations of the stages through which Europe passed to reach its present pre-eminence, most commentators now see him as struggling to make sense of the interaction between the power of reason and the opposing tendencies of ignorance and environmental challenges.[6] In his *Essaie sur les moeurs et l'esprit des nations* and his *Philosophie de l'histoire* he acknowledged that humanity had begun in a state of savagery – but noted that on the usual definition there were still savages living in Europe today.[7] The climate of Europe had prevented the initial emergence of civilization here, so the first major efforts were made in the Orient. Voltaire was anxious to

---

[4] Glanville, *The Vanity of Dogmatizing*, p. 182, and on remote communication by magnetism pp. 202–6.

[5] Mercier's *L'an deux mille quatre cent quarante* was discussed in Chapter 1; see also Bury, *The Idea of Progress*, chap. 10 and on Saint-Pierre chap. 6.

[6] On Voltaire as an exponent of stages of development see Dale, *Hegel, the End of History, and the Future*, pp. 116–17. Those who doubt his faith in progress include J. H. Brumfitt in his introduction to *La philosophie de l'histoire*, in *Oeuvres complètes de Voltaire*, vol. 59, pp. 13–78; see pp. 77–8. More generally on the Enlightenment doubts about progress see Gay, *The Enlightenment* and the same author's *The Party of Humanity*, esp. pp. 270–1.

[7] *Oeuvres complètes de Voltaire*, vol. 22, p. 23, and vol. 59, pp. 109–15; see also Voltaire's *The Philosophy of History*, pp. 26–34. This point is mentioned by Leigh, *Voltaire: A Sense of History*, whose account makes no mention of the idea of progress.

stress the antiquity and the sophistication of Chinese civilization in order to discredit the histories of those who took their cue solely from the Bible.

When Europe finally did begin to advance it enjoyed four happy episodes when a degree of progress was achieved: the ages of Greece, Rome, the Renaissance and the France of Louis XIV.[8] But between Rome and the Renaissance came the Dark Ages, and like most of his contemporaries Voltaire saw this period as a disaster in which external barbarians had brought down a culture weakened by internal decay. The age of Louis XIV represented a triumph of order and statecraft, but Voltaire was also conscious of the need to look beyond political history for the forces that actually drive progress. In his *Nouveau plan d'une histoire de l'esprit humain* he complains of the histories that only deal with kings, most of whom have been responsible for more evil than good, while ignoring the 'inventors of the arts' who have actually benefited mankind.[9] Here was a hint that science and technology had the potential to drive progress ahead in the future.

If the memory of the decline and fall of the Roman Empire undermined the Enlightenment's faith in progress, the thirty-eighth chapter of Edward Gibbon's monumental study of those events provided reassurance. Gibbon himself made no secret of his belief in progress from a primeval savagery, if only by an irregular process. The fall of Rome was perhaps the greatest irregularity, but Gibbon assured his readers that the calamity could not be repeated. Europe was not only politically more stable than Rome; it had the advantage of the latest developments of science to check any incursions. 'Europe is secure from any future irruption of barbarians, since, before they can conquer, they must cease to be barbarians.' As the gifts of modern industry and science are diffused around the world 'We may therefore acquiesce in the pleasing conclusion that every age of the world has increased, and still increases, the real wealth, the happiness, the knowledge, and perhaps the virtues of the human race.'[10]

Such reassurances encouraged the *philosophes* of the late eighteenth century to lose their fear of a future relapse and construct a theory of overall progress. The two figures who have come to represent this movement are Turgot and Condorcet – the latter famously producing his survey while in hiding from the revolutionary government that claimed to produce the reforms the previous generation had called for.

Anne-Robert-Jacques Turgot was deeply involved in the affairs of the pre-revolutionary government and had limited time to develop his thoughts on

---

[8] Voltaire, *Le siècle de Louis XIV*, in his *Oeuvres historiques*, pp. 616–18.
[9] *Oeuvres complètes de Voltaire*, vol. 27, pp. 154–7.
[10] Gibbon, *The History of the Decline and Fall of the Roman Empire*, vol. 6, p. 407 and pp. 410–11.

progress. He nevertheless produced a number of works outlining his vision of how the human race had risen from savagery to its present state. Like most of his contemporaries he accepted that there had been little advance in the arts since ancient times but argued that developments in scientific knowledge and its applications have been made whenever the few geniuses who appear in each generation have been free to apply themselves. Key steps in prehistory involved the transitions from hunter-gatherers to pastoralism and agriculture and the formation of towns. The fall of Rome was an example of how external factors could block progress, but in more recent times the spread of education via printing had reduced the forces that held things back. Turgot was convinced that there had been a linear sequence of developmental stages analogous to the growth of a plant towards its crown of flowers. The predetermined nature of the sequence ensured that various peoples could advance in parallel but at different rates up the same ladder of social evolution:

Thus the present state of the world, marked as it is by these infinite variations in inequality, spreads out before us at one and the same time all the gradations from barbarism to refinement, thereby revealing to us at a single glance, as it were, the records and remains of all the steps taken by the human mind, a reflection of all the stages through which it has passed, and the history of all the ages.[11]

This anticipates the linear model of progress that would be adopted in the nineteenth century by those anthropologists who sought to diminish the significance of cultural diversity by treating all living cultures as stages in a single process of evolution.

Jean-Antoine-Nicolas Caritat, marquis de Condorcet, was a follower of Turgot who devoted himself to reform and to the creation of a science of society based on statistics. He played a prominent role in the early phases of the Revolution but was eventually denounced and wrote his *Esquisse d'un tableau historique des progrès de l'esprit humain* while in hiding in 1794. Intended as a sketch for a more substantial work, it was still a coherent account of progress from primeval savagery, divided into ten epochs of which the last is a prediction of the future. Condorcet's purpose was to use the model that could be derived from history as a guide to what we might and should achieve:

Such a picture is historical, since it is a record of change and is based on the observation of human societies throughout the different stages of their development. It ought to reveal the order of this change and the influence that each moment exerts upon the subsequent moment, and so ought also to show, in the modifications that the human species has undergone, ceaselessly renewing itself through the immensity of the

---

[11] Turgot, 'A Philosophical Review of the Successive Advances of the Human Mind', trans. Meek in *Turgot on Progress*, pp. 41–59, quotation from p. 42; note also 'On Universal History', pp. 61–118. On Turgot and Condorcet see also Manuel, *The Prophets of Paris* and Frankel, *The Faith of Reason*.

centuries, the path that it has followed, the steps that it has made towards truth or happiness.[12]

Well aware of the obstacles to progress, he wanted to create a model of the successive stages of development that would be achieved whenever the creative faculties had the freedom to function. Keith Baker's study of Condorcet stresses his mathematical sociology and argues that the *Esquisse* should be seen not as history but as an idealized hierarchy that might in practice be followed only irregularly. Along with Frank Manuel, Baker notes the parallel between this ladder model and the chain of being. Here is an interpretation of history that allows the various civilizations of the past to be seen as stages in a unilinear hierarchy of development: 'All peoples whose history is recorded fall somewhere between our present degree of civilization and that which we still see amongst savage tribes ... so welding an uninterrupted chain between the beginning of historical time and the century in which we live, between the first peoples known to us and the present nations of Europe.'[13] Condorcet's first two periods cover the widely accepted sequence from the consolidation of tribes through to the development of pastoralism and agriculture. The remaining sequence outlines the periods of known history, focusing both on advances in science and technology and on increasing social cohesion. The fifth epoch deals with science in the ancient world, the sixth with the Dark Ages and the remainder with modern European history. Printing and the scientific revolution are hailed as major advances, and Condorcet was anxious to stress how improved knowledge of nature has led to new industries and medicines. The predictions for the future in his tenth epoch include the spread of Enlightenment values to all the peoples of the earth. A new language will be created for the natural and human sciences that will consolidate their methodologies. Unspecified improvements in agriculture will be made, and medical science will significantly extend the human lifespan.

Condorcet put great store by the extension of education to the whole population but retained the Enlightenment's view that progress depended on the creation of a more equitable social order under the control of experts. His views were soon widely publicized despite his untimely death in prison: the *Esquisse* was printed in the following year and became an inspiration for the generations that lived through the Napoleonic era and the restoration of the monarchy. It would play a role in the thinking of Auguste Comte, whose

---

[12] Condorcet, *Esquisse d'un tableau historique des progrès de l'esprit humain*, pp. 2–3, translation from *Condorcet: Political Writings*, p. 2.

[13] Condorcet, *Esquisse d'un tableau historique des progrès de l'esprit humain*, pp. 7–8, and translation from *Condorcet: Political Writings*, p. 5. On the chain analogy see Baker, *Condorcet*, p. 356, and Manuel, *The Prophets of Paris*, p. 62.

positivist philosophy served as a foundation for numerous projects to create a fairer society based on progress and a rationally imposed social order.

There was, however, another interpretation of the expectation that throwing off the yoke of ancient tyrannies would lead to a better world for all. If improvements in industry are to be the key to better living conditions, perhaps the drive for freedom should be taken as far as possible short of outright anarchy. People should be free to better themselves by their own efforts and inventions without let or hindrance from experts seeking to impose their artificial systems from above. Free enterprise would become the motor of progress in a world increasingly dependent on industry and innovation. This was a different sense of the goal towards which progress is aimed – yet one that required a similar hierarchical model of the stages by which that goal would be reached.

From Philosophical History to Utilitarianism

The Enlightenment's vision of an ideal life assumed that material well-being would be available to all, along with enough leisure to enjoy life in a cultured society. Some thinkers retained elements of Christian values: Priestley is an obvious case, and Adam Ferguson similarly retained aspects of his Presbyterian faith. But Britain also had its share of sceptics, including David Hume and Edward Gibbon. Like their Continental counterparts they shared the assumption that freedom from both ecclesiastical and political tyranny was essential to allow everyone to enjoy life and play a role in promoting further progress. Inventors and entrepreneurs could make their fortunes while providing benefits for all. They looked to the past for evidence that whenever innovators had been given their head (as opposed to being hired by the state) the provision of material goods for all had forged ahead.

Their faith in the benefits of free enterprise did, however, introduce a new dimension into the idea of human perfectibility. Everyone would have to learn to be self-reliant as well as polite to their fellows – and that would be best done in the school of hard knocks rather than through a liberal education. This might generate a materialist equivalent to the religious believers' ideal of freedom, but it constituted a very different ideology. British economists and political thinkers became utilitarians, judging everything by its value in the economic sphere and seeing the way forward in terms of dismantling state control of more or less everything except the protection of private property.

Although the goals of the British and Continental progressionists differed, they had the same level of confidence in their projected futures and devised a similar vision of the past to sustain it. Common interests in commerce and industry as the keys to a better material lifestyle led them to the same list of innovations in prehistory and a similar enthusiasm for later inventions and the

scientific revolution. There were also enough similarities between their political ideals for them to share the same assumptions about the steps in social
evolution. British thinkers thus created a model of progress very similar to that
produced by Turgot and Condorcet. Scottish economists such as Adam Smith
and philosophical historians such as Ferguson independently constructed a
ladder of developmental stages and placed other contemporary societies at
identifiable positions in the sequence below their own. The same model was
adopted by James Mill in his assessment of Indian culture. It became commonplace to assume the parallel evolution of peoples through the same evolutionary steps but at different speeds, allowing the diversity of modern cultures
to be seen as a record of the past.

British utilitarians also exploited the suggestion that an explanation for the
differing rates of progress could be found by looking to environmental factors.
Voltaire had noted that Europe had not been a suitable location for the first
steps towards civilization: they had occurred in more fertile regions to the East
and South. But once formed, these early cultures had become locked into
centralized empires that blocked further progress, so only when their achievements were transmitted to Europe did the next steps become possible. The
British extended this argument to explain why their own country was now
advancing faster than the rest of Europe. Its island location and wealth of
raw materials gave it a head start over its rivals. This argument does not seem
to have undermined faith in the existence of a predetermined sequence of
social developments, because the environmental factors were seen as
merely facilitating or hindering the transitions from one stage to the next.
They might determine some aspects of individual cultures such as the vast
extent of the ancient empires, but they did not define the steps necessary for
progress to be made.

There was little sense that the shape of future society might be influenced by
the kind of technological innovations that became available. Future technologies were still thought of as fairly predictable (better transport and health care
for instance), but there was little appreciation that they might shape the kind of
society that would emerge. The utilitarians' commitment to the model of a
predetermined ladder of progress made it hard for them to appreciate that
cultural diversity might indicate a plurality of ways in which civilizations had
advanced. Seeing history as a process constrained to advance in a single
direction blocked recognition both of past diversity and of the potential for
unpredictable environmental and technological factors to determine what a
future society might look like.

In Scotland a group of economists and social thinkers created a school of
philosophical history which viewed the past more in terms of an idealized
model of progress than a record of actual events. Adam Smith's *Wealth of
Nations* expounded the economic philosophy of the commercial age: people

are selfish and seek to improve their position, and their efforts to better themselves result in benefits to society as a whole. This 'natural identity of interests' provided the basis for the argument that even in a world ruled by selfishness an 'invisible hand' guides social progress. Smith's lectures on jurisprudence, given during the 1760s, briefly introduced the claim that social progress, if not checked by climate or repressive policies, follows an inevitable sequence of developmental stages leading towards free-enterprise capitalism. The stages are closely analogous to those suggested by Turgot: 'There are four distinct states which mankind pass thro [sic]: 1st, the age of hunters; 2ndly, the age of shepherds; 3rdly the age of agriculture; and 4thly the age of commerce.'[14] Little historical detail is provided for what seems to be an idealized reconstruction of the steps that were necessary for the process to work. Later in his lectures Smith developed the argument that Britain had forged ahead because its island location allowed it to leapfrog one step in the sequence by making the creation of a standing army unnecessary.[15]

Human ingenuity was constantly trying to provide more food for an expanding population, and as the mode of production changed societies took on characteristic forms. For Smith and his fellow economists, freedom to innovate was the key to progress, and ever more complex societies emerged as new technologies shaped the production of material goods. Adam Ferguson's *Essay on the History of Civil Society* of 1767 offered a slightly simplified version of the same developmental sequence, rolling pastoralism and agriculture into a single age of barbarism and suggesting the emergence of different concepts of property as society evolved. Ferguson invoked the analogy between individual development and social evolution: 'Not only the individual advances from infancy to manhood, but the species itself from rudeness to civilization.'[16] He was aware of the tendency for political corruption to upset the process and at one point suggested that the scheme of progress could not be absolutely uniform, because people did sometimes acquire different desires and purposes.[17]

The potential for climatic factors to influence what was otherwise a more or less automatic progress towards a commercial society was stressed in John Millar's *Origin and Distribution of Ranks* of 1771. There is always a struggle to satisfy our wants, generating a tendency to produce more goods and corresponding changes in laws and customs. Millar argued that

---

[14] Smith, *Lectures on Jurisprudence*, p. 14. For more details on the material presented in this section see my own *The Invention of Progress*.
[15] Smith, *Lectures on Jurisprudence*, pp. 265–70.
[16] Ferguson, *An Essay on the History of Civil Society*, p. 1.
[17] Ibid., parts 5 and 6 on the decay of societies and p. 122 on differing purposes.

There is ... in man a disposition and a capacity for improving his condition, by the exertion of which, he is carried on from one degree of achievement to another; and the similarity of his wants, as well as of the faculties by which those wants are supplied, has everywhere produced a remarkable uniformity in the several steps of his progression. ... There is thus, in human society, a natural progress from ignorance to knowledge, and from rude to civilized manners, the several stages of which are usually accompanied by peculiar laws and customs.[18]

Like Smith and Ferguson, Millar conceived his linear scheme of progress as an idealized sequence that emerged from the necessities of the human situation. It did not need to be justified by detailed historical evidence, and the only reason for looking at individual cultures was to investigate the reasons why some parts of the world had witnessed greater progress than others. Here Millar was clear that climate was the major factor; he mentioned the possibility of racial differences only to dismiss them. Too warm a climate favoured indolence, while Europeans benefited from harsher conditions which forced them to be more active.[19]

South of the border the idea of progress was increasingly perceived as a component of radical political thought. William Godwin uncoupled the call for reform from Priestley's religious agenda and became notorious for his denial of free will and support for anarchism. His *Enquiry Concerning Political Justice* of 1793 included the assumption that a new society would come about as the inevitable outcome of a general progressive trend by which reason gradually exerts greater influence on human affairs. He speculated that machinery would eventually eliminate the need for labour and that medicine would greatly extend the human lifespan – the latter prompting a dispute with Thomas Malthus over the dangers posed by population expansion.[20]

The utilitarians who took up the campaign against traditional privileges and in support of individualism were also concerned more for the future than for the past. Jeremy Bentham's calls for policies that would promote the greatest happiness of the greatest number were presented as radical innovations rather than the continuation of a long-standing trend. His greatest disciple was James Mill – raised and educated in Scotland before he became based in London – whose interests were eventually drawn to history by his desire to write an account of the East India Company's efforts to govern its informal empire. Mill explored both the current state of indigenous Indian culture and the means by which British legislators were seeking to improve the situation. He was thus brought to a model of progress very similar to that developed by the

---

[18] Millar, *The Origin and Distribution of Ranks*, reprinted in pp. 175–322 of Lehmann, *John Millar of Glasgow, 1735–1801*; quotation from p. 176.

[19] Ibid., pp. 177–80 on climate and race, and chap. 4 section 2 on 'The Nature of Progress of Government in a Rude Kingdom'.

[20] On progress see Godwin, *Enquiry Concerning Political Justice*, pp. 89 and 94–5 and on future technical and medical developments pp. 677–89.

philosophical historians. Convinced of the superiority of Western values, he sought to discredit the views of the orientalists who were impressed by traditional Hindu culture. Where they recognized the diversity of cultural evolution, he regarded both Hindus and Muslims as trapped at a lower stage of development. Chapter 8 of the book reflects on the overall state of Hindu civilization and is replete with references to a scale of development applicable to all peoples of the world. Owing to their environment, the cultures of India had 'made but a few of the earliest steps in the progress of civilization'.[21] In the tropics, the creation of vast empires prevented further progress.

At a more popular level, the novels of Sir Walter Scott embodied the philosophical historians' assumption that the sequence of social developments unfolded in a fixed pattern leading towards a more civilized state. In Britain, the events of the previous century showed Scotland being brought forcibly into the orbit of the more cultured society already achieved south of the border.[22] The writings of Thomas Babington Macaulay expressed the same viewpoint, although here the idea of progress driven by material causes vied with the more idealist vision of Britain as the embodiment of higher cultural values. In his review of Robert Southey's gloomy *Colloquies of Society* in 1830 Macaulay stressed the importance of private enterprise and invention rather than government in leading society towards a more perfect state:

History is full of the signs of this natural progress of society. We see in almost every part of the annals of mankind how the industry of individuals, struggling up against wars, taxes, famines, conflagrations, mischievous prohibitions, and more mischievous protections, creates faster than governments can squander, and repairs what invaders can destroy. We see the wealth of nations increasing, and all the arts of life approaching nearer and nearer to perfection, in spite of the grossest corruption and the wildest profusion on the part of rulers.[23]

Macaulay then went to India as Legal Member of Council, where his 'Minute on Education' of 1835 challenged the views of the orientalists by insisting that native society could be forced to progress only if education was presented in English rather than the vernacular languages. Here James Mill's impression of the low state of Indian culture was exploited to suggest that the only way for the country to mount the steps leading towards a civilized society was for modern values to be imposed from above. Macaulay did, at least, accept that once this process of education was complete, the Indian population would be able to govern itself effectively.[24]

---

[21] Mill, *The History of British India*, vol. 2, p. 152. On Mill and the other utilitarians see Halévy, *The Growth of Philosophic Radicalism*.

[22] See Brown, *Walter Scott and the Historical Imagination* and Culler, *The Victorian Mirror of History*, chap. 2.

[23] Macaulay, 'Southey's Colloquies', in his *Critical and Historical Essays*, p. 121.

[24] On Macaulay's time in India see Clive, *Thomas Babington Macaulay*, chaps. 11 and 2.

The suggestion that climate and geography held back the natural development of societies such as India was developed in Henry Buckle's *History of Civilization in England* in 1857. Buckle's main purpose was to show that human affairs are governed by natural law and to argue that Britain's surge to pre-eminence was the result of favourable circumstances. He insisted that progress arose not from any improvement in humanity's moral character but from advancing knowledge of the material world. John Fiske later called this Buckle's first law of progress, although the term 'law' is not used in the original. In his second chapter Buckle introduced the idea that environment regulates the extent and rate of a people's progress.[25] He suggested that once agriculture had been developed, population expanded, and in hot countries where food is easily produced this led to a stratification of society. The result was the emergence of the ancient empires and the corrupt and static state of modern India. The great empires of the past and the present were aberrations in the progress of society because they remained trapped at an early stage of development while the peoples of more temperate climates forged ahead.[26]

In the year in which Buckle published, the hopes that the Indian population could be taught European values were dashed by the rebellion that the British called the 'mutiny'. Such cultures might be out of date, but they had a hold on the people that prevented the artificial imposition of more advanced behaviour. Recognition of this point led the anthropologists of the late nineteenth century to develop a view of social evolution still based on the model of a ladder of development but assigning to each stage a kind of coherence that held it together until it was ready for the next step. This social evolutionism emerged independently of the Darwinian revolution in biology which would ultimately to demolish the idea of linear progress.

Positivism: Order and Progress

The utilitarians' vision of progress had parallels with the approach taken by French thinkers seeking to build on the Enlightenment model. John Stuart Mill, James's equally influential son who eventually repudiated his father's rather soulless approach to life, was for a time strongly influenced by Auguste Comte's ideas. As in the eighteenth century, the two approaches to the creation of a perfect future society went hand in hand for part of the way but disagreed on whether the final goal would be achieved by order imposed from above or by individuals having the freedom to innovate in a marketplace. French thinkers held to the *philosophes'* views that the perfect society would be

---

[25] Buckle, *History of Civilization in England*, chap. 4; see Fiske, 'Mr. Buckle's Fallacies', in Fiske, *Darwinism and Other Essays*, pp. 130–91, esp. p. 154.
[26] Buckle, *History of Civilization in England*, vol. 1, chap. 2, esp. pp. 70–81.

achieved only under the direction of experts who knew how to organize things. They were not always consistent on who these experts should be, suggestions including the scientists, the industrialists and students of the new science that would later be called sociology. The latter would be able to call upon insights into the laws governing human behaviour gained from the study of history.

A major problem for Condorcet's ladder model of progress had been the Enlightenment's negative image of the Dark Ages: how could an inevitable, law-governed advance have slipped back a stage, even temporarily? Saint-Simon and Comte solved the problem by suggesting that religion created the stable order needed for each stage of social development to consolidate itself before the next advance became possible. For them the Middle Ages were no longer dark, because the Church had played a necessary role by establishing a workable relationship between the spiritual and temporal powers. This insight made them aware of the role that religion would have to play in any stable society, including any future utopia. The new religion would be secular, with the experts playing the role of priests, but it would be necessary to have something equivalent to the Church so that the future society could function harmoniously. Critics thought their efforts to found a new religion only confirmed that they had lost their grip on reality.

Comte Henri de Saint-Simon's career straddled the Napoleonic era and the restoration of the monarchy. He was an eccentric figure whose literary productions combined valuable insights with implausible speculations. His name came to be associated with socialism (because he insisted that future progress must benefit the workers) and authoritarianism (because the reforms would be instituted by the rulers). Like the previous generation he was convinced that progress came only through the efforts of men of genius who gained new levels of understanding that had to be disseminated to the masses. His views on who should constitute the ruling class expanded in the course of his career, beginning with the applied scientists (he was less interested in pure science) and later including merchants, industrialists, bureaucrats and experts in the new social theories he was developing. The rule of these experts would have to be imposed through institutions that would give the people a sense of harmony equivalent to that originally propounded by Christianity. Social cohesion was the key, but it would depend on better standards of material comfort generated by developments in applied science and industry.[27]

The new understanding of society derived in part from the study of history and the law of progress it revealed. The law worked through humanity's efforts to better itself, and it was only now that it became possible to understand its

---

[27] See *Henri Saint-Simon (1760–1825): Selected Writings*, trans. and ed. Taylor; R. Manuel, *The New World of Henri Saint-Simon*; and on his followers Iggers, *The Cult of Authority* and Iggers's edition of *The Doctrine of Saint-Simon*.

complex operations and control future developments. Saint-Simon modified Condorcet's ladder of social development by suggesting that the successive stages were not simply cumulative but fell into two distinct and almost antagonistic categories, a position reminiscent of Hegel's dialectic. There were organic stages, during which new social structures were constructed, and critical or revolutionary stages, in which the previous order was challenged to make way for something better. The Hellenic period and the Middle Ages were organic, the collapse of Rome and the modern period revolutionary. A new organic period was now emerging that would lead to a final utopia in which all would be happy. Saint-Simon's followers were enthusiastic about his vision of historical inevitability, praising his view for 'the truly great and imposing sight of mankind slowly fulfilling the law to which it is subject, offering in the course of history a long sequence of corollaries linked with each other which make possible, by correct appraisal of past events, the determination of those that will follow'.[28]

Auguste Comte began his career under Saint-Simon's direction but then struck out on his own. His *Cours de philosophie positive* (1830–42) outlined his views on the development of the sciences, including his new science of 'social physics', while his *Système de politique positive* (1851–4) provided the practical applications of his ideas including the construction of a new religion to promote social harmony. His efforts to found the new religion aroused considerable scepticism and have often been regarded as a sign of increasing mental aberration, but Mary Pickering's monumental biography has shown that he had recognized the need for some emotional foundation for the system from the beginning.[29] For many of Comte's later followers, positivism came to mean a focus on observation to discover the laws of nature without recourse to theory, although Comte himself was by no means so dogmatic on this point.

Positivism was based on a model of history which supposed three stages in the development of humanity's efforts to understand the world, with corresponding social structures. Comte did not go into the same level of detail on history as Hegel and seems not to have derived inspiration from the latter's work, although he did appreciate Kant's vision of universal history. But Frank Manuel argues that there are close parallels between Comte's and Hegel's thinking. Both had a view of history that was teleological and to some extent determinist: events were tending towards a definite goal, a future utopia – although unlike Hegel, Comte was happy to predict exactly how he thought the future society would operate.[30]

---

[28] From the Saint-Simonians' lectures edited by Iggers in *The Doctrine of Saint-Simon*, p. 36.
[29] Pickering, *Auguste Comte* (3 vols.).
[30] Manuel, *The Prophets of Paris*, p. 287; see also Pickering, *Auguste Comte*, vol. 1, p. 282 and on Kant pp. 289–302.

At the beginning of the *Cours* Comte defined the three main stages of history in terms of a law of progress derived from Condorcet and Saint-Simon. 'The law is this: that each of our leading conceptions – each branch of our knowledge – passes successively through three different theoretical considerations: the theological, or fictitious; the metaphysical, or abstract, and the scientific, or positive.'[31] In the theological stage, everything is explained in terms of the activity of supernatural beings, culminating in monotheism. The metaphysical stage imagines mysterious forces inherent in nature, corresponding to the early forms of science. In the positive phase the hollowness of these concepts is exposed and knowledge becomes based on laws derived from observation (hence the later positivists' distrust of theory). There are natural associations between these methods of interpreting the world and the contemporary social structures. The theological stage encourages militarism, while the transition to positivism is linked to the rise of industrialism.[32] At the present time the sciences are gradually being transformed through the adoption of the positive method. Eventually all will work in this way, including the new social physics, which will explain society in terms of the laws governing human behaviour.

Comte noted the analogy between his system of progress through successive stages and the development of the individual to maturity. 'The point of departure of the individual and of the race being the same, the phases of the mind of a man correspond to the epochs of the mind of the race. Now, each of us is aware, if he looks back upon his own history, that he was a theologian in his childhood, a metaphysician in his youth, and a natural philosopher in his manhood.'[33] Later, Comte invoked an overall parallel between the progressive development of life on earth and the history of humanity. The advance of the animal kingdom reveals the origins of those intellectual and moral characters that have become crucial to humankind.[34] He was, however, inclined to adopt a vitalist perspective in biology and had no interest in biological transformism.

The culmination of the progressive trend will be the creation of a future utopia in which the sciences have all been modernized and their discoveries applied. This will include both the applications of the physical and life sciences and the construction of a social order based on the new social physics. Comte wanted the benefits to be shared by all, including the workers, and the religion he was trying to formulate would ensure that all were able to recognize their place in a harmonious whole. The structure would be imposed by the scientists and the experts who knew how to apply their discoveries – although Comte was sometimes suspicious of experts because of their tendency to become

[31] I have used the extracts from the translation of the *Cours* in *Auguste Comte and Positivism*, ed. Lenzer; for this quotation see p. 71.
[32] Ibid., p. 293.   [33] Ibid., p. 73.   [34] Ibid., p. 279.

over-specialized. The new state would be the final goal of human social and intellectual development. Manuel notes that Comte seems to have assumed that there would be no further technological progress once all human needs were satisfied: the race would henceforth devote itself to artistic and other enlightening activities.[35]

The *Système de politique positive* with its extension of the religious motif was not a success, and Comte's legacy in France lay more with the scientists who adopted a rigid interpretation of his emphasis on observation. It was the more radical materialists who played a role in the French contribution to the next important development of the linear model, social evolutionism. They were more influenced by ideas emanating from Germany, where Comte's work was largely discounted. In Britain, George Henry Lewes took up the positivist baton, aided by Harriet Martineau's translation of the *Cours*, while J. S. Mill – despite falling out with Comte – was probably more influenced than he would later admit.[36] There were strong positivist movements in the United States and in Central and South America, where states such as Brazil were sometimes under the control of those who wanted to see 'order and progress'.

Social Evolutionism

The extent of the ladder of social progress expanded enormously in the years around 1860. Although often linked to the impact of Darwinism, this expansion was mainly produced by a revolution in archaeology. After decades of denying the evidence for human antiquity, scholars now accepted that there had been a long 'stone age' preceding the events recorded in the Bible. Prehistory offered an additional, indeed a better, field of evidence for technological and social progress, substantiating and elaborating the early ascent from 'savagery' postulated in the schemes of Condorcet and others. There was a demonstrable sequence from the most primitive stone tools through to the discovery of bronze and later iron, a sequence widely assumed to indicate progressively more sophisticated social activity. Here was a ladder of progress even more convincing than the irregular ups and downs of recorded history.

The archaeologists who demonstrated these developments provided hard evidence for what became known as social evolutionism. This was evolution – but it was not very Darwinian because its basic model of progress was a ladder leading towards the goal of modern industrial society, not a branching tree

---

[35] Manuel, *The Prophets of Paris*, pp. 294–5. For more details on Comte's utopia see Pickering, *Auguste Comte*, vol. 3, chap. 6.

[36] See Pickering, *Auguste Comte*, vol. 1, chap. 12 on Mill and vol. 2, chap. 10 on the wider influence, including pp. 257–65 on Lewes. See also McGee, *A Crusade for Humanity* and Simon, *European Positivism in the Nineteenth Century*.

driven by divergent adaptations. The relationship to Darwinism is complex and disputed. J. W. Burrow's classic study argued that the two movements were independent, social evolutionism emerging as a solution to the problem the utilitarians had encountered when trying to modernize Indian society. It was now argued that a people had to progress through the intermediate stages to advance to the top: there was no short-cut to a higher level of social organization. George Stocking's *Victorian Anthropology* argues that there were interactions between biological and social evolutionism but accepts that the anthropologists' and archaeologists' views were based on a linear and hence non-Darwinian model.[37]

The interaction between archaeology and the emerging field of anthropology was crucial because the evidence for human antiquity came solely from the stone tools being excavated; there were few human fossils and little direct evidence of how the early humans had actually lived. Since some contemporary peoples discovered in remote parts of the world still used a stone-age technology, it was all too easy to assume that these 'savages' illustrated the early stages of social evolution: they were in effect living fossils. 'Lower' races were primitive in the sense that they were stuck at an earlier stage in the advance. Anthropologists who accepted the unity of human nature could still imagine a ladder of cultural progress which allowed the inhabitants of some geographical regions to advance more rapidly than others.

The first steps towards prehistoric archaeology were taken by Scandinavians including J. J. A. Worsaae, who established that stone tools appeared first in the record, followed by bronze and then iron. At first their work did not seem to challenge the traditional timescale, because the remains lay on the periphery of the known world. Stone tools occasionally reported from more ancient deposits were dismissed as fakes or later intrusions. The situation changed dramatically in 1858 when the geologists William Pengelly and Hugh Falconer confirmed the association of stone tools with the remains of extinct animals in Brixham Cave, Devon. More discoveries followed, and by 1863 Charles Lyell's *Antiquity of Man* argued that the human race had been in existence for tens or even hundreds of thousands of years. In 1865 John Lubbock refined the Scandinavians' system by dividing the stone age into two phases, the Palaeolithic or Old Stone Age, when the tools were of chipped stone, and following it the Neolithic or New Stone Age, when more sophisticated polished stone came into use. Here was clear evidence of technological and presumably social progress far more extensive than anything displayed in recorded history.[38]

---

[37] Burrow, *Evolution and Society*; see also Stocking, *Race, Culture and Evolution* and *Victorian Anthropology*.

[38] See Grayson, *The Establishment of Human Antiquity* and Van Riper, *Men among the Mammoths*. For my own account see *Theories of Human Evolution*, chap. 1, and *The Invention of Progress*, part 2.

Further discoveries refined understanding of the Palaeolithic as French archaeologists explored sites which revealed different styles of stone toolmaking. Since some styles were more sophisticated than others they were assembled into a linear sequence of progressive development. Gabriel de Mortillet, an archaeologist with a strong ideological commitment to materialism and socialism, used the progressive sequence he created to support the claim that progress in the modern world is inevitable. Unusually for a French scientist of the time, he was a committed evolutionist who saw the Neanderthal type of ancient humanity as an intermediate between the apes and modern humans. While admitting that Neolithic technology might have been introduced from the East, he argued that the sequence in the Palaeolithic represented a progressive refinement of skills by the indigenous people of Europe. There were six phases: Chellean, Acheulian, Mousterian, Solutrean, Magdalenian, and Azilian, named after sites in France. The primitive handaxes of the Chellean were preceded by the controversial 'eoliths' of the Prechellean, which seemed to show signs of limited human action. Mousterian tools were associated with the Neanderthals. As de Mortillet claimed in his material for the Paris Exposition Universelle of 1867, the sequence demonstrated the validity of three propositions:

> Loi du progress de l'humanité
> Loi du développement similaire
> Haute antiquité de l'homme[39]

In other words, all humans had undertaken the same path towards progress over the vast period of time now demonstrated by science.

The archaeologists were anxious to reconstruct the lives of the ancient humans who had made the tools they unearthed. In the 1860s there were few fossil hominids to go on – only the much-debated Neanderthals. The most obvious source of information on how the toolmakers had actually lived seemed to be anthropologists' studies of modern 'savages', especially those such as the native Australians who still made use of stone tools. Anthropology was just emerging as an independent discipline in need of a theoretical paradigm, which – for the more radical – meant an alliance with the idea of evolution. The cultural evolutionism that emerged combined the utilitarian faith in progress towards an industrial economy with the historicist view that progress could occur only by following a law defined by successive stages of development. Lubbock, who had close links with Darwin, forged an alliance with the school of anthropology founded by E. B. Tylor in Britain and L. H.

---

[39] De Mortillet, 'Promenades préhistoriques à l'Exposition Universelle', p. 368; see also his *Le préhistorique*. On his ideological position see Hammond, 'Anthropology as a Weapon of Social Combat in Late Nineteenth-Century France'.

Morgan in the United States. The result was an approach far more in tune with the linear model of progress than with the Darwinian branching tree. For the anthropologists, the environment could influence only the rate at which a people could progress along a scale of complexity defined by a single goal.

The subtitle of Lubbock's *Prehistoric Times* proclaimed that the subject would be illuminated by a study of the 'Manners and Customs of Modern Savages'. Noting that palaeontologists sought to understand extinct animals by studying their modern relatives, he argued that the same technique would work for archaeology. Savages were not, as some religious thinkers argued, the victims of degeneration from a civilized state. They were simply stuck at an early state in an advance that had proceeded much further in more favoured parts of the world. 'In fact, the Van Diemaner [Tasmanian] and South American are to the antiquary what the opossum and the sloth are to the geologist.'[40] A second work, *The Origin of Civilization and the Primitive Condition of Man*, was subtitled *Mental and Social Condition of Savages* and opened with the claim that studying the lives of savages would throw light on our early ancestors. It would also encourage progress in the modern world by helping us to eliminate customs we still follow for no practical purpose; these are what Tylor would call 'survivals' from an earlier time. This made anthropology a 'reformer's science'.[41]

The assumption that modern savages represent the primitive condition of humankind was also proclaimed in the title of Edward B. Tylor's 1865 book: anthropology would reveal *The Early History of Mankind*. The eighth chapter was entitled 'The Stone Age – Past and Present', again explicitly linking anthropology with archaeology. A later book, *Primitive Culture*, openly acknowledged that scholars defined the goal of progress in terms of their own society: 'The educated world of Europe and America practically settles a standard by simply placing its own nations at one end of the social series and savage tribes at the other, arranging the rest of mankind between these limits according as they correspond more closely to the savage or to civilized life.'[42] Technical inventions were driven by necessity, which resulted in a fixed sequence of development that was still continuing today:

as experience shows us that the arts of civilized life are developed through successive stages of improvement, we may assume that the early development of even savage arts came to pass in a similar way, and that finding various stages of art among the lower races, we may arrange these stages in a series probably representing their actual sequence in history.[43]

---

[40] Lubbock, *Prehistoric Times*, p. 336.
[41] For more details see for instance Burrow, *Evolution and Society* and Stocking, *Victorian Anthropology*; Adam Kuper, *The Invention of Primitive Society*.
[42] Tylor, *Primitive Culture*, p. 23.   [43] Ibid., p. 57.

Tylor also showed how other aspects of culture, including language, mythology and religion, grew by the same process of ascending scales of development leading towards the present.

As he made clear in the conclusion to *Researches into the Early History of Mankind*, for Tylor there was really only one culture exhibiting varying stages of development, and there was no real cultural diversity: 'the facts collected seem to favour the view that the wide differences in the civilization and mental states of the various races of mankind are rather differences of development than of origin, rather of degree than of kind'.[44] This cultural sequence unfolded uniformly in pattern but at different speeds in the various parts of the world. The same innovations occurred independently among peoples isolated from one another, and even when one culture borrowed from another (the process known as diffusion) this was possible only because the recipient had reached the stage where it was about to invent the new technology or belief itself. Tylor concluded with a comparison between the development of culture and the process of individual development, arguing that the savage phase is equivalent to the passage from infancy to youth, rather than from youth to maturity.[45] The latter stage was by implication that reached by the modern West.

An even more complex scale of cultural development was proposed by the American anthropologist Lewis Henry Morgan in his *Ancient Society* of 1877. Morgan had studied kinship in the Iroquois and constructed his system without reference to archaeology. He defined three basic stages of savagery, barbarism and civilization, with a host of subdivisions based on technology and social habits. Although aware that the philologists saw language evolution as a branching tree, he insisted that the stages of cultural evolution formed a 'natural as well as necessary sequence of progress' of which all but the lowest could still be seen in the modern world.[46] Morgan became known as a pioneer of the view that economic factors shape social progress, but he also insisted that invention is guided by a 'natural logic which forms an essential attribute of the brain itself', giving uniform results wherever it is applied.[47] Economic forces interact with the human mind to give a preordained sequence of advance which – wherever circumstances make it possible – leads towards the goal represented by modern industrial society.

Morgan believed that races develop their mental capacities in parallel with their level of culture, a view to which Tylor also subscribed in his later writings.[48] Neither saw the level of mental development as something innate

---

[44] Tylor, *Researches into the Early History of Mankind*, p. 361.   [45] Ibid., p. 371.
[46] Morgan, *Ancient Society*, p. 11. See Kuper, 'The Development of Lewis Henry Morgan's Evolutionism' and Trautmann, *Lewis Henry Morgan and the Invention of Kinship*.
[47] Morgan, *Ancient Society*, p. 59.
[48] Ibid., pp. 38–40 and 59; Tylor, *Anthropology*, pp. 60 and 73–4.

that would permanently limit a race's ability to advance up the scale of development, and this approach allowed them to make common cause with the biologists' views on mental development. Cultural evolutionism flourished through the later decades of the nineteenth century, and its legacy persists in the popular image of the 'savage' as a living fossil representing a primitive form of human society. Its last great exponent was Sir J. G. Frazer, whose *Golden Bough* of 1890 offered an extensive account of how early magical and religious beliefs had evolved. In a later work Frazer drew an explicit comparison between the growth of the individual mind and the evolution of culture: 'For by comparison with civilized man the savage represents an arrested or rather retarded stage of social development, and an examination of his customs and beliefs accordingly supplies the same sort of evidence of the evolution of the human mind that an examination of the embryo supplies of the evolution of the human body.'[49] Once again we see the linear model of evolution illustrated by an appeal to the goal-directed process of embryological development, with all its teleological implications.

Industrial (R)evolution

It may seem a far cry from speculations about prehistory to the industrial developments of the nineteenth century, but the whole point of evolutionism for many thinkers was to use the past as justification for their expectations of future progress. The concept of an Industrial Revolution had emerged in the early decades of the century when people became increasingly conscious of the changes that were transforming their lives as new industrial processes took over the economy. It was by no means obvious that the resulting flood of manufactured goods would be of benefit to all: at first only the well-to-do could afford them, and the workers in the factories lived lives of endless hardship. Those who envisioned a more general social progress had to hope that eventually the benefits would be spread to all, ushering in the utopia of peace and plenty that was seen as the goal of history. The key question was: how was this dissemination of material plenty to be brought about? Social philosophers from rival ideological backgrounds all sought ways of linking their programmes for change to a pattern of development that could be discerned in the past. This is why from the 1840s onwards there were endless appeals to biological evolution and the records of the human past as evidence that the future would indeed be better than the present thanks to the driving force of history.

Technological innovation was the motor of progress, but how was it to be directed for the common good? The later nineteenth century saw endless

---

[49] Frazer, *Psyche's Task*, p. 162. See Ackerman, *J. G. Frazer*.

speculations in fiction about the benefits that could be conferred by future innovations, a genre that has been called 'technological utopianism'. The most successful writers are still remembered today, Jules Verne being an obvious example. All too often, however, their speculations were constrained by expectations imposed by their current interests: the future would be more comfortable and more interesting because communication, transport and everyday household appliances would be better. Everyone would live a life of luxury as that was understood in terms of existing middle-class values, so the end of history would be a world of comfortable people free to engage in whatever leisure pursuits interested them. In William Delisle Hay's *Three Hundred Years Hence* of 1881 the human race has moved underground to leave the earth's surface free for mechanized agriculture, but the facilities offered in the subterranean cities are pretty conventional in scope. There is little indication that everyday life will be transformed beyond the obvious conveniences – a rather banal end to history. The darker side of the progress towards this utopia is evident from Hay's glorification of a process by which the non-white races who cannot participate are eliminated.[50]

These rather conventional utopian predictions tended to assume that political developments would bring about the required equalization of access to the benefits of technology. But would this come about by a gradual reform of the existing conditions, or would some more revolutionary activity be needed to force the owners of the new industries to relinquish their hold on the profits? Comte's answer to this question assumed a gradual imposition of the rule of experts, an expectation that would come to be taken seriously only in the twentieth century. In the meantime two possibilities were conceived. The most radical was Marx's call for a communist revolution to allow the workers to take control of the new industrial economy. More popular at the time, at least in the English-speaking world, was the evolutionary philosophy of Herbert Spencer.

The 'synthetic philosophy' that Spencer propounded from the 1850s onwards drew on the whole sweep of history to argue for the inevitability of progress towards more complex systems. The reforms that would bring about a more perfect society were merely the last stage in this process. The scheme drew on a vista opened up by biological evolution and the development of human societies to create a philosophy of nature based on the law of progress. There were similarities to the kind of cultural evolutionism proposed by Tylor and others via the claim that the emergence of modern industrial society was an inevitable follow-on from the earlier phases of development. For Spencer,

---

[50] Hay's *Three Hundred Years Hence* is reprinted in Clarke, ed., *British Future Fiction*, vol. 2, pp. 1–370. On predictions of future technologies see Segal, *Technological Utopianism in American Cutlure* and my own *A History of the Future*.

the motor of progress was individual competition, which forced everyone to exert themselves to the full and innovate in response to environmental challenges. This has led to his philosophy being seen as a form of social Darwinism, although in fact he argued that the struggle for existence would stimulate individuals to improve themselves, the benefits being transmitted to future generations by the Lamarckian process of inheritance. The 'robber barons' of American capitalism thought that Spencer's philosophy justified their activities, but he envisaged a world in which everyone would become perfectly adapted to the operations of the free-enterprise marketplace. This was the ultimate goal of social progress.[51]

For all his emphasis on the inevitability of social evolution, Spencer had little interest in the archaeological evidence, and his views were formulated without reference to the anthropologists' approach. The implication that a free-enterprise society is the inevitable goal of social progress was masked by an appreciation of the diversity of cultures and the divergent nature of both biological and social evolution. For Spencer the trend towards progress was inevitable, but it operated in many branches, of which only few broke through to higher levels of organization (biological or social). The line of development could be reconstructed by hindsight, but one could not have predicted the actual winners in advance. His was a hybrid of the linear and branching models of progress, and we shall return to it in discussing the impact of the Darwinian model. But for his many followers, the assumption that free enterprise was both the motor of innovation and the goal of progress seemed to point the way towards a single goal.

Marxism

Spencer's liberal followers assumed that social evolution would eventually extend the benefits of the free-enterprise system to all. For the many critics of the system it seemed obvious that the robber barons would never give up their ownership of the factories and would block any efforts to spread the wealth of material goods to all. In this case, the only recourse would be a revolution that would topple them from their position and give power to the people. The communist ideology developed by Karl Marx and Friedrich Engels provided a basis for the movement that would promote this revolution, providing it with a foundation based on yet another interpretation of social evolution as the ascent of a linear sequence of phases towards a predictable goal. Marxism would become perhaps the best-known linear model of progress, especially after it

---

[51] The classic interpretation of Spencer as a social Darwinist is Hofstadter's *Social Darwinism in American Thought*. For an expansion of the alternative position outlined here see my 'Herbert Spencer and Lamarckism'.

became the basis for the Soviet Union's bid for world-domination in the mid-twentieth century.

Here was a new form of materialism in which the economists' focus on the production of the goods necessary for life and comfort was sharpened into a revolutionary creed by exposing the necessary inequalities of the capitalist system – and showing how to eliminate those inequalities in the future. Saint-Simon's call to spread the benefits of industry to all was acknowledged, along with Hegel's emphasis on society as a dynamic entity being transformed by historical forces. The result is often seen as Hegelianism stood on its head, with the progressive sequence of social structures produced by the dialectic leading to a utopia defined by the provision of material comfort for all. This new version of social evolutionism offered a step beyond the free-enterprise utopia of Spencer to a world where all would share equally because private property had been eliminated. Earlier utopians had dreamed of establishing such a world; Marx and Engels thought they had discovered the means to achieve it by engineering a workers' revolution.

Scholars now have serious problems with the orthodox interpretation of the Marxist view of history, analogous to those they have with Hegel's views on the topic. Later Soviet ideology was rigidly materialistic and determinist, suggesting that social evolution is driven inevitably in a single direction by the rigid laws of economics:

All peoples travel what is basically the same path ... The development of a society proceeds through the consecutive replacement, according to definite laws, of one socio-economic formation by another. Moreover, a nation living in the conditions of a more advanced formation shows other nations their future just as the latter show that nation its past.[52]

There are modern efforts to present the Marxist system as determinist, but the above quotation is from the analysis by Melvin Rader, who – along with others – challenges this interpretation.[53] Human actions played a role, which is why it was necessary to be a revolutionary rather than wait passively for the laws of social evolution to do the work. As with the revisionist view of Hegel's philosophy of history, there can be no mechanical unfolding of a preordained pattern of development. Nor is it clear that Marx and Engels saw the communist future as a static utopia beyond which there could be no further development. That future might represent the end of social history, but it would leave

---

[52] From the Moscow 1963 edition of *Fundamentals of Marxism-Lenininsm*, quoted in Rader, *Marx's Interpretation of History*, p. 129.

[53] In addition to Rader, cited above, see Ferraro, *Freedom and Determination in History According to Marx and Engels* and Gandy, *Marx and History*. For more determinist interpretations see Cohen, *Karl Marx's Theory of History* and Bober, *Karl Marx's Interpretation of History*.

us free to make contributions to human experience in other areas. To many later Marxists, the Soviets' increasing use of state power to impose an order on society in the name of historical inevitability was a betrayal of the founders' vision.

Marx and Engels issued their *Communist Manifesto* in 1848, the year of revolutions, and although it went unnoticed at first, by the end of the century it had been translated into a number of languages. Marx wrote his *Critique of Political Economy* at this time, although it was not published until 1859, followed by *Capital* in 1867. Engels expanded their materialist viewpoint in his attacks on the philosophy of Eugen Dühring in 1877–8, and part of this appeared later under the title *Socialism: Utopian and Scientific*. Engels also extended their scheme of history in his *The Origins of the Family, Private Property and the State* of 1882.

The historical dimension of the programme adapted Hegel's dialectic by depicting social evolution as a series of transitions in which each stage generates within itself the conditions that would eventually spark the next upward step. Social evolution was driven by the struggles between competing social classes. The ancient world's economy was based on slavery; this gave way to feudalism. The feudal aristocracy was challenged by the rise of industry, at first organized by guilds, but increasingly in the hands of powerful individuals who displaced the aristocracy in a series of revolutions. The industrial elite, the bourgeoisie, simplified the class system by driving everyone else into the proletariat, the wage-slaves who provided the manpower for their increasingly mechanized factories. Marx recognized the bourgeoisie as a revolutionary class in its own right, which supervised a massive expansion in productive capacity unlike anything seen in previous history. But it had created its own nemesis, the proletariat, whose situation became desperate as the flood of material goods led to periodic economic crises rather than general well-being. This desperation would eventually lead to a revolution, overthrowing the class system to establish a world without private property. The conclusion to the *Communist Manifesto* outlined the key elements of the future utopia in which all would be free to express themselves.

Marx and Engels later claimed to have done for the social sciences what Darwin did for biology, class war substituting for the individualistic struggle for existence. This ignored the fact that Darwin postulated no final goal for evolution; the attraction seemed to be that natural selection was radically materialistic (the legend that Marx offered to dedicate *Capital* to Darwin has long been discredited).[54] Their writings contain passages that encourage an impression that they were advocating a non-Darwinian vision of progress

---

[54] Colp, 'The Myth of the Darwin–Marx Letter'; Fay, 'Did Marx Offer to Dedicate *Capital* to Darwin?'; Feuer, 'Is the Darwin–Marx Correspondence Authentic?'

through a preordained hierarchy. In the preface to *Capital* Marx suggested that 'The country that is more developed industrially only shows, to the less developed, the image of their future.'[55] There are occasional appeals to the analogy with the individual life-cycle, as when Engels wrote of the bourgeoisie developing in the womb of the Middle Ages, following this with the image of it arising from the burghers of the feudal period like a butterfly from a chrysalis.[56] His account of the origin of the family and private property made use of L. H. Morgan's stage theory of social development, thereby focusing attention on the latter's appeal to economic factors.[57]

Revisionist historians point out that there are also passages contradicting the impression of linearity and inevitability. In several later letters, Engels explained that the early focus on law-like economic forces was an over-simplification intended to draw attention to the most innovative part of the new approach that he and Marx were developing.[58] Social forces were influenced by a range of political and cultural factors. As with Hegel's system, the dialectic worked through human activity, which translated the law-bound forces into a complex web of interactions determining the rate of change. The situation in Russia posed questions the system could not answer. Was it possible that here a revolution might be achieved without passing through a fully developed capitalist phase? If so, the sequence of developmental stages was not rigidly predetermined.[59]

As for the future utopia, Marx and Engels may have thought it marked the end of social evolution, but they stressed that it would allow everyone the freedom to develop themselves in other fields. Curiously, one area where they do not seem to have recognized a role for creativity was technological innovation: they assumed that new techniques would be invented more or less automatically when economic development had need of them. Later Marxists tended to ignore individual creativity to see rational planning as the only way towards a workers' utopia. Despite Marx's prediction that the state would wither away, in Soviet Russia it became ever more powerful as the need to direct industrial production became acute. Lenin even seems to have hoped that the psychological conditioning developed by Ivan Pavlov might be used to

---

[55] Marx, *Capital*, vol. 1, preface, pp. 8–9.
[56] Engels, *Herr Eugen Dühring's Revolution in Science*, pp. 120 and 122; see also *Marx and Engels: Basic Writings on Politics and Philosophy*, p. 276.
[57] Engels, *The Origins of the Family, Private Property and the State*; see also the note added to the first page of the 1888 English edition of Marx and Engels, *The Communist Manifesto* (p. 79 of the edition cited). Marx and Engels's American publisher, Charles H. Kerr of Chicago, also reprinted Morgan's *Ancient Society*.
[58] Engels, letters to Joseph Bloch, Conrad Schmidt and Heintz Starkenburg; see *Marx and Engels: Basic Writings on Politics and Philosophy*, pp. 397–412.
[59] Preface to the 1882 Russian edition of *The Communist Manifesto*, p. 56 in the edition cited.

impose uniformity on the population.[60] The dream of perfecting human nature took on a new dimension as science gained the power to control the future of the race by artificial means.

As it gained influence towards the end of the nineteenth century, Marxism joined an array of theories of social evolution modelled on a hierarchical sequence of developmental stages. Whether derived from Comte's expectation of the rule of experts, Spencer's enthusiasm for free enterprise or Marx's adaptation of the Hegelian dialectic, all promised an ascent towards a materialistic equivalent of the Christian paradise as the eventual goal. All claimed that the pattern of progress revealed by the past offered a guide to the future, defining both the goal and how it was to be achieved. Each hoped to bring about the benefits anticipated in the Enlightenment, but in a different way. All survived into the twentieth century even as the whole concept of progress became more complex and the goal less clearly defined. All would face the prospect that the dream of utopia might equally well turn into a nightmare. The late nineteenth century marked the apogee in the rising fortunes of the idea of progress, but a decline would soon set in – paralleling the cycles of maturity and degeneration envisioned in the ideas themselves.

[60] See the section 'Engineers of the Human Soul' in Figes, *A People's Tragedy*, pp. 732–51.

6   End of an Era?

The twentieth century is often portrayed as the period in which faith in the idea of progress collapsed in the face of calamities such as the Great War and economic depression. Concerns about the depletion of resources and environmental degradation were already being expressed as the century began. But to focus exclusively on the pessimistic elements of twentieth-century thought deflects attention from those elements in which the hope of future progress remained active and are still with us today. Part II of this study focuses on the transformation of the idea into a more open-ended, less utopian approach to progress in which the main driving force was supposed to be technological innovation. This has its problems too, but it has allowed the technophiles to retain the expectation that life can be improved and opportunities for self-fulfilment extended in the future, as in the past. Before moving on to this very different model of progress, however, it must be noted that efforts to defend the more developmental vision of history were still being made in the early twentieth century – and indeed they are still being made today. The assumptions that the human race is the predestined end-point of progressive evolution, and that some version of utopia is the goal of social development, have remained in play even as the less structured approach to history has been consolidated.

The model of history based on an analogy with the individual's development to maturity could itself have darker implications. Freud's psychoanalytic theories drew on this source, as did Oswald Spengler's efforts to undermine the faith of those who cited the achievements of the nineteenth century as evidence for a general trend to progress. Spengler pointed out that such a rise was always followed by a decline to eventual extinction as civilizations endlessly repeated the rhythm of birth and death.[1] These pessimistic inversions of the linear model of progress are the more visible of its legacies, but its original and more positive aspects were also still at work. In the social sphere the most obvious manifestations of this optimistic vision is the continued influence of

---

[1] See Fischer, *History and Prophecy* and Farrenkopf, *Prophet of Decline.*

Marxism and its assumption that social evolution must rise through the pattern of historical inevitability towards a communist utopia. When Marxism eventually failed, there was an even more transitory effort to revive the Hegelian view of history. This rival ideology also had an equivalent in the liberal Christian churches, where faith in the ultimately purposeful outcome of history remained active. Some Christian socialists combined the insights of Marxism with an unorthodox interpretation of traditional theology. Although we see here a collection of apparently different ideological positions, they shared an underlying faith that history was advancing in a predetermined direction towards a final goal – which is why apparent enemies could sometimes find common cause.

The Goal of Evolution

The belief that there is an underlying pattern giving meaning to the apparent diversity of historical development remained active in the life sciences, though with decreasing influence. There have been continued efforts to defend aspects of the chain of being by evolutionists who insist that the human species can still be regarded as the predestined goal of evolution. This suggestion has come to seem increasingly out of place in the context of modern Darwinism, yet it still has the power to excite those who seek an overall purpose in the cosmic drama. In the early part of the twentieth century, the assumption that there was a more or less inevitable drive towards higher levels of mental and moral activity built into nature was for a time the dominant vision of how our species originated.

Palaeoanthropologists from a variety of ideological backgrounds debated rival interpretations of how the inbuilt trend operated and where the fossils now being discovered fitted in. But the whole debate was premised on the assumption that multiple lines of hominid evolution had tried to advance towards the same goal, driven by an inbuilt trend towards larger brains and higher mental functions. Our own species had come to dominance because it was the most successful candidate in this race to perfection. Other evolutionists defended the claim that the whole sweep of evolution in the animal kingdom represents an ascent towards the human level, all other forms being reduced to the status of side-branches or dead-ends. This viewpoint even managed to survive into some interpretations of the new synthetic Darwinism that began to dominate evolutionary biology in the mid-twentieth century.

In his *Descent of Man* Darwin made the radical suggestion that the growth of human intelligence was a by-product of our distant ancestors' adaptation to a new environment: they moved from the trees to the open plains, stood upright and then used their hands to begin making primitive tools. Getting a bigger brain was a by-product of an adaptive transformation in locomotion.

Most of his contemporaries preferred to believe that because our more highly developed mental faculties define our superiority over the rest of the animal kingdom, the addition of those faculties must have been the main driving force in our advance beyond the apes. The expansion of the brain must surely have begun before any mere adaptive transition to bipedalism, and indeed most thought that standing upright was an inevitable consequence of increased intelligence.

In the early twentieth century it was widely accepted that several branches of the human family tree had been ascending towards the same goal, our own being the most successful. It was still taken for granted that the driving force of this trend was the tendency to acquire higher levels of intelligence. The 'brain first' theory continued to be articulated by anatomists such as Grafton Elliot Smith. Far from seeing a bigger brain as the consequence of an adaptive shift to bipedalism, Smith thought that only the brightest of the early hominids saw the advantages of moving out onto the open plains. In his *Evolution of Man* of 1924 he wrote of Asia and Africa as 'the laboratory in which, for untold ages, Nature was making her great experiments to achieve the transformation of the base substance of some brutal Ape into the divine form of Man'. In his preface he conceded that 'in attempting to attain conciseness of expression I have used teleological phraseology in many places as a matter of convenience, and not from any idea of accepting Teleology'. In fact it is hard to avoid the impression that Smith and many of his contemporaries were still enmeshed in a world-view that took it for granted that somehow the modern human form was the predestined outcome of the evolutionary process. Anatomists and archaeologists from a variety of ideological backgrounds all shared the same assumptions.[2]

In America Earnest A. Hooton's *Up from the Ape* of 1931 endorsed Elliot Smith's claim that the adoption of bipedalism was 'not the result of environmental accident, but rather a manifestation of that superior intelligence and initiative which, inherent in the proto-human stock, determined its evolutionary destiny'. The trend towards increased intelligence had become established in certain branches of the ape stock, leading them to evolve in parallel to give rise to the races of modern humanity. Hooton had no hesitation in claiming that the races are distinct species and that the whites had advanced further than the others.[3] William King Gregory's *Our Face from Fish to Man* included a frontispiece depicting a single line of development by which the facial features gradually approached the human form.

---

[2] Smith, *The Evolution of Man*, p. 77 and pp. 19–20. See Bowler, *Theories of Human Evolution*, chap. 9.
[3] Hooton, *Up from the Ape*, p. 115 and on races as distinct species p. 395.

The popularity of the 'brain first' theory of human origins ensured that the discovery of the fossil hominid *Australopithecus* by Raymond Dart in 1924 received little attention. Unearthed in South Africa, it revealed a creature that had stood more or less upright yet had a brain little bigger than that of an ape. This was a vindication of Darwin's suggestion that it was bipedalism, not increased brain-power, that defined the human family: greater intelligence was not the driving force but a by-product of an adaptive change in locomotion. Although ignored at first, further discoveries by the South African palaeontologist Robert Broom in the 1930s confirmed the role of the Australopithecines and sounded the death knell of the 'brain first' theory. As Darwinism itself began to dominate evolutionary biology thanks to the synthesis with genetics, ideas about human origins also began to take on a more Darwinian aspect.

Curiously, Broom himself played a significant role in the attempt to defend a vision of progressive evolution in which the human form defined the goal towards which the main line of progress is aimed. He was a deeply, if unconventionally, religious man, sympathetic to the position adopted by the long line of liberal Christian thinkers who saw evolution in terms of spiritual progress. But he recognized that it was now necessary to redefine the key feature of the trend of which we are the final outcome: simply postulating a mysterious upward urge was no longer plausible. His argument drew on a point made by the neo-Lamarckian palaeontologist Edward Drinker Cope in his 'law of the unspecialized'.[4] Most branches of the tree of life represent adaptive specializations and are side-branches often ending in extinction. It is the organisms that resist the temptation to specialize, the generalists with flexible lifestyles, which preserve the hope of progress because they can add on new characters with wider applications. The argument that modern humans are the last example of this long line of generalists and hence represent the goal of evolution was expanded by Broom and then passed on to Julian Huxley and beyond. It preserved the idea of an 'end of history' implicit in earlier models of progress without implying that there has been a predetermined pattern of ascent. The 'main line' of development can be recognized only with hindsight; the fossil record alone identifies the generalists which evaded the tendency to specialize in each class. But the insistence that humans mark the end-point of the only meaningful sequence of steps in the ascent of life ensures that the notion of a goal for evolution is preserved.

Broom made his scientific reputation by studying fossils transitional between reptiles and mammals and then became well known for his discovery of the remains of fossil Australopithecines. Although these discoveries undermined the idea that the enlargement of the brain was a basic evolutionary trend,

---

[4] Cope's 'law of the unspecialized' was proposed in 1879; see his *The Primary Factors of Organic Evolution*, pp. 172–5, and Bowler, *Life's Splendid Drama*, p. 207.

Broom remained convinced that evolution exhibited a divine plan aimed at the production of humankind. In his *The Coming of Man: Was It Accident or Design?* of 1933 he argued that the branches of the tree of life consist largely of blind alleys which lead off to adaptive specializations, preventing any further development and often leading to extinction. The main line of progress is defined by the sequence of generalized forms that managed to evade the demands of adaptation and thus retained the flexibility necessary for further progressive development. The human species was now left as the only heir to this trend, progress elsewhere having ceased because of over-specialization. This 'seems to drive us to the conclusion that there was no need for further evolution after man appeared, and that the evolution of man must have been deliberately planned by some spiritual power'.[5]

Julian Huxley was one of the architects of the 'modern synthesis' of Darwinism and genetics and as such might have been expected to reject a teleological interpretation of evolution. But although he developed a philosophy of humanism, he had close links with many liberal Christians and shared their conviction that evolution must have an ultimate purpose expressed in human values. Huxley proposed a general definition of progress based on the introduction of new functions giving organisms better control of their environment. But he also wrote an essay on 'The Uniqueness of Man' in which he argued that the human species has a special place at the end-point of the most progressive sequence.[6] In search of the sources of this conviction Marc Swetlitz draws attention to his connections with Broom. A footnote in the concluding remarks of Huxley's classic *Evolution: The Modern Synthesis* criticizes Broom for supposing that there is a spiritual directing force in evolution. In fact, however, there was some similarity between their views on the topic, the main difference being that Broom was more open about the teleological implications of his approach. In the context of evolution, both Broom and Huxley saw the human species as the 'end of history'.[7]

Huxley rejected the need to invoke a divine plan, but his own view of progress was based on the claim that there was a crucial sequence of progressive steps that was necessary to bring about the emergence of human consciousness. Each step could be made only by generalized forms with the flexibility to add on completely new structures and functions, yet at each step the majority of new lines of evolution led off towards over-specialization, cutting them off from further progress. The key transitions that were necessary

---

[5] Broom, *The Origin of Man*, p. 218. [6] In Huxley, *The Uniqueness of Man*, pp. 1–21.
[7] Huxley, *Evolution: The Modern Synthesis*, p. 568. See Swetlitz, 'Julian Huxley and the End of Evolution'. Other studies of Huxley on progress include Ruse, *Monad to Man*, pp. 328–38 and 349–54, and Gascoigne, 'Julian Huxley and Biological Progress'; several contributors to the collected volume edited by Waters and Van Helden, *Julian Huxley* also address the issue.

for progress included the emergence of the amphibians onto dry land and the transition to warm-bloodedness (shown by Broom's mammal-like reptiles). More recently the primates had to move into the trees to gain visual acuity and grasping hands, but then a branch of the family had to move back onto the open plains to develop socialization, toolmaking and, ultimately, self-consciousness. The apes were now over-specialized for tree-dwelling.

> One somewhat curious fact emerges from a survey of biological progress as culminating for the evolutionary moment in the dominance of *Homo sapiens*. It could apparently have pursued no other general course than that which it has historically followed: or, if it be impossible to uphold such a sweeping and universal negative, we may at least say that among the actual inhabitants of the earth, past and present, no other lines could have been taken which would have produced speech and conceptual thought, the features that form the basis for man's biological dominance.[8]

Huxley also shared Broom's view that no other species was capable of following in our footsteps should we disappear from the earth. He may have been reluctant to openly endorse the existence of some directing force in evolution, but clearly there was only one sequence capable of achieving the goal of consciousness, and that was by definition the main line of development. Unless one were to accept that consciousness wasn't really important on a cosmic scale, the impression that it was somehow intended to come out this way was clear. Huxley would eventually throw his weight behind Teilhard de Chardin's explicitly spiritual vision of progress.

Most working biologists were suspicious of Huxley's efforts to preserve the idea of a main line of progress leading to humanity. Historians now tend to assume that he was influenced by a residue of religious faith hidden behind his philosophy of humanism. Other adherents of the new Darwinian synthesis were more comfortable with the idea of a general progressive trend in evolution but not with the claim that humans had a unique status as the end-point of the most successful line of development. A few were suspicious of the whole concept of progress, Stephen Jay Gould being an articulate critic of the claim that one organism can be defined as 'higher' than another. Gould disagreed with several components of the Darwinian synthesis, but he did champion perhaps the most radical application of its principles. His analogy of 're-running the tape' of life's development allowed him to make the claim that the number of contingencies involved would ensure that any re-run would be unlikely to produce a world anything like the one we actually know. He argued that the bizarre fossils revealed by the ancient Cambrian rocks of the Burgess shale were 'experiments' that had been winnowed out by chance, leaving only the ancestors of the living types as the lucky survivors. Even if there was a

---

[8] Huxley, *Evolution: The Modern Synthesis*, p. 569.

general tendency to progress to higher states, it was unlikely that anything like the human species would result from a re-running of earth history.

Gould's position has been challenged by the Cambridge palaeontologist Simon Conway Morris, who offers the most serious effort by a modern biologist to defend the claim that humans are the inevitable outcome of evolutionary progress. In 1998 his *Crucible of Creation* used studies of the Burgess shale fossils to show that most could, in fact, be placed within the existing phyla. Gould's vision of an open-ended Cambrian explosion was false, and so was his ultra-Darwinian model of evolution as a sequence of contingent events. Conway Morris has no time for mysterious evolutionary trends, appealing instead to the phenomenon of convergent evolution. Convergence is when two quite separate lines of evolution independently develop the same structure because both are adapting to the same way of life. Darwinians accept that this can happen, but have been unwilling to see it as a major factor because their whole approach accentuates the role of undirected variation. It was the Lamarckians of the 'eclipse of Darwinism' who were most attracted to convergence because it allowed them to argue that evolution was directed by the deliberate life-choices made by organisms.

Conway Morris is no Lamarckian, using convergence instead to limit the open-endedness of Gould's vision of how the tree of life unfolds. He argues that convergence is so pervasive that it imposes major limitations on the possibilities of adaptive innovation. Re-running the tape of history would not produce significantly different results, because the ability of living things successfully to develop new structures and functions is highly restricted. The branches of the tree of life are predictable, not because of any developmental force but because sheer practicality often ensures that there is only one way to achieve a really efficient structure. Our eyes resemble those of some cephalopods simply because that's the best way of making a functioning visual apparatus.

The conclusion to *The Crucible of Creation* spells out the wider implications of Conway Morris's position, which are then developed more fully in his 2003 book *Life's Solution: Inevitable Humans in a Lonely Universe*. He argues that one factor defining an important convergence is the advantage of developing a bigger brain. Thus although it would be silly to think of all the branches of the tree of life trying to become human, it is the case that the rise of intelligence and ultimately consciousness is inevitable in the long run. This occurs in the line leading to humanity because there is a crucial combination of convergences that boost the effect: each factor has been developed independently several times, but all have to be added together for the best result. Many creatures have developed a high level of intelligence (dolphins, for instance), and many have developed vocal communication (including birds) – but it was the combination of such factors in the line leading to humanity that gave rise to our unique mental powers.

This may not represent a 'main line' of evolution in quite the sense implied by the old chain of being, but it does suggest that it is legitimate for us to identify the line leading towards ourselves as the one that finally achieves the goal the whole process was intended to bring about. Modern humans are simply the most effective combination of these functions, and sooner or later that coming-together was bound to occur. If our ancestors hadn't achieved it, some other species would have done it anyway. Just how 'human' such alternatives would be is unclear; the most likely candidate would be another primate, but Conway Morris admits that some less closely related form might be successful. In this case we might wonder whether the similarity to ourselves would be rather superficial, although the defining features would have to be the same. The cosmic significance of the achievement is then driven home when he suggests that the conditions suitable for the emergence of life are so precise that – despite the increasing number of extra-solar planets being discovered – the earth may be the only place in the universe where the process could occur.[9]

In the battle to resist the rise of modern Creationism, Conway Morris hopes that his arguments will allow Christians to accept evolution without giving up the central pillar of their faith. Humanity, or something very much like it, is the goal of evolution and, perhaps, the Creator's central concern. If one thinks the outcome is of cosmic significance, it is still plausible to talk of a main line of ascent and even an end-point where the purpose of the whole process has been achieved. Conceptually at least there is a parallel between this position and Francis Fukuyama's claim that modern liberal capitalism is the 'end of history' defined by a Hegelian sequence of inevitable stages of social evolution. While no equivalence in philosophy or world-view is implied, both models of history are intended to show that what we see in the world today (modern humans and Western liberalism) is the intended end-product of a process that had to be followed if the world was to have any real purpose. Fukuyama backed away from his 'end of history' thesis as it became clear that the goal was by no means as clear-cut as he had implied. Conway Morris, at least, is sticking to his guns.

Spiritual Evolution

Broom and Huxley had created a more sophisticated version of the position that had long been accepted by religious thinkers who hoped to make evolution compatible with the view that humankind is the primary focus of creation. The

---

[9] Gould's claims sparked controversies among philosophers of science: see Beatty, 'Replaying Life's Tape'. The debate between Conway Morris and Gould is the centrepiece of Losos, *Improbable Destinies*, which accepts the evidence for the ubiquity of convergence but insists that there is good evidence of unpredictable developments in evolution.

chain of being had always been interpreted as a divine plan, and the analogy
between the history of life on earth and the goal-directed development of the
human embryo had appealed to liberal Christians and others hoping to retain
humanity's cosmic significance. Efforts to replace natural selection with some
more purposeful evolutionary process played an important role even in science
during the 'eclipse of Darwinism'. In 1894 Henry Drummond's *Ascent of Man*
enjoyed huge success for promoting the view that the struggle for existence
was not the driving force of evolution: instead it was the drive to cooperate that
led animals to progress, leading inevitably to the appearance of humanity's
moral character.[10]

In the early part of the twentieth century there was a determined effort by the
Modernist movement in the churches to challenge the Victorians' assumption
that science was inherently hostile to religion. Adopting a strongly teleological
approach to evolutionism played a key role in their strategy, preserving a relic
of what Eiseley called the transcendental man-centred version of progress
popular in the nineteenth century. Even agnostics with little interest in the
explicitly Christian position could still find it hard to throw off the assumption
that there was an inherently progressive trend in evolution of which we
humans are the highest product. This was an interpretation of the idea of
progress that appealed both to liberals in the churches and to some scientists
of the older generation who were still active in popularizing the older theories
of evolution.

The scientists who participated in this movement may have been regarded as
out of date by their younger colleagues, but since they wrote prolifically for the
general public they played a significant role in the wider debate. Sir Oliver
Lodge shared A. R. Wallace's belief in spiritualism and used his position as an
eminent scientist to promote opposition to materialism. His philosophy was
based on the reality of the ether, the substance that was supposed in the period
before Einstein's theory of relativity to transmit radiation through space. In the
1920s books such as his *The Making of Man* and *Evolution and Creation*
continued to defend the view that evolution was the unfolding of a divine plan
that was aimed at the emergence of beings with a spiritual component able to
survive the death of the material body on the etherial plane. Life itself was an
active agency, and the driving force of evolution was the striving of individual
organisms to improve themselves – the classic argument of the Lamarckians.[11]

Lodge's spiritualist beliefs may have been an extreme, but other scientists
wrote in support of the same approach to evolution. One of the most prolific
writers of popular-science texts in the early twentieth century was the

[10] On the response of religious thinkers including Drummond see Moore, *The Post-Darwinian Controversies*.
[11] Lodge, *The Making of Man*, chap. 6.

Aberdeen professor of natural history J. Arthur Thomson. In conjunction with Patrick Geddes he conducted a long campaign in defence of the belief that the purposeful activities of living organisms can play a role in evolution. His *System of Animate Nature*, based on the Gifford Lectures for 1915–16, followed Drummond in replacing the struggle for existence with the drive to cooperate as the directing force of evolution. Despite the side-branches, evolution exhibited a general trend towards the human species as the 'summit of the whole': we are the inevitable outcome of the tendency for the mind to acquire the freedom to choose its own destiny. More popular books such as his *Gospel of Evolution* spelled out the message for a general readership, while in a contribution to a collected volume on *The Great Design* (a follow-up to an earlier collection entitled *Creation by Evolution*) he proclaimed: 'There is something very grand in the conception of a Creator who originated Nature in such a way that it worked out His purpose: an orderly, beautiful, progressive world of life with its climax, so far, in Man, who echoes the creative joy in finding the world "good".'[12]

In America the biologist E. Grant Conklin published his *The Direction of Human Evolution* in 1921, arguing that the process was guided by something more than chance. There was a 'wider teleology' at work ensuring that there was a meaningful goal, represented by modern humans. Conklin was well aware that Lamarckism was now regarded with suspicion in the scientific community and that evolution had to be visualized as a branching tree rather than a ladder. He invoked a discontinuous version of progressionism in which the appearance of major new branches was accompanied by a burst of creativity. Struggle against the constraints of the environment was the driving force, but all too soon the initiatives ran out of steam as evolution became bogged down in specialization. There was no simple line of ascent, but a key sequence of progressive phases had eventually given rise to humankind, opening up a new arena of social rather than biological progress. Conklin insisted that no other species is currently able to progress, and none would be able to take our place if we were removed from the scene – the point that Broom would develop.[13] The conclusion to E. A. Hooton's *Up from the Ape* similarly retained the hope that there might be something in the Lamarckian theory's teleological approach. This would allow us to believe that 'the upward course of human evolution is determined by man's reaching up for higher things'. The

---

[12] Thomson, *The System of Animate Nature*, pp. 397, 565–6. *Creation by Evolution* and *The Great Design* were edited by Frances Mason; the second quotation is from Thomson's 'Introduction' in the latter volume, p. 14.

[13] Conklin, *The Direction of Human Evolution*, esp. pp. 18–22, 59, 81 and 245. See Pavuk, 'Biologist Edwin Grant Conklin and the Idea of the Religious Direction of Human Evolution in the Early 1920s'.

mechanism is 'apparently so purposive in its functioning, that it almost seems to exercise an intelligent control. We get God out of the machine.'[14]

A more sophisticated approach came from the biochemist (and later student of Chinese science) Joseph Needham, who combined High Church Anglicanism with Marxism and even had some sympathy with Herbert Spencer's evolutionary philosophy. The result was a view of history that often looks very much like Hegelianism. Needham promoted a sense of historical inevitability in lectures and essays from the 1930s and 1940s collected in his *Time, the Refreshing River* and *History Is on Our Side*, the title of the former derived from W. H. Auden's poem 'Spain': "O show us/History the operator, the/Organiser, Time the refreshing river.'[15]

One source of the religious element in Needham's vision of social evolution was Drummond's *Ascent of Man*, which had linked liberal Christianity with the claim that evolution was predisposed to develop cooperation and altruism. As a Marxist, Needham agreed that biological and social evolution were both phases of the same process, but he was more interested in the social aspect, which he interpreted as the drive towards collectivism. But this drive was also towards a more spiritual humanity:

The historical process is the organizer of the City of God, and those who work at its building are (in the ancient language) the ministers of the Most High. Of course there have seen setbacks innumerable, but the curve of the development of human society pursues its way across the graph of history with statistical certainty, heeding neither the many points which fall beneath it, nor the many more hopeful ones which lie above its average sweep.[16]

He was convinced that the process was inevitable, but like any Marxist (or good Hegelian) argued that we have to participate actively because it is the human will that drives the process.[17] Although well aware of the rise of Darwinism in biology, Needham still saw the development of society as having a definite goal, which we might eventually achieve by improving the human race: 'Evolution is not yet completed; human society has not yet attained its full development; man may perhaps now learn to control his further

---

[14] Hooton, *Up from the Ape*, pp. 600–2. For details of the progressionst evolutionism discussed in this section see my *Reconciling Science and Religion*, esp. chap. 4, and also my *Monkey Trials and Gorilla Sermons*, chap. 4.

[15] Quoted in Needham, *Time, the Refreshing River*, p. 15, and *History Is on Our Side*, p. 28.

[16] Needham, 'Metamorphoses of Scepticism', in his *Time, the Refreshing River*, pp. 7–27, quotation from p. 16. For his view of Drummond see 'The Naturalness of the Spiritual World', ibid., pp. 28–41.

[17] See the section 'Evolution and Inevitability' in his 'Integrative Levels: A Revaluation of the Idea of Progress', reprinted in *Time, the Refreshing River*, pp. 233–72, quotation from pp. 266–9. This address contains his appreciation of Spencer's evolutionism.

evolution.'[18] Needham knew that the version of Christian faith that he preferred might not survive the development of the perfect society, but he was convinced that the sense of the holy would survive to ensure that the future world-view would be both moral and spiritual.

The publications of scientists sympathetic to the traditional view of humanity as the centrepiece of creation were welcomed by the liberal Christians who hoped to modernize the faith to make it compatible with what they presumed to be current scientific theorizing. Like the increasingly marginalized biologists who shared their hopes, they wanted to retain as much as possible of the non-Darwinian thinking that had underpinned the developmental view of evolution. Lamarckism was defended against its critics, while Henri Bergson's philosophy of creative evolution based on an *élan vital* was accepted as evidence of a revolt against the materialism of the previous generation.

Drummond's views were echoed by R. J. Campbell, a Congregationalist minister whose 'New Theology' attracted much attention in the years before the Great War. He emphasized what he took to be traditional 'Christian belief in the Divine immanence in the universe and in Mankind'. Evolution was an expression of God's creativity at work in nature: 'God is ceaselessly uttering Himself through higher and ever higher forms of existence; or rather, which is the same thing, He is doing it in us'.[19] Similar views proliferated among liberal clergymen in the inter-war decades, including Charles Raven in Britain and Harry Emerson Fosdick in America. There were regular appeals to non-Darwinian theories as a means of defending the view that humanity was the goal of evolution. Raven's *The Creator Spirit* of 1927 openly endorsed Lamarckism as part of his campaign to rescue science from the materialism that had destroyed natural theology. The inheritance of acquired characteristics allowed the efforts of individual organisms to direct evolution, also ensuring that higher psychic properties would appear at key steps in the process, representing the main stages in the unfolding of the divine purpose.[20]

Towards the end of his career Raven helped to promote interest in the evolutionary world-view of Pierre Teilhard de Chardin, noting that had he known the French thinker before his recent death he would have made common cause with him.[21] Teilhard was a Jesuit priest who worked in palaeontology and palaeoanthropology: he had been involved with both the Piltdown discoveries and the expeditions to central Asia in search of Osborn's

---

[18] Needham, 'The Gist of Evolution', reprinted in his *History Is on Our Side*, pp. 121–45, quotation from p. 145.
[19] Campbell, *The New Theology*, pp. 4 and 24; see Bowler, *Reconciling Science and Religion*, pp. 224–32.
[20] Raven, *The Creator Spirit*, pp. 80–5; on Modernism in the churches see Bowler, *Reconciling Science and Religion*, chaps. 7 and 8, and *Monkey Trials and Gorilla Sermons*, chap. 4.
[21] Raven, *Teilhard de Chardin*.

'dawn man'. During this time he had elaborated a vision of cosmic evolution but had been forbidden to publish by his superiors. After his death his most important work, *Le phénomène humain*, was published in 1955 and translated as *The Phenomenon of Man* four years later. It was presented as science, and indeed the book contains modern-looking evolutionary trees, but the underlying vision was of a main line of evolutionary process designed to increase the level of consciousness in the world, with humanity as the final step leading towards an 'omega point' in which the race will coalesce into a single spiritual entity. Teilhard wrote openly of an orthogenetic trend in the primates leading towards increased cerebralization, coinciding with a wider trend in all nature towards higher levels of consciousness. His tree of primate evolution has a distinct central trunk, with the branches corresponding to the various races blending together in the future ahead of the omega point.[22]

Teilhard had tried to bring together his liberal version of Christian theology and his scientific interest in evolution. Julian Huxley wrote an appreciative introduction to *The Phenomenon of Man*, convincing his critics in the scientific community that there was indeed a quasi-religious foundation to his ostensibly humanist position. But the book sold well and reached many readers who were not formally Christians. Teilhard's philosophy attracted considerable attention in the churches but also reached laypersons, often of the younger generation, who still hoped to find some spiritual dimension in their lives. The outburst of enthusiasm for his work during the 1960s represents probably the last time in which the image of the human race as the predestined goal of evolutionary progress achieved a significant level of influence in the wider community.

Planning for Perfection

As the example of Needham and his fellow Christian socialists demonstrates, there was no necessary hostility between the Marxists' predictions of a workers' utopia and the theologians' hope for spiritual revival. In practice, though, for the vast majority the two systems seemed antipathetic, and the focus in the twentieth century was increasingly beginning to favour a utilitarian version of progress. The drive would be towards a world of peace and plenty for all, and now there was more emphasis on how to achieve a perfect society in the future, although appeals to the past for evidence of a general historical trend did not disappear altogether. There were now two rival ideologies offering the promise of future social improvements, one based on the free-enterprise tradition that had inspired Spencer, the other claiming that only rational planning imposed by experts would be able to eliminate the

---

[22] Teilhard de Chardin, *The Phenomenon of Man*, p. 200 on orthogenesis, and the primate tree on p. 213. The more conventional trees for the animal kingdom can be found on p. 137 and p. 149.

imperfections of the market-based economy. The most obvious manifestation of the latter policy emerged in Soviet Russia, and this served as a beacon for some thinkers in the West who also accepted that the market economy would have to be substantially modified if calls for outright revolution were to be stilled.

Conservative thinkers continued to insist that free-enterprise capitalism remained the only hope of ensuring eventual prosperity for all, but some began to argue that reform of the existing system was necessary, although falling short of the complete transition to communism advocated by the Soviets. The answer to social disruption, it was argued, was planning by experts who would take control of industry and the distribution network from the owners without the need for violent revolution. The sociologist Peter Wagner has coined the term 'organized modernity' to highlight the increasing confidence of technocrats who thought that the process of social evolution could be managed by experts who would direct the applications of science and technical innovation to ensure the public good.[23]

For those whose main concern was social justice, this programme encouraged the emergence of what was, in effect, a less drastic equivalent of the Marxists' vision of a technologically enhanced utopia as the goal of history. In the long run this possibility began to seem unrealistic in the face of the ongoing rush of technical innovation reshaping how society operates. In the meantime, though, the optimists retained the hope that the machine-based economy ushered in by the Industrial Revolution could be reconfigured so that its benefits were more evenly distributed. There was little talk of Condorcet and Comte (although some of Marx), but the calls for drastic reform via planning retained a sense that history still had a goal that could be achieved by one last push to overcome the tensions introduced by the Industrial Revolution. The age of the machine had the potential to ensure a comfortable life for all, but its potential had been restricted by the underlying logic of the free-enterprise system, which had restricted the benefits to a few successful entrepreneurs. If only the lessons learnt from the rise of science could be applied to the organization of society itself, the benefits of industry could be shared by all.

Europeans seeking a way out of the tensions that had generated the Great War and the revolution in Russia thus turned increasingly to the hope that planned economies would encourage social harmony and eliminate national rivalries. Perhaps the nations themselves would become redundant and a rationally organized World State would emerge as the goal of social evolution. Eminent thinkers including H. G. Wells and Bertrand Russell visited Russia and came back impressed with the Soviets' determination to organize their

---

[23] Wagner, *A Sociology of Modernity*; see also Hard and Jameson, eds., *The Intellectual Appropriation of Technology*.

society but worried that their intrusive approach would stifle individuality. Russell's *The Prospects of Industrial Civilization* of 1923 called for central-ized control of industry that might verge on socialism but would preserve freedom. Modern industry offered economic benefits, but the new society must avoid the 'worship of the machine' that had led to dehumanization. Russell imagined a world consisting of a few great states, each self-sufficient to avoid the need for conflict. Walter N. Poliakov was one of many who also called for the elimination of the profit motive in industry via some form of state control. Frederick Soddy, a pioneer in the study of radioactivity who anticipated the possibility of nuclear power, became convinced that the existing financial system had to be rationalized to allow the potential benefits of new technolo-gies to be available to all. His views were echoed by C. Marshall Hattersley, who insisted that an age of plenty was achievable – and would have to be achieved if war was to be avoided.[24]

As the skies grew darker in the 1930s the enthusiasts for planning still hoped that war could be averted and the world of plenty for all achieved. Lyndall Urwick, head of the short-lived International Management Institute of Geneva, promoted the scientific management techniques of Frederick Winslow Taylor as the guide to a solution for all the world's economic woes. A rational reorganization of society along more equitable lines was also advocated by the Irish politician Eimar O'Duffy and the British engineer A. P. Young. Ludwig Bauer offered a pessimistic view of the international situation but insisted that the move to a World Government would eliminate war. Unfortunately, none of these writers could offer any realistic means of achiev-ing their goals.[25]

H. G. Wells was the most prominent advocate of the need for a rationally planned World State, but like Soddy he also recognized that ongoing techno-logical innovation would continue to have a disruptive effect on how people live. The resulting social disruptions made the hope of a stable future utopia unrealistic even if universal plenty could be assured. Soddy had noted the potentially disruptive economic effects of unlimited nuclear energy (to say nothing of the prospects of nuclear war). Wells also began to promote a less clear-cut vision of the prospects for economic and social progress. The result was a more open-ended and unpredictable vision of the future and a new

---

[24] Bertrand and Dora Russell, *The Prospects of Industrial Civilization*; Poliakov, *Mastering Power Production*; and Hattersley, *This Age of Plenty*. Soddy's predictions for the future of nuclear energy came in his *The Interpretation of Radium*, and his later writings on economics include his *Wealth, Virtual Wealth and Debt*; see Trenn, 'The Central Role of Energy in Soddy's Holistic and Critical Approach to Nuclear Science, Economics, and Social Responsibility'.

[25] Urwick, *Management of Tomorrow*; O'Duffy, *Life and Money*; and Bauer, *War Again Tomorrow*.

interpretation of historical progress which did not depend on the assumption that there had been a necessary development towards a predetermined goal.

The United States retained a hope that its own version of liberal capitalism offered a better way to a similar goal, but here too there were suggestions that something would have to be done to control the system. Where Europeans had lost faith in Herbert Spencer's predictions, Americans continued to take him seriously, prefiguring the confidence regained after the depression of the 1930s. America's faith in progress had traditionally been sustained by exceptionalism, the belief that a new world with a boundless frontier leading to unlimited resources offered a unique opportunity now lost to Europe. Now that 'manifest destiny' was reconfigured so that it could lead the free world. The American dream was redefined as the search for happiness through material prosperity.[26] At the same time, the depression of the 1930s brought home the dangers of an unrestricted drive for profits, and there were increasing calls for some form of planned economy, in this case partially realized through President Roosevelt's 'New Deal'.

America was the birthplace of Taylor's vision of scientific management, and Henry Ford's highly organized factory system seemed to offer a model for what planning could achieve if it could be extended to the whole economy. In 1907 Upton Sinclair dedicated his *The Industrial Republic* to Wells, predicting a revolution within ten years that would bring in a government able to regulate the whole economy. The goal was the equalization of access to the flood of material goods being produced by the new industries. Jack London's *The Iron Heel* provided a fictional vision of a violent revolution against an increasingly repressive industrial elite. Sinclair envisioned a less dramatic transition to the new order, and he was by no means alone in hoping that the American economy could be transformed by a radical, but not violent transition to planning.

In the 1920s Thorstein Veblen advocated the expansion of Taylor's scientific management to the whole economy. He envisaged the rule of experts, openly describing this as a 'Soviet of technicians'.[27] Edward A. Filene wrote of 'Fordizing' the economy, this process leading to a new Industrial Revolution that would eliminate the inequalities of the first by taking the interests of the workers into account. George Soule's *The Coming American Revolution* of 1934 lamented the opposition to Roosevelt's schemes and predicted a more vigorous challenge to capitalism that would eliminate the profit motive from industry. William Trufant Foster and Waddell Catching's

---

[26] See Ekirch, *The Idea of Progress in America, 1815–1860*. In his *Monad to Man*, Ruse notes how many early twentieth-century American scientists still clung to Spencer's philosophy. On redefining the American dream see Churchwell, *Behold, America*.

[27] Veblen, *The Engineers and the Price System*, chap. 6.

The Road to Plenty offered the economists' vision of a distribution network that would allow the abundance of goods now being produced to benefit all.[28]

The economists and social reformers assumed that a world of material plenty for all could be achieved by rational planning and would be the final goal of social evolution. They envisioned a more or less static society in which people would be free to live happy and fulfilled lives. The problem was that – as Wells was starting to appreciate – planners could not predict the emergence of new technologies with the power to disrupt society, nor could they anticipate all their consequences. Sean Johnston has charted the evolution of the technocratic movement pioneered by the American engineer Howard Scott, noting how it soon became obsessed with the notion of a 'technological fix' to deal with social problems.[29] As popularized in texts such as Frank Arkright's *The ABC of Technocracy* of 1933, the movement began as yet another call for experts to be put in charge of industry and society so that the problems of the existing system could be overcome. But as Johnston points out, far from planning to achieve a clearly defined goal, the engineers increasingly found themselves having to deal with the unintended consequences of new technologies that had been promoted as offering a better life for all.

Imaginative critics used science fiction to show how both sides of the utopian dream could turn into nightmares. A World State in which all aspects of life were planned by experts would destroy individuality. In Russia Yevgeny Zamyatin's *We* parodied the Soviet dream of an artificially unified humanity spreading order to the cosmos, while Aldous Huxley's *Brave New World* offered an equally bleak picture of the perfect state achieved in the name of Ford's industrialism. Their predictions would haunt the imagination of those still hoping to achieve a new world order of the Right or the Left during the Cold War. In 1954 Jacques Ellul drew on the history of science and innovation to argue that rational planning might produce a world of plenty by the year 2000, but at a terrible cost to human nature. Once the genie of technology was out of the bottle, the world was set on a course of inevitable development towards the rule of experts. The Scientific Revolution and the French Revolution had established the trend, after which 'There was a logical and foreseeable succession of events, once the first steps had been taken.' The result would be a totalitarian dictatorship, an end of history that looks remarkably like the one predicted by Huxley.[30]

---

[28] Sinclair, *The Industrial Republic*; Veblen, *The Engineers and the Price System*; Filene, *The Way Out*; Soule, *The Coming American Revolution*; Foster and Catchings, *The Road to Plenty*.

[29] Johnston, *Techno-fixers*; see also Akin, *Technocracy and the American Dream*.

[30] Ellul, *The Technological Society*, quotation from p. 44; on the year 2000 see chap. 6, 'A Look at the Future'. Ellul's book appeared in French in 1954 and the English translation ten years later. On Ellul's influence see Jerónimo, Garcia and Micham, eds., *Jacques Ellul and the Technological Society in the 21st Century*.

Legacies of Marx and Hegel

In the inter-war years the pessimists seemed only too prescient as the West was challenged by the Soviets and by the rise of fascist ideologies with their bizarre amalgam of technological fantasy and dreams of restoring ancient empires. The triumph of the Bolsheviks in the Russian Revolution at last allowed a party drawing its inspiration from Marxism to exercise power. The model of historical development based on dialectical materialism became part of Soviet ideology, and the interpretation of Marx and Engels themselves as economic determinists, outlined in the previous chapter, became the accepted way of interpreting their work. In fact, the new model was more deterministic than its inspiration, using the image of history as the record of social development through an inevitable sequence of revolutions to bolster its hopes of progress towards a classless utopia.

Lenin had proclaimed imperialism to be the last phase of capitalism even before he was catapulted into power. When the expected revolutions in the West failed to materialize, he and later Stalin were forced to proclaim that socialism could flourish in one country, and Stalin went on to enforce it via massive industrialization and the collectivization of agriculture. This was the vision of how the future utopia would work that was offered to the rest of the world and would later become the basis for the Soviet drive for dominance in the Cold War.[31]

In the inter-war years, Marxists in the West ignored or excused the excesses of Stalinism, partly in the name of resistance to fascism. Many scientists were sympathetic to Marxism, echoing the hopes of the economists who insisted that rational planning offered the only way forward. The Marxist view of history remained influential in anthropology and archaeology. In America, L. A. White revived Morgan's evolutionary model with Marxist overtones, while in Britain V. Gordon Childe popularized a similar view. Both seem to have accepted a predetermined sequence of cultural stages though which various societies progressed. They acknowledged that there were differences between the cultures that had developed in different regions of the world, but saw these as of less significance than the parallels. Childe at least argued that Europe had advanced to the highest level, so its development represented the 'main stream' of history.[32]

---

[31] See for instance Thompson, *A Vision Unfulfilled*. There is a huge body of literature in this area, so I have kept my account brief.

[32] Childe, *What Happened in History*, p. 29; see also his *Man Makes Himself* and White, *The Science of Culture*. See Layton, *Introduction to Theory in Anthropology*, chap. 5; Harris, *The Rise of Anthropological Theory*, chap. 22; Hatch, *Theories of Man and Culture*, chap. 3; and Trigger, *A History of Archaeological Thought*, chap. 5.

If Marxist intellectuals were willing to make excuses for Stalinism, to everyone else it became obvious that Stalin and Hitler represented opposite sides of the same totalitarian coin. We have seen that there were some who still remained hopeful that the West could respond with an ideology aimed at achieving a utopia of peace and plenty without the need for a complete destruction of the existing economic and social order. In early twentieth-century France a revival of Hegelianism kept the concept of history as a progress towards a definite goal alive.[33] French intellectuals came to the subject fully aware of and dissatisfied with the development of Marxism and willing to take on board contrary views such as those of Martin Heidegger. Hegel's thought thus played a role in the emergence of radical new movements such as existentialism. Of the thinkers who constituted this revival the one who has become associated most clearly with the idea of progress towards the 'end of history' is the Russian emigré Alexandre Kojève. Although for a long time little known in the English-speaking world, Kojève's position inspired Francis Fukuyama's controversial vision of a new world order based on liberal capitalism.

Kojève wrote his Heidelberg doctoral dissertation under Karl Jaspers, focusing on the work of Vladimir Soloviev, who had developed a philosophy in which history represented humanity's efforts to work its way towards the divine. The Absolute was immanent in the world where it sought to perfect itself. He then came under the influence of both Heidegger and the Marxists, finding the latter's economic model of social progress increasingly unsatisfactory. He turned back to Hegel for an alternative, although scholars disagree over whether he distorted the original vision or worked out its underlying implications. He gave an influential series of lectures on Hegel in Paris during the late 1930s. Here he developed the thesis that led to the claim that the end of history had now been achieved.[34]

Although his thinking depended on defining the end-point towards which social progress was aimed, Kojève was unwilling to accept that the process could be represented by analogy with biological development. History was a purely human affair following no natural laws, so he made little effort to define the stages of development by which the goal was achieved. He was particularly interested in Hegel's views on the master–slave relationship and the argument that the drive for personal recognition on both sides provided the impetus for the dialectical movement towards freedom. Following the French Revolution and the reforms of Napoleon, the modern state had guaranteed personal

[33] See Roth, *Knowing and History* and Butler, *Subjects of Desire*; also Sinnerbrink, *Understanding Hegelianism*, chap. 6.

[34] Kojève, *Introduction to the Reading of Hegel*. In addition to the items cited in the previous note see Cooper, *The End of History*.

freedom for all and in this sense represented the end of history. No further social progress was possible since the Hegelian ideal of rationally justified freedom had been achieved. There might be developments in other fields, but history proper was at an end, or would be when the modern system was extended to the whole world. In the later part of his life a disillusioned Kojève worked on the project for European unity.

Kojève's thesis came to international attention when it became part of Francis Fukuyama's argument, first presented in a 1989 article and then in his best-selling *The End of History and the Last Man* of 1992. Fukuyama's starting point was the opposite of Kojève's early Marxism: it was an expression of the short-lived enthusiasm that followed the collapse of the Soviet Union in which Americans believed that their ideology could become the basis for a new world order. He saw liberal democracy and free enterprise as the best possible foundations for a society and hence as the end-point of historical progress. The element of freedom so crucial to Hegel's thinking played a key role in the argument, as represented by Kojève's appeal to the need for recognition and self-fulfilment as the driving force of human activity. Fukuyama explicitly criticized the purely utilitarian view of social interactions as 'The Coldest of All Cold Monsters'.[35] In this sense his position can be seen as the heir to Hegelianism and, indirectly, to the liberal Christian hope of re-creating paradise on earth (although now lacking the explicitly biblical image of an original paradise).

In fact, however, Fukuyama's position was a synthesis of this spiritual vision of progress with the more rationalist or utilitarian model that had emerged as a rival to it since the mid-eighteenth century. For all his criticism of self-interest as the only driving force, Fukuyama still recognized free-enterprise capitalism as the best source of the material goods we need to ensure that we can enjoy our freedom. His acknowledged the support of the RAND Corporation, which promoted the role of technological innovation.[36] His new world order thus combined freedom in both senses of the term – the Hegelian and Christian ideal of spiritual development and the utilitarians' recognition that industrial progress is an essential driving force towards better well-being.

Fukuyama's assumption that free-enterprise liberalism is the end-point of social development was linked to a revival of the traditional idea of progress as a historical trend. The end could be recognized because it was the obvious goal of a clearly defined process in time. Like both the Hegelians and the Marxists he believed that history could be 'understood as a single coherent evolutionary process' which allowed us to define societies as 'primitive' or 'advanced'

---

[35] This is the title of chap. 20 of Fukuyama's *The End of History and the Last Man*; on Kojève see ibid., part 4.

[36] Ibid., chap. 8, and on the support of the RAND Corporation p. ix.

according to their position on a scale of development. 'This evolutionary process was neither random nor unintelligible, even if it did not proceed in a straight line, and even if it was possible to question whether man was happier or better off as a result of historical "progress".'[37] Although clearly aware of the problems facing the 'last man' at the end of history, he did still believe that what had been achieved was the best of all possible worlds.

In his search for antecedents to his position Fukuyama virtually ignored Spencer – the traditional hero of American businessmen – focusing instead on Hegelianism as reinterpreted by Kojève.[38] There are two obvious and by no means mutually exclusive reasons for this choice. There is a spiritual dimension to Fukuyama's prediction of the final state of human freedom that was notably lacking in Spencer's more prosaic world-view. Hegelianism modulated with a dash of Marxism offered just the right impression that freedom offered not just material prosperity but also the hope of self-fulfilment. But is it possible that Spencer also provided a less convincing ancestry for a theory of necessary progress towards a fixed goal? We shall see how Spencer's evolutionary progressionism diluted its sense of inevitability with a recognition that there might be multiple avenues by which lines of development might achieve new levels of complexity.

Fukuyama concluded with the analogy of wagon trains starting out along a road, presumably moving westwards and hence linking Hegel's vision of history with the ideals of the American frontier. Some of these wagon trains will be ambushed by Indians or halted by other calamities, some will lose their way or even decide to set up camp along the road. But in the end most will make their way into town.[39] This was an analogy that would be recognized by any of the idealist or utilitarian progressionists of the previous two centuries. Fukuyama's ability to synthesize the idealist and the rationalist model of progress reveals the underlying parallels between their models of history: the goals and the mechanisms of change may be different, but the linear pattern underlying the process is very similar.

In the end, however, Fukuyama himself left open the possibility of further change. Following his metaphor of wagon trains fighting their way across the continent, he accepted that after reaching their final destination, some of the occupants might 'find the surroundings inadequate and set their eyes on a new and more distant journey'. In a later book he would explore the challenges of the posthuman future offered by biotechnology.[40] Looking back, we can see that some of the key figures predicting a final end to social history also left open the possibility that other forms of cultural development might take place even after the 'perfect' society had been established. Technological innovation

[37] Ibid., p. xii.   [38] There is a brief reference to Comte and Spencer in ibid., p. 68.
[39] Ibid., p. 338.   [40] Ibid. and Fukuyama, *Our Posthuman Future*.

merely added to the range of factors that continue to force change, but in this case it was obvious that the effects might well transform society itself. The image of a final goal to cultural and social evolution thus broke down under the weight of increasing recognition that evolution is not a linear process leading in a single direction, let alone towards a single goal. The utilitarian version of the idea of progress may have begun by modelling itself on the chain of being, but in arousing interest in the role of technological innovation it created an environment in which that model became irrelevant.

Part II

Towards a World of Unlimited Possibilities

7   Darwinian Visions

The linear vision of progress enjoyed its heyday in the late nineteenth century even though it was already being challenged by a rival that threatened to overthrow it. The model that defined progress in terms of the ascent of a predetermined hierarchy of forms – biological or social – was defended by acknowledging an element of diversity only to trivialize it. There were branches leading off the main line of development, but they were merely dead-ends with no significance for the overall result. Evolutionism threatened to wound the human race's pride in its supreme position by linking us to the animals. The ladder of progress salved the wound by arguing that we were still the ultimate goal that had been reached by ascending a predetermined scale of development.

The alternative viewpoint had the potential to undermine the idea of progress altogether. It insisted that there was no linear pattern in nature: the elements we perceive in the world are related in a more complex way that admits of links in many directions, not just up and down a linear scale. When translated into historical terms this more complex taxonomy could be seen as a branching tree of diversity which in principle has no trunk or main line: each branch is as valid a result of the underlying processes as any other, and there is no single goal towards which everything is tending. There can be no predetermined path of development in the cosmos, whether defined by God or by the laws of nature, because change is driven by short-term interactions in the everyday world, not by some preordained plan.

Here was an insight that threatened to undermine the preconceptions so central to European thought, the assumptions that humanity is the centrepiece of creation and that modern Western society (Christian or industrial according to taste) will be the end-point of human history. If what became known as the Darwinian ideology was pushed to its ultimate conclusion, humanity would be shown to be the tip of just one among many branches of the tree of life, and the West just one of the many cultures that have emerged around the globe. In the modern world there have indeed been some Darwinians who argue that the concept of progress is meaningless in biology, while the West's traditional sense of its own cultural superiority has been become ever more insecure.

The earliest naturalists who recognized the need to treat the history of life as a process leading to increased diversity saw this as an alternative to the idea of progress. But again it was possible to salvage something from the wreck by redefining progress to accommodate the new theories of evolution. Instead of the ascent being defined by a linear sequence of individual forms as in the chain of being, it would be measured against a hierarchy of increasing complexity and ever more sophisticated grades of organization. There were different ways of becoming more complex, so many different lines could ascend through the scale, some ascending faster and hence further than others. Because modifications were driven by short-term interactions, there would be no guarantee that any one line would advance; many lines would stagnate, and some might even degenerate. Extinction also becomes an everyday occurrence, whether due to unpredictable factors such as migration, invasion or natural catastrophe.

This would be a less rigid, more open-ended and distinctly less teleological version of the idea of progress. Because the underlying causes of change are driven by localized interactions, many developments would offer no improvements of any long-term value. They would be merely adaptive specializations useful only in a particular environment. Yet it seemed obvious that at certain points in the advance of life or of society something new and of permanent significance had emerged to define a higher level of organization. Warm-bloodedness and higher mental powers would be typical examples in biological evolution, new technologies or forms of social organization in human affairs. In some cases more than one branch in the tree of evolution would make the breakthrough to a higher level (birds and mammals became warm-blooded independently), while in others only one would become the pioneer and its descendants would block any others seeking to reach the new grade (we are the only hominid species left of the many that originally evolved).

Even in biological evolution it is easy to speak of these major new developments as inventions or innovations. The terms are obvious anthropomorphisms derived from the analogy with human creativity, yet even though the biological equivalents involve no forethought or directing intelligence they seem to have the same transformative power, causing something new and potentially exciting to come into the world. There can also be rival innovations competing for a place in the marketplace of nature, just as human inventions generate rival industries. In some circumstances an innovation may not get its chance to develop, because an alternative has already taken over the field.

The analogy with technological innovation is worth exploring as the means by which progress can be defined, and it points us towards some distinctive characteristics of the more open-ended view of progress. In the long run the changes have been cumulative and do seem to have produced results that most of us regard as meaningful, yet there is no predetermined goal towards which things are moving, and there is more potential for disagreement over the value of what has come about. It is easier to accept a relativist position in which more than one animal species or society is seen to occupy the same level of significance. Chance and contingency play a role in shaping what actually happens, so the outcome is never predictable. We can speculate in the realm of counterfactual history, wondering what would have happened if the asteroid had not wiped out the dinosaurs, or if the Confederacy had won the American Civil War. Recognizing that things need not have come out the way they did warns us that we cannot be sure what the effects of the next innovation will be.

This chapter shows how the branching tree model of progress emerged in the life sciences and, like Part I, provides a template which we can then apply to a wide range of disciplines seeking to understand human origins, prehistory and history. Here interactions between the various disciplines are often quite explicit. The analogies between the Darwinian theory of evolution and human history are striking and were already being recognized in the early twentieth century by thinkers such as H. G. Wells. There was a growing awareness that we are witnessing an acceleration of innovation to levels never seen before, with consequences for society that are unpredictable and of increasingly debatable value. Futurology would become an increasingly important component defining the meaning of progress. The sense of a fixed goal towards which history was tending evaporated, just as the chain of being was replaced by the open-ended model of evolution represented by the tree of life.

The New Taxonomy of Nature

Voltaire ridiculed the concept of the chain of being, but this was because he saw obvious gaps in the hierarchy.[1] For working naturalists there was a more fundamental problem. Whatever the chain's philosophical implications, by the late seventeenth century it was becoming obvious that it was useless as a practical guide to nature. It presented the world as a rationally ordered hierarchy with humanity at the top of the scale, but naturalists studying the relationships between species found it impossible to represent those links in the form of a linear pattern. It was hard enough with the well-known European species, but as explorers brought back new forms of life from around the globe the problems became insurmountable. They would become even worse once fossils began to be taken seriously as evidence of extinct forms of life.

Most naturalists still accepted that species were created directly by God and hoped that there was a divine order underlying the apparent diversity of forms. But there were other considerations too: what has been called the 'utilitarian

[1] Voltaire, 'Chaîne des êtres créés', in his *Philosophical Dictionary*, pp. 107–8.

argument from design' used the adaptation of structure to function as evidence of the Creator's benevolence in creating animals with structures that enabled them to live a comfortable life in the environment for which they were fitted. These adaptations came in so many forms that it was hard to believe that they could correspond to an abstract pattern such as the chain of being.

Naturalists such as John Ray wanted a practical way of describing how the various forms of plants and animals were related to one another, using observable characters visible to any collector. Ray was a deeply religious man whose *Wisdom of God Manifested in the Works of Creation* of 1691 became a classic expression of the argument from design. But he also wanted to publish a catalogue that would help naturalists to classify species, and he had no confidence in the power of human reason to fathom any meaningful order at the heart of nature. The naturalist needed to work upwards from observable characters, not downwards from a preconceived plan, linear or otherwise. Ray used the similarities between species to classify them into groups, building on common-sense relationships that we recognize when we link the dog with the wolf and the lion with the tiger. In his works on botanical classification he argued for the inclusion of as many characters as possible in the hope of creating a truly natural arrangement of plant species. Others such as Joseph de Tournefort preferred to focus on a single important character to create an artificial system that would be easier to use. Following the recent discovery of plant sexuality he chose the reproductive organs, using the structure of the flowering parts as the main guide. Neither strategy seemed likely to vindicate the chain of being.[2]

The artificial system lay at the heart of the classification popularized by the founder of modern biological taxonomy, the Swedish naturalist Carolus Linnaeus, in his *Systema naturae* of 1735. He also used the flowering part of plants to establish an artificial system that pigeonholed species according to the number of their stamens and pistils. In this system, soon expanded to cover the animal kingdom, the most closely allied species were grouped into a higher-order category known as the genus, the genera were grouped into families, and so on in a series of ever-wider categories. Linnaeus also introduced the modern binomial nomenclature, in which the first name denotes the genus and the second the individual species. The original system for plants seems to fit Michel Foucault's definition of the 'classical' approach to nature prevalent in the eighteenth century. The aim was to classify, not to explain, and there was an expectation that the activity of classifying corresponded to the real order of

---

[2] The classic biography is Raven, *John Ray, Naturalist*; see also Sloan, 'John Locke, John Ray, and the Problem of the Natural System'. For my own more detailed account see *Evolution: The History of an Idea*, pp. 41–4.

nature. That real order should be based on a rational pattern, which would have made the chain of being an ideal model if only it had worked in practice.[3]

Although Linnaeus's numerically defined categories represented a predefined structure into which the species had to be crammed, this was only part of the system: it didn't apply to non-flowering plants and certainly not to the animal kingdom. Here there was no hope of seeing a linear pattern emerging from the arrangements. The numerical straightjacket was merely a special case within a much more flexible system in which species were associated in groups within groups in an ever-wider array. In effect, Linnaeus and his followers were beginning to visualize natural relationships in two dimensions rather than the one-dimensional chain of being. Each species has similarities to many others in different directions, not just to one above and one below it in the scale. As other naturalists extended the range of characters used as the basis of classification, the idea that there might be an underlying linear arrangement was slowly marginalized. It did not die out immediately, as we saw in Chapter 2, but by the early nineteenth century it could be kept alive only by being treated as a skeleton at the heart of a much more complex arrangement.

The best way of appreciating the impact of this transition in biological taxonomy is through the range of two-dimensional representations drawn up by naturalists in the pre-evolutionary period. The collection assembled by Theodore W. Pietsch reprints over fifty diagrams from the seventeenth century to the early nineteenth century which utilize a variety of ways of depicting the complexity of natural relationships.[4] Most of these are the work of specialists trying to make sense of the similarities and differences within a single group, although a few do make an effort to show relationships more widely within the plant or animal kingdoms. Some are by eminent naturalists including Linnaeus and Buffon, but most are by specialists now long forgotten.

The simplest representation brackets the species within a genus and then uses larger brackets to indicate higher-level associations. Looked at sideways, these lists already begin to resemble a branching tree. There are true diagrams that deliberately resemble trees, although there is no suggestion that the branches indicate development over time. Others, including those by some followers of Linnaeus, show the species and genera like islands on a map of the ocean – islands of various sizes scattered irregularly over the page, some close together and others further apart. Linnaeus himself had suggested that this might be the best form of representation.[5] More formalized diagrams show a limited degree of orderliness often highlighted by lines to connect the most closely related forms. A few models use circular forms of representation, the

[3] Foucault, *The Order of Things*. See Frangsmyr, ed., *Linnaeus: The Man and His Work* and Larson, *Reason and Experience*.
[4] Pietsch, *Trees of Life*, figs. 3–57.   [5] Linnaeus, *Philosophia botanica*, para. 77, p. 27.

most bizarre being those of the quinarian system of William Sharpe Macleay, which became temporarily popular in the 1830s and 1840s. Here nature is depicted as a series of interlocking circles, each containing five elements. This was an attempt to salvage the idea that nature must be built to an orderly pattern in an age when the linear chain had become untenable. In fact, as Pietsch's collection of diagrams shows, most naturalists were beginning to suspect that no such geometrical regularity was to be found in the world.

Naturalists were, in effect, anticipating the open-endedness and lack of orderliness that would soon become characteristic of Darwinism. Diagrams depicting species as islands irregularly scattered over the ocean point towards the Darwinian image of the tree of life because they can be seen as cross-sections through the tree. A phylogenetic tree really ought to be conceived in three dimensions, two 'horizontal' to represent the diversity of forms and one 'vertical' to represent time. If we imagine a section cut through the branching tree of temporal evolution at a particular point in time, each branch will appear as a single circle in the two-dimensional display. If evolution is divergent, as the cross-section moves upwards in time, the circles will move further apart and will sometimes split into two or more sub-circles as a species divides. And if evolution is shaped by local interactions rather than some overarching divine plan, we should expect the two-dimensional maps to show an irregular distribution of the species rather than a neat pattern.

The possibility that a process of diversification in time might explain why species fall into natural groups did occur to a few observers and to the more radical philosophers of the period. They wondered whether a collection of similar species might be produced when a single original form became modified in different ways. These speculations were often on a limited scale, when specialist collectors thought about the origin of the group they worked on. As early as 1665 Robert Hooke's microscopical studies of mites led him to speculate that the various forms might have arisen from a single progenitor, diversifying by accident or by environmental influences. He even used the term 'genealogy' to denote such relationships.[6] Bolder thinkers in the following century imagined wider transformations that might, for instance, explain the whole assembly of backboned animals. Inevitably, such speculations triggered an interest in the question of what kind of process might produce this diversification of forms.

Monsters and Materialism

Long before Darwin the materialist philosophers of the eighteenth-century Enlightenment had begun to wonder about these issues. The more radical

---

[6] Hooke, *Micrographia*; see pp. 214–15 on mites. The term 'genealogy' is used on p. 207 in the context of possible wider links, for instance between crabs and insects; for details see Schneider, 'The First Mite'.

among them, including Denis Diderot and the baron d'Holbach, abandoned not only Christianity but any belief in a wise and benevolent Creator. They didn't need the concerns of naturalists and collectors to warn them that the old vision of a rationally ordered creation was no longer plausible. Perhaps natural forces were capable of producing new forms in the course of time – but the materialists doubted that there was any wider purpose to the endless variation. Their ideas occasionally anticipated aspects of later transformist theories, but some of their suggestions provided alternatives, including the possibility that complex living organisms might be spontaneously generated from formless matter. Speculations that do involve the transformation of species were conceived within a world-view that still lacked many of the elements later available to Darwin. Treating these thinkers as 'forerunners of Darwin' risks imposing modern concepts onto ideas conceived in a very different cultural and scientific environment. There were only few anticipations of the idea that nature was able to winnow out the variants to give an apparently purposeful outcome.[7]

One contribution that has attracted much attention is Pierre Louis Moreau de Maupertuis's theory of generation, presented in his *Venus physique* of 1745. Rejecting the concept of pre-existing germs, he suggested that particles within the male and female semen spontaneously arranged themselves to form the foetus. He appealed to spontaneous variations or monstrosities to illustrate the absurdity of the idea that all organisms have been preformed from the creation within either the ovum or the sperm. He used the example of polydactyly, showing how once the new character of a sixth digit had appeared, it could be transmitted to later generations. From this he went on to suggest that the appearance of the black race might be explained by a similar mutation that had appeared within the original white population. Although aware of the possibility that some environmental influence might be involved, he preferred to think that the new character had appeared more or less by accident.[8]

In his *Systême de la nature* of 1751, published under an assumed name, Maupertuis suggested that the fundamental particles might have some of the basic characteristics of life, including intelligence and memory. This would allow the origin of life on earth to come about by spontaneous generation. He also imagined the multiplication of species from an original pair by a series of monstrosities or fortuitous rearrangements within the reproductive process. 'Each degree of error would give a new species, and by the force of these

---

[7] The classic collection on this topic is Glass, Temkin and Strauss, eds., *Forerunners of Darwin, 1745–1859*. For more reliable accounts see my own *Evolution: The History of an Idea*, chap. 3, and Roger's *The Life Sciences in Eighteenth-Century French Thought*.

[8] There is an English translation of the *Venus physique* by Simone Brangier Boas, *The Earthly Venus*.

repeated departures the infinity of animals that we see today would be formed.'
Here was a clear suggestion of divergent modification over time.[9]

An more radical application of the notion of chance variations was offered
by Julien Offray de La Mettrie, best known for his materialist tract *L'homme
machine*. In his *Système de l'epicure* of 1750 he retained the old idea that all
life develops from pre-existing germs, but wondered how the first forms would
have been produced on the primeval earth since there would have been no
parents to receive them. The germs would have had to develop as best they
could, producing a process of trial and error in which vast numbers of imperfect forms would appear only to die. Viable forms of life would have been
achieved only after a vast period of time: 'Through what an infinity of
combinations must matter have passed before arriving at the one which results
in a perfect animals. Through how many others before the generations arrived
at the perfection which they have today.'[10] La Mettrie thus managed to turn his
vision of a world governed by chance into a theory of gradual progress with a
superficial resemblance to evolutionism.

An even more materialistic application of Maupertuis's ideas came from the
pen of Denis Diderot. He was already thinking about the implications of
monstrosities in the conclusion of his *Lettre sur les aveugles* of 1749, where
he imagined the blind mathematician Nicholas Saunderson speculating about
the origin of life on earth. He suggested that when life was first spontaneously
generated it appeared in an array of randomly produced forms, most of which
would have lacked the organs needed for survival. Living things achieved
viability only by trial and error. In his later *De l'interpretation de la nature*,
Diderot speculated that an original prototype of all the animals had gradually
diversified to give all the species we see today. Structures such as the limbs
were modified in various directions to give, for instance, the human hand and
the hoof of a horse. The modifications are clearly purposeful but there is no
sense that the environment plays a direct role in shaping the adaptive
variations.[11]

Acquiring Adaptations

Diderot never lost his fascination with monstrosities, and in the notes later
printed as his *Eleménts de physiologie* he asks, 'Why should not man and all
the animals be but types of monsters a little more durable?'[12] But elsewhere in

---

[9] Maupertuis, *Système de la nature*, section 45, in his *Oeuvres*, vol. 2, pp. 148–9 (my translation).
[10] La Mettrie, *Système de l'epicure*, in his *Oeuvres philosophiques*, vol. 1, p. 235 (my translation).
[11] Diderot, *Lettre sur les aveugles*, in his *Oeuvres philosophiques*, pp. 121–2; *De l'interpretation de la nature*, in *Oeuvres philosophiques*, pp. 186–8.
[12] Diderot, *Eléménts de physiologie*, p. 209 (my translation).

this and other works he recognizes another possibility, that the efforts made by less perfect forms of life seeking to cope with their environment might gradually modify the species. For this to be possible, the adaptive modifications made by the individual organism would have to be transmitted to the offspring so that they could accumulate over the generations – the process of the inheritance of acquired characteristics. This mechanism of adaptive evolution, long regarded as the only alternative to natural selection, is usually known as 'Lamarckism' – although as we have already seen, for Lamarck himself it was only a secondary process that modified the progressive ascent of life.

In his *Rêve de d'Alembert* Diderot's dreamer imagines a world in a constant state of flux. Complex forms of life can be produced by spontaneous generation at any time, and the endless generation of monstrosities ensures that no species remains fixed. There is also, however, the suggestion that the efforts of the organism can modify its structure in ways that can accumulate to give adaptations.[13] As a philosopher rather than a naturalist, Diderot saw no need to restrain his speculations within the constraints of observation. His vision of a world without rational order could accommodate many different possibilities for change, and the only sense of progress was that the earliest forms might sometimes lead to more complex adaptations.

Diderot was impressed by the work of the comte de Buffon, whose monumental *Histoire naturelle* began publication in 1749 and continued over the next three decades. Although he was once hailed as a pioneer of evolutionism, modern studies suggest that Buffon's view of the history of life had few similarities to the modern one.[14] In his *Les époques de la nature*, a supplement to the *Histoire naturelle*, he imagined a total replacement of the earth's original species following what we would now call a mass extinction caused by climatic change as the planet cooled down. Both the old and the new populations were produced by spontaneous generation from unorganized organic matter, very much as the materialists supposed. But unlike Diderot he insisted that the end-product was predetermined. Exactly the same array of species would be generated on each of the planets in the course of its history, almost as though the viable forms were inherent in the laws governing interaction of organic matter. There were precisely thirty-eight quadruped types allowed by these restrictions on the activity of nature.

Buffon came closer to a limited version of evolutionism in a chapter 'De la dégénération des animaux' of 1766, where he suggests that the originally

[13] Diderot, *Rêve de d'Alembert*, in his *Oeuvres philosophiques*, p. 310.
[14] For my own analysis of Buffon's thinking see my 'Bonnet and Buffon'. My views are in general agreement with those expressed by Roger in his *Buffon: A Life in Natural History*. There is a new English translation of *Les époques de la nature*, which was originally supplementary vol. 5 of the *Histoire naturelle* published in 1778.

created forms would become modified as they migrated to different parts of the earth.[15] The term 'degeneration' should not be taken too literally: Buffon was merely indicating that the original form was distorted. The crucial point, however, is that the extent of the modifications was limited and could never transcend the basic form defining the species. The lion and tiger were still big cats and could never become anything else. Buffon had long disagreed with Linnaeus on the nature of species, and what he was now suggesting was that the species recognized in the latter's system were merely well-marked varieties within a single type defined by the genus or perhaps even the family. These thirty-eight types were the real species, and although they were endowed with limited flexibility their basic forms were permanently fixed.

One of the first true evolutionists was the chevalier de Lamarck, who worked briefly with Buffon at the start of his career. He eventually came to accept that living forms (he did not believe in the reality of species) can be slowly transformed over many generations, and he is thus sometimes presented as a forerunner of modern theories. We have seen, however, that this interpretation has been challenged by modern scholars who argue that Lamarck's theory was based on different principles. He accepted that living forms can be modified through the process we now call 'Lamarckism'. The two laws of adaptive change proposed in his *Philosophie zoologique* claim first that animals can modify their structures when they acquire new habits to cope with changed environments, and second that such acquired characteristics can be passed on to future generations and thus accumulate to give significant changes in the species. But the main agent of change was a force that drove successive generations steadily up the chain of being. Adaptation was a disturbing factor that distorted the lines of development as they ascended, producing diversity but no actual branching of the lineages.[16]

If we are looking for the originators of the modern way of explaining evolution, Lamarck himself has to be crossed off the list. During the 'eclipse of Darwinism' in the late nineteenth century critics known as neo-Lamarckians did adopt the inheritance of acquired characteristics as an alternative to natural selection in a theory based on the model of an ever-diversifying tree of life. Even then, the Lamarckian mechanism was also associated with the recapitulation theory and models of linear development, which were in some respects more reminiscent of Lamarck's own views

A contemporary of Lamarck who came closer to the idea that evolutionary progress is best seen as a by-product of the adaptation is Erasmus Darwin. As the grandfather of Charles Darwin, Erasmus – like Lamarck – is recognized as

---

[15] 'De la dégéneration des animaux' was a chapter in vol. 14 of Buffon's *Histoire naturelle*.

[16] See especially Hodge, 'Lamarck's Science of Living Bodies'; for other discussions of Lamarck see the section on his work in Chapter 2.

a pioneer of evolutionism despite the fact that his expositions are less substantial (some are in the form of poetry). Although he relied on the inheritance of
acquired characteristics, he did see adaptive modifications as the means by
which a common ancestral form can be diversified into many different
branches. He also saw progress to higher levels of organization as a by-
product of the constant efforts made by living things to adjust to their environment, not as a separate force driving towards a predetermined goal. Unlike
Lamarck he was also prepared to accept that some species have gone extinct in
the course of time.

Darwin was a medical doctor whose only exposition of his ideas in prose
came in the chapter on reproduction in his *Zoonomia* of 1794–6. There are also
references to transmutation in his poems, especially *The Temple of Nature*. The
development of life on earth was presented as an expression of the Creator's
power, with the fecundity of nature ensuring the ability of living things to
challenge their environment and develop new structures and functions. He was
influenced by David Hartley's belief that the human soul would acquire
permanent modifications in the course of its life. For Darwin, this could be
translated into a wide process by which all living organisms could adapt
themselves to their environment. 'All animals undergo perpetual transformations which are in part produced by their own exertions in consequence of
their desires and aversions', he claimed, and 'many of these acquired forms or
propensities are transmitted to their posterity'.[17]

Darwin thought that all living things came from a common origin, an
'original living filament' which had diversified in the course of time.
Common descent from a single ancestral form explained the underlying
similarities uniting the members of each class. The overall trend was always
towards higher levels of organization. The various types had branched off from
the original filament 'which THE FIRST GREAT CAUSE endued with ani-
mality, with the power of acquiring new parts, attended with new propensities,
directed by irritations, sensations, volitions and associations; and thus possess-
ing the faculty of continuing to improve by its own inherent activity, and of
delivering down those improvements by generation to its posterity, world
without end'.[18] Perhaps even the plants shared the same original ancestry, as
implied in this poetic description of the development of life on earth:

> ORGANIC LIFE beneath the shoreless waves
> Was born and raise'd in Ocean's pearly caves.
> First forms minute, unseen by spheric glass,
> Move on the mud or pierce the watery mass;

---

[17] Erasmus Darwin, *Zoonomia*, vol. 1, pp. 502–3.The most recent study of his life and work is
Fara, *Erasmus Darwin*.
[18] *Zoonomia*, vol. 1, p. 505.

These, as successive generations bloom,
New powers acquire and larger limbs assume;
Whence countless groups of vegetation spring,
And breathing realms of fin, and feet and wing.[19]

As animals strove to adapt themselves, some at least would develop improved intelligence too, so humanity would be merely the highest expression of a more general evolutionary trend towards progress. Darwin saw adaptive change as being able to do far more than merely modify existing structures: it could produce entirely new ones with associated new functions. He did not use the term 'invention' or 'innovation', but as a member of the Lunar Society he was in touch with the inventors and industrialists who were transforming Britain at the time, and his views may have been inspired by parallels between his vision of the history of life and our own innovative powers.

Fossils, Extinction and Divergence

Erasmus Darwin accepted that species had become extinct in the past, anticipating a major change in opinions on this issue. By the 1840s knowledge of the more ancient geological strata had yielded an enormous array of bizarre forms, including the dinosaurs, and the history of life on earth was beginning to take on the appearance we recognize today. For anyone trying to formulate a theory of evolution to explain these developments, the extinct forms posed a problem that could be solved only by accepting that there were multiple lines of development, some of which had come to an end long ago. For Charles Darwin these would be branches on the tree of life that had become dead-ends, leaving others to carry on and diversify.

Fossils were not the only factor threatening the linear model of development. Georges Cuvier's reconstructions of extinct species were aided by his knowledge of comparative anatomy, and his understanding of the internal structure of the various animal types convinced him that they could not be arranged into a hierarchical sequence. Each type displayed a wide range of variant adaptations to different ways of life. In 1828 Karl Ernst von Baer applied a similar approach to embryology, showing that the embryos of each type gradually become more specialized in the course of their development. Embryos are never identical to adult forms of 'lower' species, making the linear model of recapitulation inadmissible. It was becoming clear that the best way of visualizing the relationships between animal forms was the image of a branching tree, with each branch specializing in its own particular way. Before

---

[19] Darwin, *The Temple of Nature*, lines 295–302.

long students of the fossil record were showing that the same process of divergent specialization was typical in the history of each major type. Cuvier and von Baer did not accept evolution, yet between them they were piecing together the evidence for one of the major innovations of Darwinism – the concept of evolution based on a tree-like pattern of divergence.

Early debates on the significance of fossils have been widely studied, but here we are concerned with the rapid change of opinion prompted mainly by Cuvier's research in the years around 1800.[20] His papers, collected as *Recherches sur les ossemens fossiles* in 1812, described a wealth of new species, some quite unlike anything alive today. Cuvier also helped to establish a correlation between the fossils and the relative age of the strata in which they were deposited: the older the rocks, the greater the difference between the ancient forms and those of today. By the 1840s an outline of the history of life on earth had emerged, including the 'Age of Reptiles' and preceding periods from which only fish and invertebrates were known. This sequence formed the basis for the belief that life had advanced step by step from the simplest forms to the highest – the human species. But the sheer variety of fossil species, as well as the fact that most had become extinct, made it impossible to see this progress as a simple ascent of a linear chain of being. The diversity of forms within each class suggested that other forces must be at work. It would be left for another Darwin, Charles, to show how extinction could be fitted into a theory of evolution based on the constant diversification of populations.

Cuvier provided another factor that undermined the chain of being and paved the way for Darwinism, his new approach to classification based on the internal structure of organisms. The external similarities once used to create links in the chain of being were now exposed as superficial and misleading as to real affinity. Underlying similarities associated with the possession of a backbone allowed fish, reptiles and mammals to be grouped into the vertebrate type or 'embranchement'. Cuvier identified three other distinct but equally fundamental types: the molluscs, the articulates (segmented animals) and the radiates (those based on circular symmetry such as sea urchins). Later naturalists had to multiply the number of types because Cuvier had used the radiates as a dumping ground for a number of less well-known forms. This only drove home the point that the types could not be arranged in a linear pattern; they were simply different basic forms upon which an animal could be constructed. The ever-increasing array of fossils confirmed the extent of diversity.

Although many retained the hope of identifying a 'main line' of ascent within the vertebrate classes, it was obvious that this was buried within a fundamental diversity within the classes and types. In 1848 Richard Owen's

---

[20] On Cuvier's work see Coleman, *Georges Cuvier, Zoologist*, and on the fossil debates Rudwick, *The Meaning of Fossils*.

On the Archetypes and Homologies of the Vertebrate Skeleton argued that the type was unified in the sense that the anatomist could discern a basic underlying pattern of organization (the archetype), all the living forms being modifications of that pattern in various adaptive directions. The archetype was an idealized abstraction recognized by the anatomist seeking unity within the diversity of God's creation. For Darwin, the archetype would be replaced by the common ancestor from which all later vertebrates had diverged in the course of evolution.

The new way of thinking about relationships was driven home by developments in embryology pioneered by Karl Ernst von Baer. In 1828 his *Über Entwickelungsgeschichte der Thiere* showed that the linear model of development that had underpinned the simple form of the 'recapitulation theory' was false. The human embryo does not pass through stages corresponding to the adult fish, reptile or lower mammal on the way to maturity, so the lower forms cannot be seen as merely immature versions of the highest. Every embryo begins in an undifferentiated state and then gradually adds the characters that define its type, its class and so on down to the more superficial adaptive features of its own species. Embryological development is a process of specialization, not the ascent of a linear scale, which means that the adult form of one species cannot be equivalent to an embryological stage in another.

Von Baer's embryology suggested that to gain an overview of the whole pattern of developments within the animal kingdom it was necessary to think in terms of a branching system in which many different lines branch apart from a common starting point. Those working on fossil species realized that this model served as an analogy that helped them to understand the succession of forms in geological time. Although there were still many gaps, in some cases there were now enough fossil species to show a succession of forms within a particular family. In every case where this was possible, the earliest were the very generalized and were succeeded by increasingly specialized members of the group. This point was stressed by William Benjamin Carpenter and by Richard Owen, the latter arguing that for each group the earliest members were closest to the undifferentiated archetypical form, from which a variety of lines of specialization branched out through time. Owen still believed that the human form was the most perfect expression of the vertebrate type and must thus represent the goal of creation, but he now recognized that this progressive sequence was buried within a complex branching tree. His work would provide Darwin with an important line of evidence in support of his theory of divergent adaptive evolution.[21]

---

[21] See Bowler, *Fossils and Progress*, chap. 5, and Ospovat, 'The Influence of Karl Ernst von Baer's Embryology, 1828–1859'.

Darwin and Progress

There were two critical questions facing the naturalists of the mid-nineteenth century: how to reconcile the evidence for diversity with the assumption that there was a progressive trend in the history of life leading to humanity and – most controversially – how to explain how the succession of new forms was generated. The idea that a central trunk could be perceived within the branches of the tree of life would remain active well into the twentieth century, at least among popular representations. Darwin would challenge this version of the tree, but he also proposed a highly radical explanation of how what we now call evolution actually worked. Most efforts to deal with this question fudged the issue by imagining mysterious trends that some saw as the direct expression of the Creator's plan. This was still the case for Robert Chambers's transmutationist tract *Vestiges of the Natural History of Creation*, which triggered a bitter controversy when published in 1844. The only suggestions for a purely natural mechanism of species change were Lamarck's theory of the inheritance of acquired characteristics and the idea of sudden mutations caused by disturbances to the growth of the embryo. The latter – later dubbed the theory of the 'hopeful monster' – was suggested by the materialist French anatomist Geoffroy Saint-Hilaire in the 1820s but had also remained highly controversial.[22]

Darwin came to the question of species change from a very different background. The story of how his work in natural history led him towards the theory of natural selection has been told so often that we need only sketch in the relevant points. His voyage on HMS *Beagle* (1831–6) brought him face to face with biogeographical and other lines of evidence that forced him to question the fixity of species and look for a theory that would explain their relationships in terms of natural modifications. The first crucial step in his speculations – recorded in notebooks now scoured by historians – led him to the concept of divergent evolution driven by the fragmentation of populations due to migration and geographical barriers, followed by gradual adaptation to the different conditions encountered by each group. The classic example seen by Darwin himself was provided by the Galapagos archipelago, where populations derived from South American ancestors had split into groups of related species on the individual islands.[23]

[22] See Appel, *The Cuvier–Geoffroy Debate*, and on British radicals naturalists of this period Desmond, *The Politics of Evolution*.

[23] Biographical studies include Browne's two volumes *Charles Darwin: Voyaging* and *Charles Darwin: The Power of Place*; Desmond and Moore, *Darwin*; and my own *Charles Darwin: The Man and His Influence*. For a guide to the literature see my *Evolution: The History of an Idea*, chap. 5. On Darwin and progress see Ruse, *Monad to Man*, chap. 4.

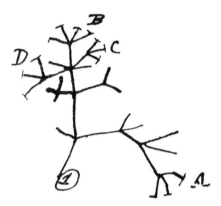

Figure 7.1 Branching tree drawn by Darwin while developing his theory.
There is no effort to imply a privileged direction of change.
From Darwin's B Notebook, p. 36; see *Charles Darwin's Notebooks* (1987), p. 180

From the start Darwin was concerned with diversity. Already in a diagram
scribbled in his notebooks we see the image of the tree of life, whose branches
diverge constantly and which is really a bush rather than a tree because it has
no central trunk. There is no suggestion of a main line of development towards
a single goal (see Figure 7.1). The tree metaphor was also being developed by
philologists studying the evolution of languages, who were confronted by a
very similar process of divergence in time. Species, like languages, can be
classified into groups because the individual elements making up each group
have diverged from a common ancestor and still share many of its basic
features, with later modifications superimposed.[24] Darwin was searching for
a general theory that would explain these relationships, but it would be a
theory founded on the need to recognize the detailed and historically unpredictable events governing how populations became divided and forced to
move apart in order to adapt to different environments. The open-ended and
opportunistic nature of the process was well illustrated in the case of the
Galapagos – islands hundreds of miles out in the Pacific that had originally
been colonized by small numbers of animals and plants accidentally transported from the mainland.

Darwin now had to wrestle with the question of how populations might be
changed to adapt to new environments. He knew of the Lamarckian mechanism and always accepted that it might play a limited role, but did not think it
adequate to explain the facts he was accumulating. Perhaps inspired by his

[24] See Chapter 8 and for more details Alter, *Darwinism and the Linguistic Image*.

grandfather, he was fascinated by the process of generation (sexual reproduction) and the possibility that this might play a role in the formation of new characters. Working with animal breeders, he gained an awareness of the extent of individual variation within every population, an insight confirmed by his extended study of the barnacles that made his reputation as a systematic naturalist. Individuals differ among themselves, and entirely new characters appear from time to time and are preserved in subsequent generations, the process we now explain in terms of genetic mutations (although Darwin's own theory of heredity was based on different principles). The crucial point was that reproduction included a creative element, although what was produced did not seem to have any relevance for the species' needs – which is why breeders can produce their bizarre varieties of dogs, pigeons, etc. Darwin's critics would claim that his theory was based on chance or accidents. In fact he knew that there must be some causal process involved, but the results were multifarious and unpredictable.

How could a process that seemed only to blur the boundaries of the species become the basis for directional changes that would adapt the population to its environment? The breeders again provided the clue, because they modified species by picking out individuals showing a new character suited to their requirements and preventing the others from reproducing, the process of artificial selection. Darwin now had to conceive of a natural process that could replace or mimic the breeders' ability to choose which animals could reproduce, and he found it in Thomas Malthus' study of population expansion. Malthus had written his *Essay on the Principle of Population* to refute the claims of eighteenth-century optimists such as Condorcet and Godwin. He insisted that there was a natural tendency for the population to increase exponentially, inevitably taking it beyond the level that could be sustained by the food supply. No amount of social reform could evade this force of nature, so all hopes of progress and happiness for all were an illusion.

It is one of the great ironies of the history of ideas that Darwin was able to use this pessimistic view of life as the basis for a new, if harsher, version of the idea of progress. He realized that in the constant 'struggle for existence' (Malthus's term) caused by population pressure, any individual which by chance acquired a character conferring some slight adaptive benefit would be better able to survive and breed. Harmful variants would be eliminated. Here was a process of natural selection that would constantly tend to modify any population in a direction that would make it better adapted to its environment. Occasionally these adaptations would be genuinely new structures with new functions supplying benefits far beyond their original purpose. As Giuliano Pancaldi has pointed out, Darwin himself recognized the parallel with human inventions, which suggests that he was aware that here was the basis for a view of evolutionary progress based on occasional innovations rather than a built-in drive towards a predetermined goal.[25]

Darwin would spend the rest of his life refining his theory, publishing the *Origin of Species* in 1859 – twenty years after he first conceived it. The implications for the idea of progress were profound but not easily worked out. After all, natural selection produced only features adapted to a particular way of life, and Darwin was well aware from his barnacle work that adaptation did not necessarily require more complex structures and might even result in degeneration. His insistence on divergent evolution only increased when he realized that, by analogy with the 'division of labour' in human manufacturing, natural selection would tend to drive species further apart even in a stable environment because greater specialization was more efficient. There could be no main line of progress towards humanity because every branch in the tree of life is developing in its own unique way: ancestors can be sought only in the fossil record, not among living species. The only rare exceptions are those 'living fossils' so well adapted to a stable environment that they have never been superseded, hardly good evidence for an inherent progressive trend.

The critical reaction to Chambers's *Vestiges of the Natural History of Creation* had made Darwin wary of implying that evolution represented a trend towards 'higher' forms. In a letter to his botanical confidant J. D. Hooker in 1858 he admitted that his views on 'highness' and 'lowness' were 'eclectic and not very clear'. At best one might take 'highness' to mean a greater differentiation from the basic characteristic of the group – a definition which clearly allows for many different forms to become higher.[26] But if these modifications were merely adaptive, in what sense would this correspond to the more intuitive sense we have that some groups are more highly organized in a general way than others?

To retain an element of progress Darwin would have to define it in a way that recognized the ability of many different kinds of organization to advance each in their own way. Humans might be the highest form so far evolved, but many other types of organization had advanced well beyond the primitive origins of life (an issue Darwin did not want to dwell on). Successively higher levels of organization have emerged even though there is no main line of development and evolution is not aimed towards a single goal. The only diagram provided in the *Origin of Species* was an effort to explain the divergence of species and was very clearly intended to counter any suggestion that the tree of life had a central trunk (see Figure 7.2). Yet Darwin ended his book with an appeal to the reader which insisted that in the long run evolution did generate higher organisms:

Thus, from the war of nature, from famine and death, the most exalted object which we are capable of conceiving, namely, the production of higher animals, directly follows.

[25] Pancaldi, 'Darwin's Technology of Life'.
[26] Darwin to Hooker, June 1854, in *The Correspondence of Charles Darwin*, vol. 5, p. 197.

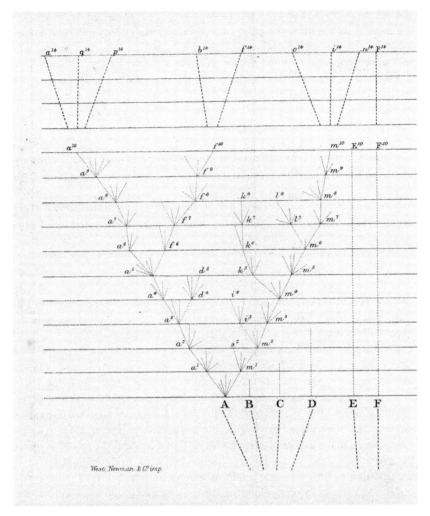

Figure 7.2 Part of the tree diagram used by Darwin in *On the Origin of Species* (1859), between pp. 116 and 117, to illustrate how evolution branches out in many directions. Again there is no implication of a main line of development.

There is a grandeur in this view of life, with its several powers, having been originally breathed into a few forms or into one; and that, whilst this planet has gone on cycling according to the fixed law of gravity, from so simple a beginning endless forms most beautiful and most wonderful have been, and are being, evolved.[27]

---

[27] Darwin, *On the Origin of Species* (1st ed.), p. 490.

Significantly, this draws attention to the diversity of advanced forms without mentioning that humans might be considered higher than any others. It would, however, be the status of humankind that would concern the critics who insisted that his theory reduced the process to a chapter of accidents.

From his concluding statement we might expect Darwin to take up the position adopted by Herbert Spencer, assuming that all or most lines of evolution inevitably tend to advance towards higher states. Just before the passage quoted above he does try to present his theory in this light: 'And as natural selection works solely by and for the good of each being, all corporeal and mental endowments will tend to progress towards perfection.' How was this to be squared with the fact that many well-adapted creatures are still lowly organized, and that adaptive specialization did not always seem to lead to higher complexity? In another letter to Hooker in 1858 he returned to the notion of 'highness' and extended it into something he called informally 'competitive highness'. What he meant by this was the observation that species from Eurasia introduced into Australia and New Zealand sometimes displaced the equivalent native species. Darwin suspected that because the larger continent offered a more challenging set of environments, its inhabitants had been forced to become more efficient in their specialization for various ways of life. But he was also well aware that specialization, however efficient, was not necessarily a route to a generally higher level of organization.[28]

For Darwin's followers the crucial question centred on the origin of what were conventionally taken to be the 'higher' groups – the amphibians, reptiles and mammals – and also the appearance of humans. Apart from the latter, Darwin paid little attention to these transitions. In a letter to his geological mentor Charles Lyell in 1860 he did discuss the origin of mammals, rejecting the view that the marsupials were an intermediate stage linking the egg-laying monotremes to the higher placentals. He preferred a branching model in which the marsupials were an entirely separate development from the placentals. On the question of why the new grade of organization had developed from the reptiles he offered no opinion, but an aside at the end of the letter implies that there was no inevitability about the transition. He asks what would happen if the modern mammals were somehow wiped out: would they be 're-evolved' from the reptiles in due course? He suggests that after a vast period of time a new and higher class would emerge – but it would not be the same as the mammals of today and might even be something more highly developed. In this brief foray into counterfactual thinking, Darwin confirms that his vision of

---

[28] Darwin to Hooker, 31 December 1858 in *The Correspondence of Charles Darwin*, vol. 7, pp. 228–31. The term 'competitive highness' is not used in the *Origin of Species*, but on the displacement of the Australian and New Zealand species see p. 116 and pp. 201–2 (1st ed.).

progress does not include the mounting of a preconceived pattern of development leading towards humans.[29]

Darwin's reticence on what would later become key areas of debate can be understood in terms of his interests and his style of argument. He was never anxious to participate in the effort to reconstruct the history of life on earth from the fossil record, preferring to leave this to followers such as Huxley and Haeckel. Moreover, his whole world-view was shaped by his commitment to Lyell's 'uniformitarian' geology in which all changes were supposed to take place slowly over vast periods of time. Theirs were theories of *continuous* change, with no room for the relatively abrupt transitions postulated by the rival 'catastrophists'. Neither wanted to admit anything resembling what we now call mass extinctions, preferring to see the apparently abrupt nature of some transitions as the result of the 'imperfection of the geological record'. Any suggestion that the apparently sudden expansion of a class might be the result of a unique breakthrough in evolution, or the abrupt elimination of a rival group, was unthinkable. Darwin wanted to preserve an element of progressionism but knew that progress was by no means the most common product of adaptive specialization. However important in the long run, it was a by-product of adaptive evolution, not the main driving force. Recognizing this point made it necessary to identify the unique circumstances that made each significant advance possible. The appearance of major new structures and functions was something that occurred only on rare occasions. Even if it was achieved gradually, the consequences could be dramatic. Evolution might be progressive as well as divergent, but it was also episodic, opportunistic and unpredictable. Darwin opened the door to this recognition, but it was his followers who would pass through the door to complete the transformation.

Spencer and Progress with Divergence

Darwin published his *Origin of Species* in 1859, and the following decade saw a rapid conversion of the scientific community to acceptance of evolutionism – the event we now call the Darwinian revolution. But Darwin's biology was not the only factor involved, for the growing popularity of Herbert Spencer's evolutionary philosophy was also influential. Their theories were often seen as complementary; after all, Spencer coined the term 'struggle for existence', and his focus on individualism led to his political views being labelled as 'social Darwinism'. Both were identified with the idea of progress, and there were certainly similarities between their positions on this issue. Both

---

[29] Darwin to Lyell, 23 September 1860, in *The Correspondence of Charles Darwin*, vol. 8, pp. 377–80. Note the branching tree diagram showing the divergence between marsupials and placentals.

abandoned the linear model of the chain of being, recognizing that evolution was a process of diversification that could not be seen as an advance towards a single predetermined goal (although some of Spencer's followers did not appreciate this point). But there were also differences, Spencer seeing progress as a far more active agent than Darwin did.

For Spencer progress was the result of an inherent tendency for nature to drive organization towards increasing levels of complexity. Evolution consisted of many branches all becoming more complex in different ways, some faster than others. One branch of biological evolution, that leading towards humans, had advanced faster and hence further than any other, although Spencer showed little interest in specifying why that one branch had been accelerated. Darwin explored diversity not as an inherent tendency linked to progress but as a consequence of the inevitable division of species by environmental factors. This made him more aware than Spencer of the contingency of evolution, the possibility that the actual courses taken by diverging populations were unpredictable, given our inability to identify in advance the various natural events that might come into play.

Spencer insisted that evolution involved a tendency to move towards increasing levels of complexity, both of the individual organisms and of the overall ecosystem (to use a modern term). The immediate driving force was the necessity for organisms to adjust to their environment by either the Lamarckian process or natural selection, but he took it for granted that this was normally achieved with an accompanying increase in complexity. Animals would automatically tend to become more intelligent, for instance, as they strove to adapt. For Spencer degeneration was an exception: lines of evolution normally progressed and environmental factors might lead some to progress faster than others. He missed the point, expounded by his American follower John Fiske, that there seemed to be certain critical points where evolution had fairly abruptly jumped to a new level.

Spencer began to formulate his social philosophy in the 1850s (see Chapter 8), soon extending this into a comprehensive evolutionary worldview. In 1855 his *Principles of Psychology* already explained the faculties of the human mind as products of evolution. In an 1857 essay 'Progress: Its Law and Cause' and in his *First Principles* of 1862 he argued that evolution is an inevitable consequence of the basic laws of nature. There is a constant tendency for structures to move from homogeneity to heterogeneity, that is, towards higher levels of complexity. In theory, progress was thus defined as an open-ended process: there was no single goal, and many different forms could become more complex, each in their own way.[30]

---

[30] 'Progress: Its Law and Cause' is reprinted in Spencer, *Essays, Scientific, Political and Speculative*, vol. 1, pp. 1–60. There is an extensive secondary literature on Spencer, including

Two years later *Principles of Biology* moved into the territory already highlighted by Darwin's theory, confirming Spencer's commitment to a branching model of evolution. He was primarily a Lamarckian, but he accepted natural selection as a secondary mechanism of adaptive evolution, coining the phrase 'survival of the fittest' to denote the process. The driving force of change was the individuals' constant need to adapt themselves to an ever-changing environment: like Lamarck, Spencer believed that characters acquired as a result of the adults' struggle with the environment could be passed on to their offspring and would accumulate to change the species. Like many Lamarckians, Spencer took it for granted that the efforts made by the organism would increase its mental powers, so that intelligence would always tend to increase. Since environments were never uniform, individuals would sometimes be led to adapt different strategies of adaptation, which would lead to a diversification of species and, to use the modern term, an increasingly complex biosphere. He was aware of von Baer's embryological findings and appreciated that the main groups of animal and plants could not be arranged into a linear hierarchy.

Darwin's work in areas such as biogeography and the study of migration was of little concern to Spencer: such details were far less important than the underlying trends derived from first principles. Nor is there anything to suggest that Spencer had recognized the problem posed by the fact that there was good evidence for discrete episodes in evolution when entirely new levels of organization had appeared and flourished. These were what Fiske called the 'critical points', and they included the appearance of the main animal types, new classes such as the mammals and the dramatic increase in human brain development. For Spencer, these transformations were inevitable consequences of the fact that some branches might advance faster and further than others, but they did not pose any threat to his confidence in the inevitability of progress. They were quantitative variations in the rate of progress, not qualitatively different from what went on through the regular process of divergent evolution.

An example of how this way of thinking about progress could be applied can be seen in the 'law of brain-growth' proposed by the palaeontologist Othniel C. Marsh. Marsh's discoveries of fossil mammals in the rocks of the American West were described by T. H. Huxley as 'demonstrative evidence of evolution' – his sequence of horse fossils vindicating the theory of continuous specialization. Although widely regarded as a Darwinian, Marsh was also an enthusiastic supporter of Spencer's philosophy, and the law of brain-growth

---

Francis, *Herbert Spencer and the Invention of Modern Life* and Taylor, *The Philosophy of Herbert Spencer*; see also Francis and Taylor, eds., *Herbert Spencer: Legacies*, which contains my own 'Herbert Spencer and Lamarckism'.

proposed in his *Introduction and Succession of Vertebrate Life in America* in 1877 fitted exactly into the Spencerian model in which many different lines progress towards higher levels of organization. He showed that in all of the lines of specialization he had recognized in the American mammals, the brain tended to increase in size. Any species that lagged behind soon became extinct: 'In the long struggle for existence in Tertiary times, the big brains won, then as now.'[31] Despite the Darwinian language, it is doubtful that Darwin would have felt comfortable with such a comprehensive law of progress, and later discoveries showed that Marsh's neat lines of development had to be broken up into irregularly branching trees. Like Spencer, he showed no interest in the question of why only one or a few lines of reptilian evolution had produced the first mammals, nor in why one particular line of mammalian evolution had progressed to the human level of mentality. All such developments were merely individual expressions of the general trend towards progress.

Life's Splendid Drama

In the half-century following the debate sparked by the *Origin of Species* major efforts were made to reconstruct the history of life in evolutionary terms. Darwinists including Ernst Haeckel and T. H. Huxley used morphological, embryological and fossil evidence to trace the sequence in which new and more advanced forms had been introduced. Some topics, including the origin of the vertebrates, had to be tackled using comparative anatomy and embryology, leading to the construction of rival theories which because the fossils were lacking could not be tested. At the end of the century William Bateson would abandon this project in disgust and go on to become one of the founders of genetics. But in other areas the fossil record was being extended by a host of new discoveries, and the effort to trace out 'phylogenies' (Haeckel's term for evolutionary sequences) increasingly depended on this evidence.

There is some debate among historians over the scientific value of this project. Michael Ruse dismisses the whole episode as second-rate science of no permanent significance.[32] The title of this section is borrowed from that of my own book replying to this claim (in turn taken from the palaeontologist W. D. Matthew) and arguing that Ruse's position focuses too closely on the process by which natural selection was turned into a professionally accepted theory. Evolutionism is also a historical science, and reconstructions of the

---

[31] Marsh, *Introduction and Succession of Vertebrate Life in America*, p. 55. On Marsh's discoveries, including Huxley's reaction, see Rudwick, *The Meaning of Fossils*, chap. 5, and Bowler, *Fossils and Progress*, chap. 6. For Marsh's links to Spencer see Bowler, 'American Paleontology and the Reception of Darwinism'.

[32] Ruse, *Monad to Man*, chaps. 6 and 7.

past were the most visible aspect of the theory for the general public. Note that this was not a project inspired solely by Darwin's theory of natural selection, for biologists who favoured non-Darwinian theories made important contributions. They may still have been inspired by the developmental viewpoint, but they recognized that evolution is an open-ended process in which major developments can be triggered by environmental challenges. They played a role in constructing a world-view that was Darwinian in the broader sense, ensuring that when the selection theory was eventually synthesized with the new science of genetics, it could be fitted into a view of the history of life that no longer saw it as the unfolding of a pattern aimed at the production of the human form.

The consequences of these developments for the idea of evolutionary progress were profound. In an 1862 address Thomas Henry Huxley had rejected the claim that the fossil record displayed progress towards higher types, endorsing Darwin's theory precisely because it did not imply inevitable progress. Only after reading Haeckel did he begin reconstructing ancestries and take an interest in the origin of new classes.[33] Some of his work focused on the origin of the birds from the reptiles, a major development that was clearly separate from the one that led to the mammals. Two branches of the tree of life had independently become warm-blooded in completely different ways. The record of progress was so complex that it was impossible to see the human species as the goal of evolution, and the very criteria by which progress could be judged would have to become more sophisticated.

Darwin's focus on the tendency for species to become specialized for a particular way of life had built on trends already recognized in the fossil record. There was an increasing recognition of the fact that specializations were often blind alleys as far as overall progress was concerned: the end-products were so committed to a particular lifestyle that there was no possibility of them switching to another, let alone of advancing to some entirely new form of organization. It was the neo-Lamarckian palaeontologist Edward Drinker Cope who formalized this point in his 'law of the unspecialized', which proclaimed that new classes could emerge only from forms that had retained a more generalized structure and lifestyle.[34] Cope also recognized that highly specialized species left themselves vulnerable to extinction if the environment changed to eliminate their ecological niche. As Julian Huxley

---

[33] Huxley's 1862 address is 'Geological Contemporaneity and Persistent Forms of Life'; see also his later 'Palaeontology and the Doctrine of Evolution'. Both are reprinted in his *Discourses: Biological and Geological*, pp. 272–304 and 340–88 respectively. On Huxley's phylogenetic work and other developments in this field see my *Life's Splendid Drama*.

[34] Cope, *The Origin of the Fittest*, pp. 232–3; see Bowler, *Life's Splendid Drama*, pp. 207–8, 334 and 428.

and others would later insist, it was only because at least some members of each class did not specialize that further progress became possible.

Admittedly, Cope saw each specialized trend as a predetermined pathway taken by several parallel lines, although even the more Darwinian O. C. Marsh tended to depict the trends as linear patterns. One of the most important factors tending to undermine Cope's developmental viewpoint was the tendency for more fossil discoveries to reveal complex sequences that could be represented only by irregularly branching trees; the modern horse, for instance, was not the goal of its family's evolution, and similar forms had appeared earlier in the record only to go extinct. By the end of the century even popular books on the history of life included diagrams showing just how complex and irregular the evolution of most classes had been.[35]

Another point recognized by Darwin was that some adaptive specializations led to actual degeneration, as when parasites lose their organs of locomotion. This was driven home by E. Ray Lankester in a popular book published in 1880. Lankester argued that it was a general principle that any form adapting for a sedentary lifestyle would degenerate. Evolution progressed only when organisms were challenged or stimulated by the environment, and any attempt to evade this pressure would lead to a decline. Later on Lankester would become a friend of the young H. G. Wells, who would apply this rule to the future of the human race in his story 'The Time Machine'.[36]

Lankester's focus on the struggle against the environment fitted the ideology of 'social Darwinism' (including Spencer's Lamarckian version) but left open the question of why there seemed to be only a few cases where a positive response had led to an entirely new level of organization. Evolution might be continuous, but there did seem to be rare instances where it jumped fairly rapidly to a new level. Lyell and Darwin had insisted that geological change is always slow and gradual, but the rival 'catastrophist' school had always taken seriously the evidence for abrupt changes in the sequence of geological formations. Fully fledged catastrophism had now declined in popularity (to be partially revived by the modern theory of asteroid impacts), but most geologists remained convinced that there had been several relatively sudden transformations of the earth's crust interspersed with long periods of stability. By around 1900 it was widely accepted that episodes of mountain-building served as punctuation marks in the earth's history and hence in the development of life. The rapid decline of the great reptiles at the end of the Mesozoic was a real event, prefiguring the modern notion of mass extinctions.

[35] Many examples are reproduced in Pietsch, *Trees of Life*.
[36] Lankester, *Degeneration: A Chapter in Darwinism*; see my *Life's Splendid Drama*, pp. 153–4, 164 and 431–2.

Palaeontologists were also increasingly inclined to think that the emergence of new conditions and new opportunities following a geological upheaval had provided exactly the kind of challenge that might stimulate the appearance of some new level of organization. Cope's disciple Henry Fairfield Osborn suggested that the mammals had evolved to occupy the harsher conditions of upland regions. Like most of his contemporaries he also saw the rapid expansion of the mammals as the exploitation of an opportunity presented by the dinosaur's decline. It had to be admitted, though, that the early mammals had remained insignificant for a long period until the elimination of the dominant class occurred. It was Osborn who coined the term 'adaptive radiation' to denote the sudden expansion of mammalian forms at this point – for all that he thought their subsequent development to be predetermined by orthogenetic factors.[37]

There was now increasing use of the metaphor of the 'rise and fall' of groups in the fossil record, analogous to human empires. Geologists became aware not only of discontinuities in the rate of activity within the earth's crust but also of evidence for major geographical transformations (although the theory of continental drift became widely accepted only in the mid-twentieth century). These too could have significant and in some cases abrupt impacts on populations, as when the joining of the isthmus of Panama allowed North American species to flood into the southern continent, exterminating many of the indigenous inhabitants. These events again triggered the rise of some forms and the decline of others, encouraging the view that the history of life was episodic, representing what Richard Swann Lull called the 'pulse of life'.

Thus time has wrought great changes in earth and sea, and these changes, acting directly or through climate, have always found somewhere in the unending chain of living beings certain groups whose plasticity permitted their adaptation to newly arising conditions. The great heart of nature beats, its throbbing stimulates the pulse of life, and not until that heart is stilled forever with the rhythmic tide of evolution cease to flow.[38]

Most attention focused on the pulses that triggered the rise of an entirely new class with a higher level of organization. It was appreciated that only the less specialized members of one class had the flexibility to progress to another level, although in some cases more than one group might have this capacity. Birds and mammals had both acquired warm-blooded constitutions, and it was widely believed that the primitive mammals that had evolved during the Age of Reptiles had emerged independently from the ancestors of those that eventually rose to dominance. Even here, the giant mammals that Osborn

---

[37] See Osborn's *The Age of Mammals*; on their origin as a response to upland conditions see p. 58 and p. 98.

[38] Lull, *Organic Evolution*, epilogue 'The Pulse of Life', pp. 687–91, quotation from p. 691. On these metaphors see my *Life's Splendid Drama*, chap. 9.

studied from the early Tertiary seemed to have died out fairly rapidly to be replaced by the modern families.

The crucial question was how and why the founders of a new class had been able to develop entirely new structures and functions. Critics of Darwinism had argued that entirely new structures could not develop, because the intermediate stages were not viable – what is the use of a limb that is neither a leg nor a wing? Lankester's friend Anton Dohrn had one answer to this: in some cases a structure that had developed for one purpose turned out to be useful for another. Jaws had developed from the gill-arches of fish, and feathers had evolved as insulation before being adapted for flight. Evolution was opportunistic, making use of whatever was available, and it sometimes transformed simple structures developed for one purpose into complex ones useful for something with more potential.[39]

Another insight was the recognition that the variations introducing new characters did not necessarily arise from modification of the adult structure. The developmental view of evolution adopted by the Lamarckians rested on the assumption that acquired characteristics were added onto the original adult form, which allowed earlier phases to be recapitulated in the growing embryo. In his 1861 book supporting Darwinism, Fritz Müller pointed out that variations might also be produced if development were switched into a new channel at an earlier point in the sequence, in which case the subsequent stages of the original sequence would be lost. In the Darwinian scheme, variation was best seen as a distortion of development, not an addition to it.

In the early twentieth century this approach was extended by the Russian morphologist Alexei Sewertzoff, who saw variation in the course of ontogeny as a process that would allow a species the freedom to develop entirely new structures and functions. This would be progress without a predetermined goal, defined by the production of higher levels of organization rather than by a particular structural target. Sewertzoff clearly distinguished between adaptive specialization within one level of organization and the rare and unpredictable transitions that generated new levels. His views were extended by I. I. Smallhausen, who studied the origin of the terrestrial vertebrates. Through Smallhausen's influence on Theodosius Dobzhansky this sense of progress as a rare event brought about by processes significantly different from those governing normal adaptations entered the thinking of the next generation of Darwinists.[40]

---

[39] Dohrn advanced this idea in the context of his controversial theory of vertebrate origins; see the modern translation of his *Der Ursprung der Wirbelthiere* as 'The Origin of Vertebrates and the Principle of Succession of Functions'.

[40] See Levit, Hossfeldt and Olson, 'The Integration of Darwinism and Evolutionary Morphology' and the same authors' 'From the "Modern Synthesis" to Cybernetics'. For more detailed on Müller and the critique of recapitulationsim see Gould, *Ontogeny and Phylogeny*.

The process by which the new function was perfected was often described as one of experimentation, as though nature was somehow groping its way towards the new possibility. Thus Alfred Sherwood Romer described a group of early mammals as 'an early "experiment" on the part of the insectivore stock in the creation of a herbivorous form' – unsuccessful in this case because they soon became extinct.[41] This was certainly anthropomorphic language, but it made evolution seem more opportunistic or serendipitous, less the unfolding of a preordained pattern. Whatever the implicit teleology, there was a definite sense that the forces at work had no clear guideline: the advantages were there to be exploited but evolution didn't know how best to achieve them. When applied to the main upward steps this allowed something of the old vision of progress to be preserved, but the process would have to be seen as less regular and less predictable, more in tune with a theory which postulated 'random' variation as its raw material.

An even stronger element of serendipity can be seen in Romer's discussion of the origin of the amphibians. This was normally presented as the starting point for the triumphant 'invasion of the land' – a key step in the advance of vertebrate life. Romer argued that far from being developed to allow the conquest of the land, air-breathing and limbs had evolved in a desiccated world where fish desperately struggled to get from one drying pool to another.[42] Only later would their descendants use the new functions to move permanently onto the land. The progressive step was made possible only by a combination of circumstances that was unpredictable on the basis of previous trends in evolution.

Another indication of the new approach is provided by the emergence of the first uses of counterfactual thinking by evolutionary biologists. Counterfactualism – imagining a world that is different from ours as the result of some slight change of circumstances in the past – makes sense only if history is seen as a record of contingent events, not the unfolding of a predetermined sequence. Romer adopted this approach when he noted that the early Tertiary had been dominated by giant birds as well as primitive mammals and asked: 'What would the earth be like today had the birds won and the mammals vanished?'[43] The same way of thinking would soon be extended to the origins of the human species, as we shall see in the next chapter.

The Modern Synthesis and Progress

A whole range of ideas associated with the old developmental version of evolutionism now began to lose credibility. Genetics exposed the weakness

---

[41] Romer, *Vertebrate Paleontology*, p. 278.
[42] Ibid., p. 105. Romer was drawing on ideas suggested earlier by Joseph Barrell and others; see my *Life's Splendid Drama*, chap. 5.
[43] Romer, *Man and the Vertebrates*, p. 98.

of the evidence for Lamarckism, orthogenesis and the recapitulation theory, all key elements of the developmentalist programme. The 'eclipse of Darwinism' ended as natural selection emerged as the only credible explanation of how genetic variation could be channelled to shape a population's evolution. Studies of mutations confirmed Darwin's belief that the variation existing within populations is generated by mechanisms that are not directed by the organisms' needs, or indeed by any predetermined trend. Adaptation was, after all, the main driving force of evolution – but it had to work with an array of genetic variation which was, in effect, unpredictable.

This last point enabled exponents of the new synthesis to present Darwinian evolution as a process that could occasionally transcend the immediate requirements of adaptation to come up with something really new. As the paleaontologists had already realized, most evolution led to adaptive specialization, but it was possible for entirely new structures and functions to emerge under certain unusual circumstances. The suggestion that natural selection could be seen as a creative process analogous to human inventiveness had been made much earlier by William James. His intention was to challenge the Spencerian view that evolution worked according to predictable laws by introducing an element of unpredictability equivalent to genius. James became an enthusiastic supporter of Henri Bergson''s philosophy of 'creative evolution' because it made the same point against Spencerianism.

Widely dismissed nowadays as an alternative to materialism based on the notion of a mysterious life-force, the *élan vital*, Bergson's philosophy was really an attempt to rethink the relationship between psychology and evolution. The *élan* was a metaphor to help us understand why science finds it difficult to come to grips with the process of transformation, and it was in this context that it was able to stimulate the imagination of some Darwinians. Bergson transformed attitudes towards teleology in evolution by promoting a model in which life groped its way upwards but had no predetermined goal. Biologists who played key roles in the foundation of modern Darwinism, including Julian Huxley, were inspired by the idea of *creative* evolution, arguing that natural selection was a materialistic process capable of generating the innovations that Bergson envisioned. Although welcomed by those hoping for a reconciliation between science and religion, Bergson's vision was also able to stimulate the emergence of a theory that revived all the old fears of Darwinian materialism. Progress now had to be recognized as a by-product – even if an inevitable one – of a process based on 'random' variation that was not primarily aimed at the production of higher levels of organization.[44]

---

[44] See James's essay 'Great Men and Their Environment', in his *The Will to Believe and Other Essays in Popular Philosophy*, pp. 216–54. Bergson's *Creative Evolution* was published in an

The revival of interest in natural selection began in the 1890s when August Weismann transformed thinking on heredity with his concept of the 'germ plasm' – a material substance located in the chromosomes of the cell nucleus that was responsible for transmitting characters from parent to offspring. In effect, this anticipated the role we now assign to DNA, and its impact was crucial because it forced biologists to consider the possibility that heredity and embryological development could be treated as distinct processes. Traditionally they had been seen as a single, integrated effect, which is why the Lamarckians could assume that characteristics acquired late in development could be transmitted to the next generation. Weismann insisted that the germ plasm was insulated from external forces: the flow of information was one-way, from the germ plasm to the developing embryo but not the reverse. Acquired characters could not be inherited, and this left natural selection based on germinal variations (due to internal causes) as the only way of directing evolution along adaptive channels. The decades around 1900 witnessed an intense debate between neo-Darwinians who saw natural selection as the sole cause of evolution and the supporters of the various non-Darwinian alternatives. Studies of wild populations showed that they did indeed contain a range of variation for each characteristic just as Darwin had claimed, and even demonstrated small-scale effects of selection on the range.[45]

A new factor was introduced by the 'rediscovery' of Gregor Mendel's laws of heredity, proposed in 1865 but long neglected because they were based on experiments with artificially bred varieties of the garden pea. In his samples, variation consisted of discrete character differences rather than a continuous range (the peas were either tall or short, for instance, not distributed around an average height). These discrete characters bred true over many generations, following what became known as Mendel's laws of inheritance. The increased focus on heredity at the end of the century led several biologists to perform similar experiments, leading to the recognition in 1900 that these long-neglected laws might hold the clue to a revolution in the understanding of the field. In Britain, William Bateson abandoned his work on the origin of the vertebrates to take up the experimental study of inheritance, published a translation of Mendel's work, and coined the term 'genetics' to denote the new science. The geneticists took up Weismann's claim that the hereditary factors they studied could not be modified by external forces and thus continued the assault on Lamarckism.

English translation in 1911; on his influence see Herring, 'Great Is Darwin and Bergson Is His Prophet' and the discussion in Chapter 10.

[45] There are many accounts of the revolution leading to the emergence of genetics, including my own *The Mendelian Revolution* and *Evolution: The History of an Idea*, chap. 7.

They did not, however, take up the cause of Darwinism, because they were now convinced that the continuous variation seen in many wild populations was caused by temporary environmental factors that were not inherited. One of the original rediscoverers of Mendel's laws, Hugo De Vries, studied the production of discrete new characters in the evening primrose and proposed his 'mutation theory' in which evolution was driven by the sudden production of new forms due to apparently random reconfigurations of the germ plasm. This was widely hailed as a new alternative to Darwinism, a theory in which evolution was indeed based on unpredictable innovations – but they were sudden rather than gradual, and seemed to involve no adaptive benefit (although De Vries himself realized that harmful mutations would soon be eliminated by natural selection).

It soon became apparent that the substantive characters studied by De Vries were not the product of new germinal factors: the evening primrose was a complex hybrid that occasionally threw off rearrangements of existing forms. Detailed experimental work in America led by Thomas Hunt Morgan used the fruit fly *Drosophila* to establish the existence of true genetic mutations and to show that the new characters they produced bred true according to Mendel's laws. The Morgan school also provided an explanation of these laws by showing that the genes were structures arranged along the chromosomes of the nucleus. Many of the mutations they studied were quite small, and the organisms carrying them bred within the existing population. Instead of founding entirely separate varieties or species as De Vries believed, they simply added to the variation of the original population. It now became possible to see that far from representing an alternative to Darwinism, genetics might actually explain the existence of the range of variation within a population upon which natural selection could act. If a number of different genes affected a particular characteristic, their discrete effects would overlap to create the appearance of a continuous range. The ultimate origin of the variation would be the process of mutation, which was multifarious and undirected just as Darwin had supposed.

The crucial question was: could natural selection act on the assembly of genetic characters in the population to modify the range of variation in a direction that provided adaptive benefit? The answer was supplied by the creation of population genetics, using statistics to analyse the balance of genes in large groups and to assess the impact of selection if some genes conferred a greater chance of reproduction. The result was, in the title of R. A. Fisher's classic book of 1930, *The Genetical Theory of Natural Selection*. Two years later J. B. S. Haldane's *The Causes of Evolution* gave a popular account of the new Darwinism, while in America Sewall Wright used somewhat different mathematical techniques to reach similar conclusions. The selection theory could now provide a satisfactory account of adaptive evolution, using the undirected variation generated by mutations as the raw material. The once rival theories of genetics and Darwinism had been brought together.

The mathematics of population genetics were incomprehensible to many field naturalists, but the Russian émigré Theodosius Dobzhansky provided information about the new theory in a format that they could apply in his *Genetics and the Origin of Species* of 1937. In the early 1940s studies by Ernst Mayr and Julian Huxley showed how the new theory could be used to explain a wide range of observations in the wild. George Gaylord Simpson extended the argument into palaeontology, building on the growing consensus that the evidence once used to support orthogenesis and other theories of directed evolution was unreliable. The recapitulation theory – once a mainstay of the old developmental viewpoint – was no longer taken seriously, because mutations produced distortions of the growth process, not additions to it.[46] Huxley's book *Evolution: The Modern Synthesis* gave the new theory its name, and from this point onwards the Darwinian synthesis dominated the thinking of evolutionists.

There have certainly been controversies, but these have never undermined the commitment of most biologists to some form of Darwinism. The most recent development is the emergence of evolutionary developmental biology or 'evo-devo', which seeks to challenge the claim that the genes offer a blueprint for the organism that cannot be significantly affected by environmental factors. Enthusiasts see this as an avenue through which something resembling the old Lamarckian mechanism can again be seen to play a role. Even so, their ideas suggest only a means by which the organism's development might shape the opportunities for natural selection to respond to environmental pressures. Long-term evolution still depends on changing the genetic structure of the population, and the overall process is still opportunistic and open-ended; there is no return to the old developmentalism.[47]

As Michael Ruse has shown, the founders of the modern synthesis were almost all committed to the belief that evolution is progressive, but it is important to stress that most of them thought that progress now had to be defined without the implication of a predetermined goal.[48] If natural selection is the driving force, evolution must be primarily based on adaptation and specialization, which is why we see such a diversity of living forms in the fossil record and the present day. Extinction is a constant threat if environments change rapidly, with the result that whole groups can be replaced by

[46] De Beer, *Embryology and Evolution*; see Rasmussen, 'The Decline of Recapitulationism in Early Twentieth-Century Biology'.

[47] For my own account of the modern synthesis see *Evolution: The History of an Idea*, chap. 9. More detailed studies include Mayr and Provine, eds., *The Evolutionary Synthesis* and Smocovitis, *Unifying Biology*.

[48] See Ruse, *Monad to Man*, chaps. 11–14.

others in the course of time. If there is progress it will occur on rare occasions when a transformation works by establishing a new level of organization with improved general functions. Such progressive steps may yield the founders of an expanding network of descendants, often triggering the abrupt replacements that define geological epochs.

The problem with any system based primarily on adaptive divergence is that it can deflect attention away from the element of progress, or even be taken to render the idea meaningless. One biologist who emerged as a critic of strict Darwinism in the 1980s was Stephen Jay Gould, who argued against both the extreme focus on adaptation and the idea of evolutionary progress. He seized on the fact that many 'lowly organized' forms have survived for vast periods of time to argue that the conventional definitions of 'highness' were illusory. There was no evidence that complexity guaranteed survival: this was as much a matter of luck as of better organization. In his *Wonderful Life* of 1989 he illustrated this point through his interpretation of the bizarre creatures revealed in the Burgess shale fossils, the best evidence available of the 'Cambrian explosion' when multi-celled animals abruptly appeared on the earth to kick-start a new phase in evolution. He insisted that many of these creatures could not be fitted into any of the known animal phyla; they were totally distinct forms that had gone extinct – and his view was that they had died out not because they were unfit, but because survival was a matter of luck.

Few modern evolutionists would accept Gould's anti-adaptationism. Indeed we have ever-increasing levels of evidence confirming how populations respond to challenges from their environment. The evidence even confirms (from a very different perspective) another one of Gould's arguments, originally presented as yet another challenge to conventional Darwinism. Gould's theory of 'punctuated equilibrium' was presented as an alternative to Darwinian gradualism, postulating rapid changes in small populations exposed to strong environmental stress. But whereas Gould thought that such punctuations would be rare events, we now know that everyday processes affecting small populations routinely produce dramatic changes in their genetic make-up. Evolution can indeed proceed much more rapidly than the original Darwinian paradigm imagined – yet it still works by natural selection acting on genetic mutations.[49]

Innovation and Purpose in the New Darwinism

The new Darwinism replaced the chain of being with a branching model of taxonomic relationships, requiring naturalists to accept that there were many

---

[49] For a recent account of these developments see for instance Losos, *Improbable Destinies*.

different ways in which life had acquired complex organizations and more sophisticated functions. Examples of very different forms with the same levels of ability to control of their environment were widely available. Some cephalopods show high levels of intelligence and can manipulate objects with tentacles. Colonial insects are also able to modify their environments in significant ways. In a biology textbook published in 1927 Haldane and Huxley gave a table illustrating the grades of organization reached by different groups in which molluscs and insects both reached grades equivalent to lower vertebrates (humans, of course, were the only species to reach the top level).[50] When applied to a theory of evolution, these points imply that progress can be measured against an abstract scale and has occurred in many of the separate branches of the tree of life. Bergson himself had used the insects to illustrate the diversity produced by 'creative evolution', although he insisted that their instinct-driven behaviour imposed limitations that were eventually transcended by the more flexible form of individual intelligence developed by the vertebrates. The Darwinists now argued that natural selection was as innovative as Bergson's imaginary *élan vital*, but they had to address the question of whether the element of progress they wished to retain allowed for anything resembling the old idea of a cosmic purpose for life.

Efforts to retain the idea of progress led to definitions that could be applied to different forms of organization: Julian Huxley wrote of improvements in the 'machinery of living' which opened up new environments and opportunities. Mental powers also increased, which is why we find obvious value in the achievements of evolution and can see our appearance as a final step in the overall advance of life. Huxley accepted the image of occasional breakthroughs in which evolution had produced a new level of organization, seeing these as illustrations of the creativity which Bergson had envisaged. For Huxley, Bergson was the poet of evolution, but scientists had to recognize that his mysterious life-force was really only a metaphor for the ability of natural selection to innovate. In an essay published in 1923 he had acknowledged Bergson's thought but made a connection that would become commonplace in the new version of progressionism: natural selection was capable of something equivalent to human invention and innovation. In effect, evolution was equivalent to our technological developments: 'what could be more striking than [the] parallel between the rise of the mammals to dominance over the reptiles, and the rise of the motor vehicle to dominance over that drawn by horses?'[51]

---

[50] Haldane and Huxley, *Animal Biology*, fig. 77 between pp. 236 and 237.

[51] Huxley, 'Progress, Biological and Other', in *Essays of a Biologist*, pp. 17–61, quotation from p. 41. See Herring, "Great Is Darwin and Bergson Is His Prophet'.

The significance of this comparison is that it was now increasingly accepted that technological innovation was the main driving force of change in human society. And as we shall see, it was acknowledged to be unpredictable: experts were constantly being confounded by the appearance of things they had deemed to be impossible. At first, Huxley himself seemed to accept this implication. In *The Science of Life* of 1931, which he wrote in collaboration with H. G. Wells, he presented progress as a 'tangled skein' and, significantly, saw humanity's emergence from the trees as a key step in our evolution at a time when the Australopithecines were still not recognized as human ancestors. In a section asking 'Is There Progress in Evolution?' Wells and Huxley wrote that 'Life experiments and discovers; Nature selects; everlastingly the old is surpassed by the new and fitter.' They also implied that humanity was by no means the only vehicle by which evolution could achieve further progress: 'But a check in the advance of *Homo sapiens* is not necessarily the end of progressive evolution.'[52]

These are all insights that would become widely accepted as the modern Darwinian synthesis emerged, making it all the more remarkable that Huxley himself reined in his sense of the unpredictability of evolution and ended up defending the view that humans are indeed the only viable form by which mental and social progress can continue. His humanistic philosophy was intended to provide a substitute for the emotional needs satisfied by traditional religion, and he seems to have become concerned to give our species a cosmic significance that could be sustained only if we were the unique end-product of evolution. We have seen how he turned to Robert Broom's explicitly teleological view that progress has taken place only through a series of breakthroughs originating in unspecialized types, and that each of these steps narrowed down the opportunity for further development. In the end, the human form was the only one by which the advance to consciousness could be made, and without us evolution would remain unfulfilled – hence Huxley's enthusiasm for Teilhard de Chardin's mystical vision of spiritual progress.

Two other founders of the evolutionary synthesis retained more conventional religious beliefs than Huxley, but both were more willing to see an element of unpredictability in the outcome. R. A. Fisher shared Huxley's enthusiasm for Bergson's views and saw the element of indeterminacy implied by Darwinism as crucial for allowing humans the freedom to achieve the task that God has assigned to them.[53] Theodosius Dobzhansky retained his Russian Orthodox faith and shared Huxley's appreciation of Teilhard's mystical view

---

[52] Wells and Huxley, *The Science of Life*, pp. 477 and p. 480. For further reading on Huxley's progressionism see the discussion of his later views in Chapter 6, notes 6–8.

[53] On Fisher's views see Ruse, *Monad to Man*, pp. 295–303, and on his enthusiasm for Bergson my *Reconciling Science and Religion*, p. 154.

of progress, although he agreed that it could not be seen as scientific. In his *The Biology of Ultimate Concern* he addressed what he called 'The Teilhardian Synthesis' and argued that the image of evolution 'groping' its way upwards was informative: there had been 'a succession of trials and errors, some of them resulting in inventions' by which progress had been achieved. He went on:

The history of life is comparable to human history in that both involve creation of novelty. Both proceed by groping, trial and error, many false starts, being lost in blind alleys failures ending in extinction. Both had, however, also their successes, master strokes, and both achieved an overall progress.

If this was a process designed to achieve a certain goal, it was incredibly clumsy and unpredictable. A biologist studying the earth's inhabitants at the start of the age of mammals could not have anticipated the emergence of humans. If there was life on other worlds, it would be unlikely to contain any form resembling humans.[54]

Most of the founders of the modern synthesis had lost all religious faith yet retained some hope that evolution had a meaningful purpose in human terms. J. B. S. Haldane was a Marxist who believed that technological innovation must play a role in improving living conditions. The textbook he wrote with Huxley included a chapter on progress and used one of Huxley's analogies to suggest that the balance between predators and prey could lead to a situation resembling and arms race in which the search for better weapons and armour leads to improvements in both. Adaptive evolution in this form was equivalent to human technological innovation. Progress to higher grades was less predictable, and in his *Causes of Evolution* Haldane insisted that such steps were only an occasional feature in the history of life. A popular essay entitled 'Man's Destiny' expressed concern that if we did not take control of our own evolution we might become extinct, but a jocular addition makes it clear that he did not see us as the only possible form of intelligent life: 'If this happens, I venture to hope that we shall not have destroyed the rat, an animal of considerable enterprise which stands as good a chance as any other of evolving towards intelligence.'[55]

Haldane's remark suggests that a well-established species could serve as the springboard for further progress, undermining Huxley's claim that no replacement for humanity was possible. Several of his contemporaries challenged the 'law of the unspecialized' and insisted that specialization was not a barrier to

---

[54] Dobzhansky, *The Biology of Ultimate Concern*, pp. 120–8, main quotation from p. 125. On his wider views see Adams, ed., *The Evolution of Theodosius Dobzhansky*.

[55] Haldane, 'Man's Destiny', in his *The Inequality of Man and Other Essays*, pp. 142–7, quotation from p. 146. See also Haldane and Huxley, *Animal Biology*, pp. 237 and 248–9; Haldane, *The Causes of Evolution*, p. 152–3.

progress, as long as it was not too extreme. Gavin De Beer's embryological studies led him to support the theory of 'paedomorphosis', which supposed that a species could actually throw off specialized adult characters and begin new variation from an earlier stage of development.[56] The necessity for generalized ancestors was also questioned by the palaeontologist G. G. Simpson, who endorsed the idea of progress even though he acknowledged the difficulty of defining that progress actually was, especially in the context of 'dominant' forms of life. In his *The Meaning of Evolution* of 1949 he included chapters on the opportunism of evolution and on progress, insisting that the latter occurred in many branches. In *This View of Life* he challenged the evolutionary mysticism of writers such as Teilhard and the comte de Noüy and insisted that progress occurred only through occasional unpredictable 'breakthroughs'. There was no single way forward: think, for instance, of kangaroos replacing quadrupeds in Australia.

The most blatant challenge to the claim that evolution can have a predictable outcome came in Stephen Jay Gould's *Wonderful Life* of 1989. Gould was hostile to many aspects of the new Darwinism, as we saw above. But he was a good Darwinian in one crucial respect, in his stress on the contingency of evolution. We have already encountered his position as the antithesis of Simon Conway Morris's claim that the extent of convergence in evolution means that its outcomes are fairly predictable. Gould insisted that the sheer complexity of the interactions involved in shaping the earth's populations over time makes it impossible to predict the course of evolution from its past trajectories. *Wonderful Life* introduced the analogy of imagining a 're-running' of the tape of evolutionary history from its starting point – and Gould insisted that each time the tape was run the outcome would be different. The present situation, with humans at the top of the scale, was just one among a vast number of potential outcomes because at each major step in the process slight differences in either environmental or biological forces might switch the outcome from one potential trajectory to another. This particular issue will re-emerge in the next chapter when we consider the question of whether or not humanity (or something like it) must necessarily evolve on any earth-like planet.

Gould's arguments generated some controversy among philosophers of science. Did he intend to imply that there is an absolute element of contingency in history, i.e. that an earth identical to ours in every respect (down to the last atom) at the start would nevertheless witness a different pattern of evolution? If we allow for quantum indeterminacy this might well be the case, since mutations occur in individual molecules and it would thus be possible for a different train of mutations to emerge in the course of time, offering different

---

[56] De Beer, *Embryology and Evolution*, pp. 95–6.

opportunities to adaptive evolution. I suspect that what Gould actually meant was that if, as a thought experiment, we imagine a world with just a tiny difference in its starting configuration, would we expect that difference to result in a completely new pathway of evolution? Gould thinks it would, while Conway Morris argues that minor changes cannot affect the inevitability of what must emerge. On the latter view, even if there is an element of contingency, the substantive outcome is predictable. For Gould, contingency means that the end result is uncertain: since evolution depends on the interplay of a vast number of independent causal chains, predicting the outcome is for all intents and purposes impossible.[57]

This element of contingency was integral to the Darwinian synthesis, although few shared Gould's antipathy to the element of progress. Most Darwinians accept that there *has* been progress towards higher levels, and humans are the highest product of that trend so far produced – but the ascent has been irregular and unpredictable, and it might easily have had different outcomes. In one sense, the main foundation of the new way of thinking is acceptance of the possibility that we humans are not the inevitable outcome of the development of life on earth, a point explored in the next chapter. The most recent popular account of the latest research on the issue by the evolutionary biologist Jonathan Losos concedes Conway Morris's point that convergence – the emergence of identical responses to environmental challenge in different species – is more prevalent than the previous generation of Darwinists assumed. But there is evidence that even when identical populations are exposed to identical conditions, each may develop different innovations to cope. The element of contingency is still there.[58]

In the end, the most important focus for the debate over the purposefulness of evolution was the appearance of humankind at the end of the process. Was it also the goal of the process, as Huxley wanted to believe, or at least an indication of the end-point that evolution was supposed to achieve? Haldane's suggestion that the rat might evolve a high level of intelligence if the human race were to disappear highlighted an issue that would be actively debated by many of his colleagues as the Modern Synthesis was consolidated. If the human species is indeed the product of an unpredictable sequence of adaptive innovations, is there any sense in which we can be regarded as the expression of some ultimate purpose in nature? Or does innovation itself have to be seen as the only factor that can define progress?

---

[57] On the conceptual issues involved see Beatty, 'Replaying Life's Tape'.
[58] Losos, *Improbable Destinies*.

8   The Uniqueness of Humans

The most crucial application of the new way of thinking about evolutionary progress was to the issue of human origins. Even in the late nineteenth century, many still believed that progress was automatic and had a privileged direction aimed towards the human form. Sooner or later something like humanity as we know it today was bound to emerge. The full implications of Darwinism tended to undermine this faith. If evolution really was a process undirected by any long-range trends and merely the outcome of a complex web of adaptive responses to environmental challenges, then the discontinuous events that initiated the great advances in the history of life were undirected too. One might expect that in the long run there would be a sequence of significant innovations in life's ability to dominate the environment, but the exact ways in which these innovations are made would depend on individual circumstances, so the new structures and functions would also be unpredictable. Applying this insight to the origins of the human species would undermine one of the most fundamental components of traditional Christianity, the belief that humans were created in the image of God and hence that our appearance in the world was inevitable.

Full recognition of this point came only in the twentieth century, although Darwin himself offered a pioneering suggestion on the topic. Eventually the palaeontologists who recognized the episodic and unpredictable nature of evolutionary progress began to apply the lesson to human origins, just as new fossil discoveries confirmed a key point in Darwin's own suggestion that an adaptive transformation had triggered the expansion of the human brain. By the middle of the twentieth century the new way of thinking had consolidated into a widespread belief among the pioneers of modern Darwinism that no observer of the previous episodes in the history of life could have predicted this particular development. Had circumstances been different at any point in the history of life, we would not be here today. Even if one supposed that something equivalent to the human level of intelligence and awareness would eventually appear, it might be something very different from ourselves. One illustration of the growing popularity of this inference can be seen in the willingness of science-fiction authors to write about non-human alien civilizations. Such scenarios make sense only if one accepts that evolution has no predictable goal and that if it occurs on other worlds than ours the outcomes will be very different.

Darwin on Human Origins

Darwin had little interest in the efforts made by his followers to reconstruct the history of life on earth. He was far more concerned with working out the general principles that governed adaptive evolution, focusing especially on its implications for the diversity of life. One of the few cases in which he did seek to provide an explanation for a particular breakthrough in evolution was the crucial one of human origins. Most early evolutionists saw the expansion of our mental powers as the driving force, in effect presenting humans as just a special case of Marsh's law of brain-growth. Darwin realized that this approach begged the question of why the line leading to modern humans had advanced so far when the great apes – fellow members of the primate family – had failed to benefit from the trend. It was necessary to specify some unique event that had affected our ancestors but not theirs, creating the special circumstances in which extra brain-power became beneficial.

In his *Descent of Man*, Darwin suggested that it was our upright posture, not our bigger brains, that originally marked out our ancestors as different. He noted that standing upright would have been an advantage if a group of primates had moved from their traditional habitat in the forests onto the open plains. He believed that this was most likely to have taken place in Africa, challenging a widespread assumption that Asia was the centre of origin for progressive types. Standing upright freed the hands from the demands of locomotion and created the opportunity to use them to manipulate the environment and make primitive tools. The use of the hands was thus 'partly the cause and partly the result of man's erect position'.[1] He seems to have assumed that toolmaking would have encouraged the development of a higher level of intelligence. The apes had stayed in the trees and remained merely intelligent animals.

Many other factors were involved, of course, including the development of language and the social instincts. But the suggestion that standing upright was the key to a breakthrough that made all the other advances possible shows that Darwin was beginning to recognize the significance of questions most of his contemporaries had not thought to ask. The details of his argument may not have stood the test of time, but he was right to predict that the event may have

---

[1] Darwin, *The Descent of Man*, vol. 1, p. 144, 2nd ed., p. 53. On the African origin of humans see vol. 1, p. 199, 2nd ed., pp. 55–6. For a wider discussion of his views see my *Theories of Human Evolution*, pp. 157–9.

taken place in Africa, and that the earliest hominids had walked upright before the expansion of the brain began. Darwin was proposing what we now call an 'adaptive scenario' that explains a particular evolutionary advance not as a special case of a general trend but as a unique event in which a species exploits the environment in a new way. In this one instance he pioneered a way of thinking that would emerge as his followers sought to understand why the fossil record seemed to display a series of key innovations in the history of life.

Rethinking the Goal

The possibility that the evolution of intelligence might have come about differently was raised in 1895 by the sociologist Lester Frank Ward, who argued that if any other animal species had developed mental powers equivalent to our own it 'would have had the same rank and secured for that race the same mastery over animate and inanimate nature'. Here was a suggestion that intelligence did not have to come in a human form, although Ward still saw an erect posture as a consequence of our increasing intelligence, not as its cause.[2]

By the turn of the century there were signs that the wider implications of Darwin's approach were starting to be appreciated more generally. The palae-ontologist William Diller Matthew addressed a question that Darwin had asked Lyell over seventy years earlier: how would evolution proceed if humans were removed from the world? Matthew believed that in the end something equivalent to humanity would emerge but it would be 'some super-intelligent dog or bear or glorified weasel'. We have seen that J. B. S. Haldane later answered the same question by favouring the rat as a candidate. Matthew extended the theme by asking what would happen if all the mammals disappeared, concluding that the reptiles would eventually produce intelligent descendants, but they would be quite unlike ourselves both physically and mentally.[3] There is progressionism here in that it is taken for granted that the higher functions will re-emerge, but it is progress in a very different form from an ascent of the chain of being. Matthew saw progress as open-ended and unpredictable, at least in its details. In the world of divergent evolution driven by adaptive opportunities, we cannot be sure which of the many branches will actually move to higher levels in the scale of organization.

Support for Darwin's hypothesis on human origins came in 1917 from the palaeontologist Richard Swann Lull, who linked it to the belief that innovations were triggered by environmental challenges. He suggested that our ancestors were driven out onto the open plains by a diminution of the forests.

---

[2] Ward, 'The Relation of Sociology to Anthropology', pp. 243–4.
[3] Matthew, *Outline and General Principles of the History of Life*, p. 234. For the Darwin–Lyell interaction see below, Chapter 7, n. 28, and for Haldane's remark n. 54.

They were thereby forced to walk upright and only then found that they could use their hands for toolmaking. Other palaeontolgists took up the theme, and since it was still widely assumed that humanity had evolved in central Asia it was argued that the uplift of the Himalayas had isolated our ape ancestors to the north, where they had been forced to adapt to gradual deforestation.[4] The implication was that without this division of the ape populations by a newly emerged geographical barrier, the evolution of modern humanity might not have occurred.

The growing sense among evolutionists that there was nothing inevitable about the emergence of humanity in its current form began to percolate into popular culture. Novelists moving into the area of what we now call science fiction began to imagine that evolution on other worlds might produce intelligent aliens – but they would not necessarily be humanoid. Victorian novelists had certainly written about life on other worlds, but tended to assume that the dominant race would be humanoid – and probably advanced beyond our own capacities. But if the new vision of evolution was to be taken seriously, it would seem highly unlikely that life evolving on another planet would take the same course as on earth. Even if we assume that the emergence of intelligence is inevitable in the long run, it would be most unlikely that it would have a human form.

In an appendix to the 1904 edition of his *Man's Place in the Universe*, the venerable Darwinist Alfred Russel Wallace used the contingency of evolution to argue that the results could never be duplicated elsewhere. Wallace was actually convinced that there was no life on other worlds, but this was an additional factor allowing him to insist that even if there was, it would not include humans. The chances of all the events in the sequence that led to humanity being duplicated, even on a world with similar conditions, were infinitesimally small.[5] This open-ended view of evolution was subsequently forgotten when Wallace wrote his *The World of Life* with its revival of the belief that humanity is the goal of creation.

It was H. G. Wells who drove home the point that extraterrestrials might be truly alien. His *The War of the Worlds* – which first appeared as a serial in 1897 – described an invasion by ruthless Martians with superior technology, depicting them as monstrous creatures with tentacles. *The First Men in the Moon* has the explorers encountering a race of ant-like creatures, again with a high level of technology. A 1908 article 'The Things That Live on Mars' imagines vaguely bird-like creatures with tentacles instead of hands. An

---

[4] Lull, *Organic Evolution*, pp. 672–3; see Bowler, *Theories of Human Evolution*, pp. 176–81.
[5] Wallace, 'Appendix: An Additional Argument Dependent on the Theory of Evolution', in his *Man's Place in the Universe* (1904 ed.), pp. 326–36.

illustration depicts them living in a futuristic city of skyscrapers.[6] No one reading these pieces could remain comfortable with the old image of progress aimed inevitably at humankind. Evolution might be progressive and lead inevitably to intelligence and high levels of technology, but the way that goal would be reached would be determined by a complex web of interactions. In 1921 Matthew made this point explicit in a popular article, concluding that if there was life on other worlds 'it probably – almost surely – would be so remote in its fundamental character and its external manifestations from our own, that we could not interpret or comprehend the external indications of its existence, nor even probably observe or recognize them'.[7]

The Modern Synthesis and Human Uniqueness

Because few authorities took Darwin's hypothetical sequence of events ser- iously, when the first fossil *Australopithecus* was found in South Africa by Raymond Dart in 1924 it was widely dismissed as of little real significance. Although having human-like dentition it seemed to have stood upright despite having a brain no larger than an ape's. This was a combination incompatible with the prevailing assumption that expansion of the brain was the key step initiating the emergence of the human species. Only in the 1930s, when more Australopithecine fossils were found by Robert Broom, did it become apparent that here was evidence that the founders of the human family had indeed become bipedal before they began to gain new levels of intelligence. There was no automatic trend towards brain growth that had simply operated more efficiently in our ancestors. The origins of humanity lay in an adaptive transformation based on a move into a different environment: the apes remained apes because they did not leave the trees. The implications of these developments could be fully appreciated only when the overall philosophy of evolutionism became truly Darwinian. Evolutionary progress is not produced by built-in trends; it is the outcome of contingent events, and it might very easily have come out differently, just as Gould insisted with his metaphor of re-running the tape of history.

As the modern Darwinian synthesis was consolidated, a consensus on the main steps in the origin of humanity emerged. The Australopithecines repre- sent the first step in the separation from the apes, while *Homo erectus* (combining the Java and Pekin hominids once treated as distinct genera) was

---

[6] See Dick, *The Biological Universe*, chaps. 5 and 7. The illustration from Wells's 'The Things That Live on Mars' is Dick's frontispiece. On the earlier speculations see Crowe, *The Extraterrestrial Life Debate, 1750–1900* and Hillegas, 'Victorian "Extraterrestrials"'. 'The Time Machine' is reprinted in *The Short Stories of H. G. Wells*, pp. 9–103.
[7] Matthew, 'Life on Other Worlds', p. 241.

the first truly human form, with a larger brain and the capacity to use tools and fire. It persisted largely unchanged for a considerable period before the first modern humans appeared, again in Africa, and spread out over the rest of the world. Ernst Mayr at first argued that only a single hominid species could exist at any one time (to avoid competition), but it soon became apparent from further fossil discoveries that the human family tree contained several branches, although only one has survived into the present. In effect, our evolution exhibits exactly the same pattern of divergence and extinction as can be seen in almost every major group in the fossil record. The much-debated Neanderthals are now seen as close relatives of modern humans, possibly giving some idea of what the very first specimens of *Homo sapiens* would have looked like.

One consequence of these moves was the discrediting of older ideas about the evolution of the modern human races that had persisted among physical anthropologists into the early twentieth century. The claim that several different branches of the primate family had independently evolved into the modern races was incompatible with the new model in which even the Neanderthals could be seen as merely an extinct race of *Homo sapiens*. Ranking the races (living and extinct) in a hierarchy defined by the ascent from the ape to the modern European seemed equally outdated. The biologists who created modern Darwinism were able to use genetics to undermine the claim that the human races were distinct species with separate origins. Race was now reduced to a less clear-cut recognition of limited genetic diversity within a widely dispersed global population. These developments in biology were driven by the increased sensitivity to the issue created by the Nazis' horrific applications of the older hierarchical definitions of race.[8]

Debates continued to rage over the factors that led to the expansion of human intelligence. In the 1960s Sherwood Washburn revived an old suggestion that hunting with weapons marked a key change defining modern human behaviour. The importance of hunting had already been stressed by Raymond Dart, who had depicted the Australopithecines as brutal killers to create an impression of humanity evolving with inherently violent instincts. The 'anthropology of aggression' was later popularized by writers such as Robert Ardrey in his *Territorial Imperative*. The claim that we have built-in aggressive instincts might seem like a revival of 'social Darwinism' but in fact owed little to the new evolutionary biology, which eventually developed its own slightly more subtle messages about behaviour through E. O. Wilson's sociobiology and Richard Dawkins's notion of the 'selfish gene'. The image of 'man the hunter' had in any case soon been challenged by the advocates of the

---

[8] See Barkan, *The Retreat of Scientific Racism*, although this study tends to ignore the informal continuation of prejudice within science.

view that gathering, an occupation primarily associated with women, was more important. Many Darwinians now preferred to assume that the key features of modern humanity are defined not by genetically implanted instincts but by the behavioural flexibility generated by our capacity for learning.[9]

That the emergence of modern humans constituted the last in a sequence of progressive steps within vertebrate evolution was still taken for granted by most of the new generation of Darwinians. Their model of progress was, however, defined by the insight that evolution is an opportunistic process, with significant innovations being rare and unpredictable. In such a model the human species might be the highest form reached so far, but to present the sequence of breakthroughs that defined the steps towards humanity as the 'main line' of development could be justified only by hindsight. Other outcomes might have been possible, as Gould stressed, in which case we would not be here on the earth today. But if evolution is progressive, at least in the long run, this might imply that it would be capable of generating higher levels of organization and mentality by a different route. Some other outcome would then be its crowning development.

These implications were developed in other areas, including speculations about extraterrestrial life and its possible evolutionary trajectories. Matthew's suggestion that if there were alien life forms they would almost certainly be very different from ourselves was taken up by the leading palaeontologist of the Modern Synthesis, George Gaylord Simpson. In his *This View of Life* he explored the theme in a chapter entitled 'The Nonprevalence of Humanoids', in which he insisted that to imagine the human form to be the inevitable outcome of progress was to reinstate the determinism that Darwin had shown to be implausible. There was no main line of evolution, and the path to humanity passed through many points where a slight change of circumstances would have produced a different outcome. 'If that causal chain had been different, *Homo sapiens* would not exist.'[10] Although dubious about the popular belief in the existence of extraterrestrial life, he thought it worth asking whether evolution elsewhere in the cosmos would duplicate humanity. It might, he thought, produce 'humanoids' in the general sense of something with intelligence equivalent to ours, but that these beings would be biologically similar to us was highly unlikely.

Ernst Mayr adopted an even more sceptical view of the search for extraterrestrial life on the grounds that Darwinian evolution made it unlikely that even the human level of intelligence could be reached elsewhere. Yet he was not

---

[9] On the concept of 'man the hunter' see Haraway, *Primate Visions*, chap. 8. More generally on these issues see for instance Ruse, *Monad to Man* and my own *Evolution: The History of an Idea*, chap. 10.

[10] Simpson, *This View of Life*, chap. 13, quotation from p. 267.

against the idea of progress itself. In his history of biological thought he noted that Darwin had been criticized for denying progress because he had insisted that there was no inherent progressive principle or law built into evolution. Progress was possible, but it was not law-like – and certainly not linear – because it occurred through occasional innovations. He argued that a 'new structural, physiological, or behavioral invention made in any gene pool can lead to evolutionary success and thus to progress as traditionally defined'.[11] But in addressing the question of extraterrestrial life he challenged the view that progress towards higher levels of intelligence would inevitably reach the human level. The last steps in the process had required an extremely unlikely combination of circumstances. This would mean that even if life was evolving on some other planet, the chances of it developing to a level where we could communicate were vanishingly small.[12]

Mayr complained that it was the physical scientists supporting the SETI (Search for Extraterrestial Intelligence) project who tended to adopt a determinist position in which the progress of life must inevitably develop into something resembling human technological civilization. To be fair, their speculations may only have included the more general notion of 'humanoids' as specified by Simpson, creatures equivalent to ourselves without being physically identical. Some life scientists were prepared to go along with this view. In 1958 Loren Eiseley's *The Immense Journey* provided a popular survey of the new evolutionism which concluded with the view that evolution might produce intelligence of some form elsewhere, but nothing like our own:

But nowhere in all space or on a thousand worlds will there be men to share our loneliness. There may be wisdom; there may be power; somewhere across space great instruments, handled by strange manipulative organs, may stare vainly at our floating cloud wrack, their owners yearning as we yearn. Nevertheless, in the nature of life and in the principles of evolution we have had our answer. Of men elsewhere, and beyond, there will be none forever.[13]

Carl Sagan was a leading supporter of the SETI project whose commitment was not based on any assumption that intelligent aliens would resemble ourselves. In 1966 he suggested that although aliens anatomically similar to humans were unlikely, the emergence of our intellectual equivalents might be 'a pervasive evolutionary event'. His *Dragons of Eden* made the point more explicitly, noting that although the precise sequence leading to humans on earth would not be duplicated elsewhere, there would be equivalent pathways to a similar result. Evolution must encourage 'a progressive tendency toward

---

[11] Mayr, *The Growth of Biological Thought*, pp. 531–4.
[12] Mayr, *Toward a New Philosophy of Biology*, pp. 67–74; see also his 'The Search for Extraterrestrial Intelligence'.
[13] Eiseley, *The Immense Journey*, p. 162.

intelligence' because 'smart organisms ... leave more offspring than stupid ones'. Sagan's hugely successful *Cosmos* of 1980 was similarly optimistic, but did include the caveat that the final transition to a level of intelligence capable of maintaining a technological society might be less likely to occur.[14]

The 1980s saw a number of conferences organized to debate the question of extraterrestrial intelligence in which the issue of progressive evolution was discussed. A meeting of the International Astronomical Union in 1987 included a symposium asking 'Is Intelligence an Inevitable Evolutionary Trait?' Rival views were expressed, including concerns that the final stages leading to anything equivalent to human consciousness were dependent on unlikely combinations of events. Mayr used this argument against the SETI programme in another symposium organized to debate the issue. Evolutionary biologists seemed happy with the idea that intelligence up to a certain level might be inevitable, but tended to side with Mayr on the unlikelihood of development to the human level.[15]

A related topic that was raised several times in these debates was the suggestion by Dale A. Russell of a counterfactual history of terrestrial evolution based on a scenario in which the asteroid impact that wiped out the dinosaurs did not happen. He imagined a dinosaur that had evolved in a direction that might have led to a high level of intelligence, illustrated by a widely depicted model of a creature that was vaguely humanoid in appearance. Russell himself, however, suspected that the creature could not have progressed through to the human level of intelligence.[16] The time was picked up in the other area where such issues were explored: science fiction. In 1984 Harry Harrison's novel *West of Eden* was set in a world dominated by intelligent reptiles (evolved from mosasaurs, not dinosaurs). They use biological technology to control the world and have very different social lives and morals from humans. Unfortunately, the implication that a different form of intelligence is possible is partially obscured by the scenario in which a close equivalent to modern humans has developed from the New World monkeys in an isolated North America.

The more extensive exploration of this theme in science fiction and related literature is in speculations about aliens from elsewhere in the universe. In a sense, the debate over whether or not evolution is automatically directed towards the human form is encapsulated in the division between two schools

---

[14] Sagan, *The Dragons of Eden*, p. 230; see also Shklovshii and Sagan, *Intelligent Life in the Universe*, p. 359, and Sagan, *Cosmos*, pp. 296–8, 300–1 and 314.

[15] See Billingham, ed., *Life in the Universe*; Marx, ed., *Bioastronomy – The Next Step* (proceedings of the 1987 conference); Regis, ed., *Extraterrestrials: Science and Alien Intelligence*; and Zuckerman and Hart, eds., *Extraterrestrials: Where Are They?* (containing Mayr's paper).

[16] Russell, 'Speculations on the Evolution of Intelligence in Multicellular Organisms'; an image of the humanoid dinosaur can be found in Ruse, *Monad to Man*, p. 528.

of thought (or imagination): will the extraterrestrials be Little Green Men or Bug-Eyed Monsters? The LGM school reflects the belief that something resembling the human form will evolve elsewhere (although the humanoid aliens so prevalent in accounts of UFO visitations are grey rather than green). Many movies and television series have worked with vaguely humanoid aliens, but in the early days this was probably more a consequence of the fact that they had to be played by human actors. The alternative BEM approach started out from H. G. Wells's totally alien Martians and was so prevalent in the early science-fiction magazines that it became a source of derision among critics. In the late twentieth century stories about totally alien aliens (ones so different that it was hard for us to interact with them) became much more sophisticated. A good example is David Brin's 'uplift' series of novels, including his *Startide Rising*. Here the galaxy is filled with civilization, some friendly, some aggressive and some completely weird. Such realms of the imagination became possible only when the old belief in evolution aimed at a particular goal had been replaced by a more open-ended view of what had happened even here on earth.

Brin's novels include an extra twist on the story of evolution. He imagines that the majority of intelligent races in the galaxy have been artificially 'uplifted' to full consciousness by visitors from an already existing civilization. Humanity is unusual in having evolved to this status naturally, and we are taken seriously by the extraterrestrials only because we ourselves have uplifted the chimpanzees and dolphins to something approaching our own level of intelligence. This scenario endorses Mayr's position, in which the last stages in the progress towards humanity rest on an unlikely combination of circumstances, but also points us towards another area of concern: the possibility of using our understanding of biology to control and direct the future of ourselves and perhaps other species. If the major steps in natural evolution have been the result of events analogous to human inventions, why should we not use our new technologies to invent new directions for an artificial progress in the future?

9 Branching Out
The Evolution of Civilizations

The Darwinian reconfiguration of evolutionary progress as open-ended and unpredictable has close parallels with developments in thinking about human history. The two areas sometimes interacted directly, as when the diversification of languages traced by philologists inspired Darwin's tree of life or biologists saw the upward steps in evolution as inventions. Darwinians tended to define progress in terms of life's increasing ability to control material nature. Their focus on the material wellbeing of individual organisms directly paralleledthat of the emerging economic philosophy of utilitarianism. Living things have 'invented' many different ways of dealing with nature, and by the end of the nineteenth century it was becoming apparent that the same point applied to human civilizations. Focusing on the role played by innovation opened up a new version of progress in which the past had to be seen as a branching tree of cultural diversity. Future developments too were rendered unpredictable by the force of innovation.

By the early twentieth century a new vision of human progress had emerged, drawing on two originally incompatible trends in the interpretation of the past. One was the growing willingness of some Westerners to accept that other cultures were fundamentally different from their own, but might have equivalent moral value. The diversity of cultures that had appeared in different parts of the world was equivalent to the various advanced forms of life on different branches of the evolutionary tree. This made it impossible to treat the West as an illustration of the goal towards which all other societies were progressing, if only at a slower rate.

Attention now focused on the question of why Europe and later North America had acquired the power that allowed them to dominate the rest of the world. As a result there was a growing conviction that technological innovation was the driving force of social progress. But it was difficult to see this force as being constrained by a rigidly defined trend towards a goal represented by the West. Some historians began to argue that there had been a unique breakthrough in the West's development, analogous to the sudden appearance of dominant new classes in the history of life. The present situation was not the inevitable outcome of history: it was the result of a sequence of opportunistic initiatives, some of far greater consequence than others. The creation of an industrial economy that gave the West its current superiority was the most significant of these breakthroughs.

The Diversity of Invention

New attitudes towards ancient and non-Western cultures now allowed them to be seen as having intrinsic values of their own rather than as stages on the way to their becoming modern in the Western sense. This initiative could easily turn into a reaction against the whole idea of progress, especially as the link between cultural and biological evolution had now been shattered. Although some anthropologists rejected any form of evolutionism, the extended historical dimension created by archaeology made it hard to escape the conclusion that all advanced cultures had emerged from stone-age origins. Cultural progress had to be acknowledged, but in a divergent, not a linear form. Relativism didn't have to mean a rejection of progress, only a rejection of the model of progress based on the assumption of Western superiority. Even those who though that the West (or its Soviet derivative) represented a more sophisticated level of development could acknowledge the value of different cultures – and wonder what might happen as they came to grips with the rise of modern industrialism.

It was now taken for granted that the West had gained its superiority thanks to its achievements in science and technology. Earlier materialists had seen these as the product of a more or less predetermined sequence of innovations steadily improving our control of nature: Europeans had simply moved on to the next step earlier than anyone else. But if cultures could invent different ways of flourishing in the world, the sequence could no longer be seen as inevitable, and the eventual triumph of the West had to be the result of some unique factor in its history. Human inventiveness was recognized as a source of divergence throughout cultural evolution, but in recent centuries it seemed to have moved to a more intense level of activity – but only in one branch of the evolutionary tree.

The latest industrial developments were the product of a new force promoting innovation and creativity which Alfred North Whitehead dubbed the West's 'invention of invention'. Here was another parallel with Darwinism: the population was exposed to a competing flood of variations in how it sought to control the world – although unlike genetic mutations all of these were introduced with the assumption that they conferred some benefit. The proliferation of new inventions, each with its own implications for society, generated a sense of uncertainty which made it hard to predict the future course of progress. There was the constant impression that things could go in many possible directions. The diversity that had driven cultural evolution was being mirrored, but within a single culture experiencing an orgy of innovations that offered many different futures.

Future progress would not be a predictable, step-by-step ascent; it would be driven by a Darwinian competition between rival inventions that would be decided by certain techniques gaining dominance over their competitors. The structure of society itself would be shaped by the winners. By the start of the twentieth century a sense that society had entered a phase of uncertain progress driven by the flood of new technologies was becoming commonplace; these were what the historian Philip Blom called 'the vertigo years'.[1] Later decades have seen the emergence of life-changing technologies thrust upon us at an ever-increasing rate, often preceded by periods of uncertainty when no one was sure which would succeed. The technophiles still see this as progress towards a richer society offering everyone greater levels of personal development. The goal is no longer a static future utopia, but an ever-expanding world of opportunities. Almost inevitably, though, their optimism is challenged by pessimists who see the downside of what is being produced.

Curiously, one area that did not participate in this transition was academic history. It was the archaeologists and physical anthropologists whose broader historical range preserved some element of progressionist thinking even as the divergent nature of cultural development was recognized. Where we do see the popularization of a new view of historical progress is in the efforts of technophiles such as H. G. Wells – who was well aware of the new Darwinism – to promote an integration of their hopes for the future with a revised interpretation of the past. It is no accident that Wells was also one of the founders of science fiction, a literary genre that focuses on the future and has tended to assume continued technological expansion but with a variety of possible outcomes.

Movements on the fringes of academic history have also contributed, most obviously the growing interest in counterfactual thinking and the emergence of a discipline focused on the history of science and technology. Counterfactual history doesn't necessarily ally itself with the idea of progress, but it certainly reflects a sense that the sequence of events can be deflected into different channels, sometimes by quite trivial events. The history of science inevitably charts technical progress and examines its consequences. Although strongly focused on the modern period, it does look back to ask questions about why the system of invention and discovery we now take for granted did not emerge, or became sidetracked, in other cultures. The fact that this system emerged only in a very particular social environment highlights the element of contingency in the new vision of progress.

[1] Blom, *The Vertigo Years.*

This chapter outlines how a new and less deterministic vision of human progress in the past emerged, starting with some tentative challenges to the linear model of progress in the eighteenth century and rising to a culmination in the early decades of the twentieth. The next chapter will move on to the parallel transformation in thinking about the present, ongoing state of social change and about the consequences of further instabilities in the future. My argument is that we need to recognize a link between the two phases. It can hardly have been a coincidence that anthropologists and historians began to represent the progressive steps in the development of human civilizations as a branching tree with a multitude of outcomes at the same time as thinkers from a wide range of disciplines were waking up to the emerging popular fascination with rapid social change. The fact that the latter process was routinely seen as a consequence of technological innovation, not of some inbuilt social momentum, encouraged historians to think of the past in a non-deterministic way. Key events in the past could also be seen as the consequence of technological innovations that would have been as unpredictable to any contemporary observer as the open-endedness of current technological innovation seems today. The idea of progress became more utilitarian just as it became less deterministic – which is why it also became easier for those suspicious of the applications of technology to challenge the assumption that the process represented a progression towards some ideal future.

Histories of Others

The Enlightenment view of the development of civilization created by Voltaire and Condorcet did not go unchallenged at the time. Giambattista Vico's *Scienza nuova* (*New Science*) of 1725 offered a different perspective on history, arguing that civil societies were generated in a law-like manner by human needs and activities. All societies followed a cyclic pattern of development passing through three phases in which culture was based successively on religious, mythological and rational grounds. There was no overall pattern linking the various civilizations, even though all had ultimately emerged from savage origins. As Bury notes in tracing Vico's later influence, he pioneered the view that each culture should be understood on its own terms, not as a step towards some ultimate goal represented by modern Europe.[2]

Vico's thought was largely ignored until the following century, although it may have had some influence on the thinking of the baron de Montesquieu, who wrote on the rise and fall of Rome but became best known for his *L'esprit*

---

[2] Bury, *The Idea of Progress*, pp. 268–71. See also Fisch's introduction to *The New Science of Giambattista Vico* (a translation of the 3rd ed. of 1744). Book 4 presents Vico's system of the phases through which each nation's history passes.

des lois of 1748. He too wanted to understand the forces that govern the functioning of societies, but his approach did not depend on the establishment of historical sequences. Montesquieu was interested in the conditions that govern how people come together to form an organized society, with an emphasis on environmental factors. He postulated a number of basic types of society – republics, aristocracies, monarchies and despotisms. Each was a product of a certain kind of physical and geographical environment that was due to the effects of climate and the soil on the fibres of the human body which determine temperament and behaviour, thereby predisposing the population towards certain kinds of relationship.[3] Montesquieu offered no theory of progress, but his environmentalism would be taken up by those who looked for an explanation of why the peoples of some areas had advanced up the scale of civilization further than others. The same principles could equally well be used to support the view that societies emerging in different environments would be fundamentally different.

One of the severest critics of the Enlightenment view of progress was Johann Gottfried von Herder, who acknowledged the work of both Vico and Montesquieu. To some of his later admirers Herder was a pioneer of cultural relativism and a champion of nationalism verging almost on racism, although his emphasis on the *Volk* stressed cultural and linguistic factors rather than biological ones. His vision of society included a spiritual dimension that would be recognized by Hegel while admitting a role for technical and industrial innovation. He denounced Europeans' assumption of their own cultural super- iority and their mistreatment of those peoples they considered inferior. Each culture – past and present – should be evaluated on its own terms. If there was a divine purpose in history it was being worked out through our own efforts and would be fulfilled through diversity, not by the achievement of a single goal. Yet there are inconsistencies in Herder's writings which allow parts to be read as a reformulation of the conventional view of European superiority. Perhaps he is best seen as a pioneer of the model of progress that allows development towards higher levels to operate along many different pathways.

Herder's most effective attack on Voltaire's approach was his youthful *Auch eine Philosophie der Geschichte zur Bildung der Menschheit*, variously trans- lated as *This Too a Philosophy of History for the Formation of Humanity* and *Yet Another Philosophy of History*. An early introduction to this complains about the focus on the 'wretched and mistreated question' of whether the world has improved or worsened. The main text begins with a traditional account of the rise of civilization through periods of boyhood, youth and manhood. But this is soon subjected to a sustained attack which argues that nations cannot be

---

[3] Montesquieu, *The Spirit of the Laws*, esp. books 14–18.

judged good or bad by European standards and that even savage societies can provide a culturally satisfying life. Accounts of 'the universal progressing improvement of mankind' are dismissed as novels that no one believes. The rise of the West has generated more evil than good, most evidently in our treatment of other peoples. Europe is but a tiny part of the world, and its history could very easily have turned out differently (an early hint at the possibility of counterfactual history). There is a divine purpose in the world, but it will be worked out through diversity, not through the triumph of the West.[4]

Herder's best-known work is his *Ideen zur Philosophie der Gechichte der Menschheit* of 1784–91, originally translated as *Outlines of the Philosophy of the History of Man*. This includes many of the arguments critical of European attitudes made in his earlier study but has a wider scope which allows much room for inconsistencies. As Frank E. Manuel notes in his introduction to a partial English translation, some sections read almost like an exposition of the interpretation of progress originally attacked.[5] There is an outline of the development of life on earth which, as Arthur O. Lovejoy pointed out, envisions a progressive series of creations culminating in humanity.[6] Book 7 discusses the role of climate in shaping culture, and book 8 makes the point that all cultures are capable of providing humans with a happy life. Book 14 presents a more progressionist viewpoint, suggesting that there is a divine purpose in history and that humanity has advanced through degrees of civilization. Herder also notes that improvements in technology and industry have played a role in this development.[7] Later books outline something very much like the conventional view of the rise of civilization from the ancient world onwards, although Herder rejects the idea that the sequence was predetermined. If Rome was supposed to provide a bridge by which the treasures of the ancient world were transmitted to the present, it was the worst bridge that could have been contrived.[8] The conclusion makes the Baconian point that inventions such as the compass, gunpowder and printing played a major role in promoting the rise

---

[4] I have used the translation as *This Too a Philosophy of History for the Formation of Humanity* in *Herder: Philosophical Writings*, ed. Foster; see pp. 268–71 for the early introduction and pp. 272–358 for the 1774 text; the material cited is on pp. 298, 314–20, 339 and 358. This edition also prints several letters making Herder's opposition to European expansionism clear. See also *J. G. Herder on Social and Political Culture*, trans. Barnard, which translates the piece as *Yet Another Philosophy of History*. On Herder's thought see Barnard, *Herder on Nationality, Humanity, and History* and Clark, *Herder: His Life and Thought*.

[5] Manuel in his introduction to Herder, *Reflections on the Philosophy of the History of Mankind*, pp. xx–xxi. This is a modified and abridged version of the original English translation of 1800.

[6] Lovejoy, 'Herder: Progressionism without Transformism'.

[7] Herder, *Reflections on the Philosophy of the History of Mankind*, pp. 93–6.

[8] Ibid., book 14, p. 267.

of Europe. Climate and geography created the environment that made this possible, but the result was to raise Europe to a rank above all other cultures.

Herder thus presents a complex vision of progress combining several of the themes we are tracing. His willingness to see a divine purpose in history links him to the idealist viewpoint, yet he acknowledges that Europe's pre-eminence arises in part from its technological innovations. His emphasis on cultural diversity ends up being integrated into a more sophisticated version of the conventional view. If there is an overall divine purpose to the scheme, it will be achieved through diversity, not by reaching a single goal. His scheme points the way towards a more flexible form of progressionism in which many different cultures can be seen to have achieved something of value. Progress is genuinely divergent, even if one branch can be seen to have advanced further than others.

Prehistoric Branching

The nineteenth century saw the emergence of new areas of research throwing light on the earliest phases of cultural progress. We have seen how physical anthropologists debated the origin of races, while archaeologists unearthed relics of the stone age and philologists charted the evolution of languages. The latter area played a major role in creating the branching model of evolution taken up by Darwin and his followers. In principle any theory of divergent biological evolution implies that the human races must have separated from a common ancestor, which should have made room for a theory of divergent *cultural* evolution. Indeed, it was at first assumed that each culture was unique to a particular race. Although evolutionism encouraged a move towards a branching model of progress, in practice its impact was much delayed.

The models of the past adopted by the early archaeologists and anthropologists tended to deflect attention away from the study of initial origins and to focus instead on the parallel development of separate races and cultures through a hierarchy of stages leading up to the present. If divergence from a common ancestor had occurred, it was so far back in the past that it was beyond the scope of investigation. Races and cultures were independent entities following similar paths of development but at different rates. Apparent diversity was dismissed as evidence that some had reached different levels of the one unified developmental sequence. The replacement of one culture by another was usually attributed to invasion rather than local evolution.

It was the philologists who recognized the branching nature of language evolution, although they did not see the process as progressive. Philology emerged from the study of ancient Sanskrit and its relationship to the whole family of Indo-European languages. Soon there were attempts to reconstruct the historical development of all the major language groups, often identifying them with particular racial types. As Max Müller insisted in his *Lectures on the Science of Language*, it was essentially a geneaological science, tracing how diverse modern languages had descended from ancient roots. Long before the races of Europe had moved into the continent 'there was a small clan of Aryans, settled probably on the highest elevation of Central Asia, speaking a language, not yet Sanskrit or Greek or German, but containing the dialectical germs of all'.[9] Single deep roots such as the original Aryan tongue gave rise to an ever-diversifying group of descendant languages, a process that was soon being represented in the works of F. W. Farrar, August Schleicher and others in the form of tree-like diagrams. As the historian Stephen Alter has shown, these diagrams were used as a model for representing historical developments by Darwin and Haeckel, creating the evolutionary trees that became the most popular way of depicting the development of life on earth.[10]

For the Darwinians, the tree diagram provided a convenient way of expressing the belief that evolution was both divergent and progressive: the multiplicity of branches allowed one to see how there could be different ways of becoming more complex, and to admit that not every branch continued to progress. But for the philologists themselves there was no implication that the branches could progress to higher states. Müller and Farrar were conservative thinkers, the latter associated with the liberal Anglican view that history is the unfolding of a divine plan. Müller saw the whole process of language evolution as one of decline from an original purity: 'On the whole, the history of all the Aryan languages is nothing but a gradual process of decay.'[11] The British polymath William Whewell included philology in his discussion of what he called the 'palaetiological sciences' (those that deal with reconstructing the past) and argued that it showed a slowing-down of the rate of change. The fundamental divisions between the language groups took place in the deep past, while the later ones had led to only trivial modifications.[12] He did not realize that if branching occurs at a more or less constant rate it is inevitable that those lines that separated a long time ago will have moved further apart than those that separated only recently.

Philology thus played a double role in the emergence of a new idea of progress. Originally it presented divergent evolution as a degenerative process, yet its metaphor of the branching tree provided the inspiration for Darwin to create one of the most enduring icons of evolutionism. For the Darwinists divergence was not a sign of degeneration: it was a process that could be

---

[9] Müller, *Lectures on the Science of Language*, p. 199.
[10] Alter's *Darwinism and the Linguistic Image* reprints many of the classic branching diagrams.
[11] Müller, *Lectures on the Science of Language*, p. 222.
[12] Whewell, *Philosophy of the Inductive Sciences*, vol. 1, pp. 677–9.

incorporated into a more sophisticated model of progress by being made open-ended, irregular and unpredictable. Alter notes that by the early twentieth century evolutionists were no longer making comparisons with the process of language development. Philology itself changed: some of its early reconstructions were challenged, and there was a wider appreciation of the possibility that people could learn or borrow from other languages, a process with few parallels in animal evolution.

The New Human Sciences

Towards the end of the nineteenth century the linear progressionism of anthropologists such as Tylor and Lubbock began to unravel. Instead of trying to rank other societies into a single hierarchy, anthropologists began to recognize the differences between them as a genuine diversity in the ways humans could organize themselves. Even Herbert Spencer seems to have anticipated this trend, for all that his early followers thought that he was advocating inevitable progress towards free-enterprise individualism. Cultural relativism transformed the human sciences, but it also impacted on the general public's view of the West's position in the world. Books and exhibitions revealed the diversity of cultures from around the world, and although they were often described using the rhetoric of imperialism, they did create a sense of the 'otherness' of different ways of life. In the early twentieth century artists began to take inspiration from what had hitherto been dismissed as 'primitive' traditions, heralding a decline in the West's self-confidence.

Some anthropologists and sociologists rejected the whole idea of evolution, although their real target was the linear progressionism that saw the West as the only way forward. Now that archaeology had demonstrated the emergence of all cultures from stone-age origins, it was hard to deny that there had been an advance towards more sophisticated ways of life, at least in the long run. There were debates about the origins of the first civilizations, some arguing for a single breakthrough spread by diffusion around the world. But better understanding of the ancient cultures of South and Central America suggested the independent development of agriculture and urban living with outcomes significantly different from those of the Old World. Cultural relativism made it less easy to see the West as the goal of progress, but it also focused on technological innovation as the force that made the development of civilization – and the West's recent surge ahead – possible.

The American businessmen who hailed Spencer as the advocate of 'progress through struggle' assumed that his social evolutionism predicted the emergence of a society of perfectly adjusted self-reliant individuals. His early work was, perhaps, easy to interpret in this way. He began with little interest in archaeology and focused mainly on how to adapt his own society to the emergence of an industrialized economy. His assumption that feudal militarism was a primitive state from which Europe had now emerged gave the impression of an overall progressive sequence. He was well aware of the divergent nature of biological evolution, and his overall philosophy stressed the inevitability of progress from homogeneity to heterogeneity in both individual organisms and what we would now call ecosystems. The development of intelligence and aptitude was shaped by efforts to adapt to the environment, leading to the differentiation of both races and cultures. Although it was possible to define a hierarchy of social complexity, there could be differences between groups occupying the same rank in the scale.

In the first volume of his *Principles of Sociology* of 1876 Spencer proclaimed that social development represented a realm that had to be understood on its own terms; this was 'super-organic evolution'. The interactions between individuals generated functions that could not be explained solely in terms of psychology. He challenged some of the assumptions that underpinned the kind of evolutionary anthropology championed by Tylor. Modern savages were not the equivalent of stone-age primitives, because both degeneration and progress could occur as societies struggled to cope with their environments. Advances to higher levels of complexity were actually quite rare, and progress was something that could be observed only in the long run. In the third volume of 1896 archaeological evidence was used to show just how slow the earliest advances in stone technology had been. Now, towards the end of his career, Spencer explicitly proclaimed the divergent nature of social evolution: 'Like other kinds of progress, social progress is not linear but divergent and redivergent.' In this area too, evolution was a branching tree, not a ladder.[13]

Spencer's use of the term 'super-organic' to describe culture seems ironic given that it was precisely a sense that culture could not be explained in terms of biology and psychology that led some early twentieth-century anthropologists to reject the whole idea of cultural evolution. In 1917 A. L. Kroeber published a combative paper making this point under the very title 'The Superorganic'. Many anthropologists and sociologists lost interest in the origin and development of cultures, concentrating instead on how they allowed a population to function in their environment. But this is not the whole story: many who sneered at nineteenth-century evolutionism were concerned mainly to reject the model of progress in which the West represented the goal towards which all societies were developing. They retained an interest in the historical development of societies and cultures and were, in effect, still dealing with 'evolution' under a different name. If it was called 'historical particularism' this only stressed that the development was branching rather than linear. Since

---

[13] Spencer, *Principles of Sociology*, vol. 3, p. 325, and on stone-age cultures p. 322.

all cultures had less complex origins in the stone age, this was still evolutionism, but in a form more closely aligned to the Darwinian model than the simple hierarchy constructed by the previous generation of anthropologists.

The newly professionalized social sciences certainly defined themselves against the social Darwinism and hereditarianism that saw human behaviour as determined by race and genetic inheritance. Cultural evolution is more Lamarckian than Darwinian because innovations made by individuals can be transmitted to future generations by learning. But Lamarckism is a mechanism of adaptation and could explain how populations exposed to different environments might diverge. Unlike animal species, cultures that had become separated could nevertheless borrow from each other and even merge. But this did not mean that it would be impossible to construct an evolutionary model of cultural development: cultures clearly have diverged from ancient origins, and if they occasionally merge this merely adds a new dimension to the process by which new branches are created. (We now know that even in biology there are mechanisms that can transfer genetic material from one branch of the tree of evolution to another.)

Nineteenth-century cultural evolutionists had assumed that the emergence of a complex civilization was inevitable in any population not held back by unfavourable conditions. A very different perspective was promoted by the advocates of 'diffusionism', who insisted that civilization was an unnatural state of affairs that had arisen only once and been spread from that centre around the world. The palaeoanthropologist Grafton Elliot Smith outlined this position in his *The Evolution of Man* of 1924 and extended it in books such as *Human History*. He argued that warfare had emerged only when large-scale agriculture based on irrigation – along with centralized authorities and urbanization – had led to the establishment of empires bent on conquest. This unnatural innovation could have only occurred once, in the highly unusual environment of the Nile valley when 'some man of genius' recognized the value of irrigation.[14] The Egyptians had disseminated this way of life around the world, first to Mesopotamia, then to India and China, and finally even to the New World. This was hardly a theory of progress, but it did perceive a role for contingency in the origins of one great invention – at the price of abandoning any sense that innovation was an ongoing process.

There was, however, a more sophisticated form of diffusionism that played a significant role in the emergence of cultural relativism. This acknowledged that there were real differences between the cultures that had emerged around the globe, even if they did occasionally spread their influence over a wider area.

---

[14] Smith, *Human History*, p. 272; see also his *The Evolution of Man*, pp. 125–6. On the link between Smith's views on the origin of humanity and the origin of civilization see Bowler, *Theories of Human Evolution*, pp. 215–18.

There was a greater appreciation of the independence and value of non-Western cultures, a refusal to treat them as merely immature versions of what progress might achieve. Despite their suspicions of evolutionism, anthropologists began to focus on diversity rather than hierarchy. The potential to explain this variety as the product of a historical process was there, and given the availability of archaeological evidence for a prehistory in which there had been no complex societies this could be exploited only by admitting a form of progress in which multiple branches of the human family had acquired more complex social structures and cultural practices.[15]

The work of Emile Durkheim and Max Weber was central to the emergence of the new science of sociology, but anthropologists and archaeologists were also influenced by efforts to replace the old linear evolutionism with a geographical dimension that recognized regional differences. Friedrich Ratzel, whose name became associated with the later misuse of the term *Lebensraum*, created a human geography that defined the major regions of the earth in terms of racial and, more importantly, cultural identities. Tylor himself wrote a brief introduction to the English translation of Ratzel's *History of Mankind* (1896), indicating that he was by no means unwilling to admit some cultural diffusion. The book surveyed a series of cultural areas recognizing the peoples of America and the Pacific, the lighter race of Central and South America, the Negro races and the 'Cultured Races' of the Old World as distinct entities. Despite the assumed superiority of the latter, the independence of the other regions was acknowledged.

Ratzel had been influenced by Ernst Haeckel, but his later position was not intended as a contribution to race science. Nor was it explicitly based on a religious agenda, although the idea of distinct cultural circles was soon taken up by a Viennese school of Catholic scholars led by Wilhelm Schmidt, whose diffusionism defended the view that the inspiration for cultural development in the Old World came from an original divine revelation. Nevertheless, the scheme did allow for the independence of other regions and defined levels of development applicable to each. Marvin Harris notes that some American scholars defended Schmidt despite his openly religious agenda because they were no longer interested in treating their field as a *science* of human behaviour – identified with linear evolutionism.[16]

In 1911 the British anthropologist W. H. R. Rivers reported that he was unable to fit his studies of the people of Oceanea into a linear pattern. He turned to diffusionism, although in a form less extreme than that adopted by

---

[15] On these developments see for instance Harris, *The Rise of Anthropological Theory*; Hatch, *Theories of Man and Culture*; Smith, *The Fontana History of the Human Sciences*; and Trigger, *A History of Archaeological Thought*.

[16] Harris, *The Rise of Anthropological Theory*, pp. 389–90.

Smith. British anthropology now turned its back not only on evolutionism but also on any real concern with the historical development of cultures. It was left for archaeologists such as V. Gordon Childe to promote an interest in the rise of culture. Strongly influenced by Marxism, Childe popularized the idea of progress driven by technological advances from the Palaeolithic through to the rise of the great ancient empires. He still saw the sequence leading through to European civilization as the 'main stream' of progress, but he was now willing to accept that there were significant differences between the ways in which cultures developed in the various regions of the Old World. Identifying successive levels of technological sophistication did not preclude the recognition of variation in the ways in which the culture of each level could be expressed.[17]

In America the situation was more fluid. The leading figure in the move to reject linear evolutionism was Franz Boas, who trained the first generation of professional anthropologists. They insisted on recognizing the independence of cultures and on rejecting the efforts of those who sought to treat the white race or Western civilization as the highest form of human progress. Some of Boas's students lost all interest in the effort to trace the historical development of cultures, but Boas himself refused to abandon the general idea of evolution provided that it was restructured to allow for multiple branches of development. His 'historical particularism' traced the unique events shaping the emergence of the different cultures. The next generation of anthropologists were more willing to recognize diversity even though they identified successive levels of sophistication. Leslie White and Julian Steward debated the relative significance of evolutionary parallels and geographical differences. White, like Childe, applied a Marxist perspective which focused on the similarities that inevitably emerged between peoples striving to improve themselves through technological innovation. Steward's concept of 'multilinear evolution', expounded in his book of 1955, placed more emphasis on the differences that emerged as societies advanced to successive levels of development.[18]

A collection of essays edited by Marshall Sahlins and Elman Service in 1960 highlighted the re-emergence of evolutionism in anthropology, but shows how difficult it was for the field to address the issue of divergence and progress. Lamenting what they perceived to be the complete rejection of evolutionism by the previous generation, the editors suggested that the current

---

[17] See Rivers, 'President's Address, Anthropology Section' and also Slobodin, *W. H. R. Rivers* and Langham, *The Building of British Social Anthropology*. Childe's books referred to above are *Man Makes Himself* and *What Happened in History*.

[18] See for instance the essays collected in Boas, *Race, Language, and Culture* and Steward, *Theory of Culture Change*. In addition to the historical surveys noted above see Cravens, *The Triumph of Evolution*.

revival was in part inspired by the centenary of the publication of the *Origin of Species* the previous year. They were now prepared to recognize parallels between biological and cultural evolution, drawing on the Modern Synthesis in biology to make the comparison. Julian Huxley was cited approvingly for his claim that the two areas should unite around the theme, but his efforts to modify the new Darwinism to retain the claim that humans represented the final point of development gave rise to some confusion, since this was the one area where Huxley had gone out on a limb and had reintroduced an element of the linear model of progress.[19]

Sahlins himself distinguished between 'specific evolution', which traced the development of individual cultures and followed the phylogenetic or branching-tree model of the biologists, and 'general evolution', which defined the levels of development through which societies pass. In principle this hierarchy was based on efficiency in exploiting and controlling the natural world and could be ascended by completely different cultures. Unfortunately Sahlins followed Huxley by providing a diagram of evolution with a central trunk leading towards humankind. Perhaps as a consequence of using this model as the basis of comparison, his analysis confuses rather than clarifies the role of diversity, since it accepts that cultures can advance independently through the scale of development, yet still implies that the sequence is more or less predetermined. Linear evolution is dismissed as vulgar Marxism, while Tylor and Morgan are promoted as early proponents of multilinear evolution. It seems that anthropologists were prepared to accept a certain level of diversification in evolution, but the legacy of the linear view in which all branches tend to move upwards in parallel was proving hard to throw off.[20]

The issue of independent invention and parallel development was especially relevant for the interpretation of native American cultures, especially those of South and Central America, which were routinely compared to the ancient empires of the Old World. The extreme diffusionists saw these as derivatives of Egyptian civilization, although the independent character of the American cultures had long been recognized. In 1966 Robert Adams's *The Evolution of Urban Society* presented a detailed comparison of the civilizations developed in ancient Mesopotamia and in Mexico, noting the underlying parallels required by the acquisition of agriculture and urban centres but also stressing the cultural differences. Although there was a 'common processional core' in the developments, they were '*not* the reenactment of a predetermined pattern but a continuing interplay of complex, locally distinctive forces whose specific forms and effects cannot be fully abstracted from their immediate geographical

[19] Editors' introduction to Sahlins and Service, eds., *Evolution and Culture*, pp. 1–11; see also the foreword by Leslie White, pp. v–xii.

[20] Sahlins, 'Evolution, Specific and General', pp. 12–44. The diagram is on p. 17.

and historical contexts'.[21] Here prehistory was at last interpreted in terms of technological progress, which nevertheless led to significantly divergent results as cultures evolved in isolation.

In the 1970s and 1980s there were limited efforts to exploit the latest developments in neo-Darwinism as models for cultural evolution. More recently there has been a wider effort to use the more general Darwinian framework as a way to understand long-term developments in culture. Alex Mesouli argues that cultural evolution is Darwinian, not neo-Darwinian: like Darwin's original formulation of his theory it allows for other mechanisms than natural selection. He argues that it is possible to use phylogenetic analysis to reconstruct the branching tree of cultural developments, contrasting this approach explicitly with the ladder of development used by late nineteenth-century figures such as Tylor and Morgan. Although mostly concerned with gradual change, Mesouli's analysis does allow for sudden and dramatic breakthroughs equivalent to those defining the major episodes seen in the fossil record.[22]

Outlining History

Archaeology had extended the range of known history and confirmed the primitive origins of society. Progress was admittedly not continuous, but the sequence of development seemed obvious. Even those theories of history based on the dialectic supposed that the interruptions were necessary preludes to the emergence of the next step. The twentieth century saw the emergence of a new interpretation of human progress paralleling that created by modern Darwinism. Episodes of real progress were seen as the exception rather than the rule. There was no inevitable trend, let alone a law of development, because progressive steps were occasional by-products of processes that did not require an increase in complexity. Degeneration was as likely as progress, and the only reason there had been a cumulative advance was that the rare innovations sometimes became so dominant that total reversion was impossible. Each of those innovations was the result of a unique and unrepeatable combination of circumstances, making the overall pattern of development unpredictable. History became a series of contingent events with diverse outcomes that only rarely advanced things to a new level.

One generalization that did seem inescapable was that the pace of change had accelerated. Archaeology showed that stone-age cultures had persisted for

---

[21] Adams, *The Evolution of Urban Society*, pp. 173–4.

[22] For the earlier episode see Cavalli-Sforza and Feldman, *Cultural Transmission and Evolution* and Boyd and Richerson, *Culture and the Evolutionary Process*. Mesouli's *Cultural Evolution* gives several examples of the phylogenetic approach; see for instance the tree-like diagram of stone tool developments on p. 93.

vast periods of time before the next steps were taken. When agriculture and urban civilizations eventually emerged, their inception was relatively sudden and seemed to have taken place in various locations where the conditions were exceptionally favourable, The rise of modern Europe was the result of a new approach to scientific discovery and technological invention. Far from being the resumption of a general trend after the temporary lapse of the Middle Ages, the linking of science and industry was a major step forward, ushering in a new era of progress analogous to the emergence of the human mind itself. It had let loose a storm of social and cultural changes whose future could hardly be predicted. The rise of Europe was the product of a unique combination of social and geographical factors affecting just one of the many branches of cultural evolution. Some of those other branches had had their own periods of opportunity and in some cases had fed innovations into the new entrepreneurial culture, but they were now sidelined.

This new interpretation of the rise of the West has now become widely accepted through the work of Joel Mokyr and others, but it has its origins in the early twentieth century.[23] It was adopted in some broad-brush historical writings that bypassed the work of specialist historians to bring a more complex idea of progress to the ever-wider reading public. Of the authors who specialized in presenting synthetic overviews none was more influential than H. G. Wells, whose *Outline of History* was published in fortnightly parts in 1920 and then issued in a variety of book formats which sold like hot cakes and made Wells a small fortune.[24] *The Science of Life*, which he wrote with Julian Huxley to publicize the new biology and the Darwinian synthesis, was a follow-up to this project. The *Outline of History* itself began with the development of life on earth, drawing on Wells's friendship with the Darwinian zoologist E. Ray Lankester. It then moved on to the emergence of humanity, to the creation of the first civilizations and – in ever more detail – to the stream of history culminating in the Great War and its aftermath. Here was the broadest possible canvas on which to depict the progress of life and thought, but the picture Wells painted was of a sequence far removed from the inevitable developmentalism of previous efforts.

Lankester had written an essay on evolutionary degeneration which Wells had drawn on for his depiction of the future of humanity in *The Time Machine*. While *The Outline of History* was in preparation Lankester wrote a commentary on J. B. Bury's book about the idea of progress in which he contrasted Spencer's vision of evolution directed inexorably towards the production of the 'ideal man' with the Darwinian model of a branching tree of life. It was true

---

[23] See Mokyr, *The Gifts of Athena* and *Culture and Growth*.
[24] See Skelton, 'The Paratext of Everything'. General studies of his life and thought include Batchelor, *H. G. Wells*; Corren, *The Invisible Man*; and West, *H. G. Wells: Aspects of a Life*.

that some branches of the tree had undergone degeneration, but most had witnessed significant progress, defined as an 'increase of structure and capacity'. Lankester himself argued that human history showed the same kind of open-ended progress, but contrasted the rapid developments following the appearance of the human mind with the slow and irregular advances by which biological evolution had reached that point. In his 1905 Romanes lecture, Lankester described humanity as 'Nature's Insurgent Son'.[25]

Wells's interpretation of human history paralleled the new Darwinism in several respects. He recognized the diversity of races and cultures, stressing the sequence of events leading to the rise of Europe but acknowledging the independence of other cultures and their contributions. His account represented progress as episodic in character, arising from occasional innovative breakthroughs rather than a continuous (if sometimes interrupted) trend. It included a major role for environmental factors and adaptive opportunities. It also implied that the rise of the West was something more than just another of those rare innovations: it was a leap into a new phase in which change would be more rapid and would allow unlimited opportunities for further development. Wells had his own ideas about how global society could be improved, but he also knew that technological innovation was – like Darwinian evolution – creative and hence unpredictable. The past showed us how contingent the process of development had been, but Wells expanded the debate about progress to include the uncertainties of the future. He had turned to history for background, but his real interest was in trying to imagine the opportunities and perils to come.

Nineteenth-century efforts to imagine a utopian future such as Bellamy's *Looking Backwards* focused on social improvements. Wells too was capable of utopian visions, but he was also aware that technical innovations would make it impossible to achieve a perfect and stable society. He was not the first to link a history of human progress with speculations about future technologies. Winwood Reade's *Martyrdom of Man* of 1872 portrayed the history of humankind as a struggle to control the environment all too often thwarted by superstition, ending with the prediction that three inventions would transform future life: a new source of power, aviation and the artificial production of food. He also imagined travel to the stars and the creation of new worlds.[26] But Reade skipped over recent history and made no effort to present the modern era as a new phase of progress; this was a Spencerian approach with little room for discontinuities.

---

[25] Lankester, 'Progress!', reprinted in his *Great and Small Things*, pp. 74–82 (it was originally written as a magazine article). 'Nature's Insurgent Son' is the title of chap. 1 of his *The Kingdom of Man*, based on the Romanes lecture. See also his *Degeneration: A Chapter in Darwinism*.

[26] Reade, *The Martyrdom of Man*, pp. 513–15.

Wells's more Darwinian account represented a new interpretation. Book 1 of *The Outline of History* covers the development of life through geological time and includes a chapter on natural selection. The advances are described in a deliberately episodic manner, stressing the abrupt appearance of new types after long periods of adaptive stagnation. The original version has a chapter title 'The Invasion of Dry Land by Life' to drive home the impression that here was an innovation which transformed the whole situation.[27] Environmental stress is a stimulus to evolution, so there can be no clear-cut line of progress: a visitor from another world would have been unable to predict the next progressive step on the basis of what had gone before. Such a visitor would have found the Devonian mudfish, for instance, 'a very unimportant side-fact in that ancient world of great sharks and plated fishes . . . a poor refugee from the too-crowded and aggressive life of the sea'. But the mudifsh nevertheless 'opened the narrow way by which the land vertebrates rose to predominance'.[28] The rise of the mammals to dominance is presented as something that could not have happened until the great reptiles had been eliminated, probably by climatic change. Here Wells exploits the palaeontologists' episodic view of the advance of life to drive home the message that history involves contingencies with unpredictable consequences.

The same message emerges in the treatment of the emergence of humankind. Wells emphasizes the climatic stress of the Ice Ages and presents the evidence for the existence of several different forms of early hominids. He uses diagrams of branching trees to depict the evolution of the modern races and languages.[29] After tracing the immensely slow development of Palaeolithic cultures, he follows Elliot Smith in arguing for a single Neolithic culture diffused over the whole world. But his treatment of the origins of agriculture and urban societies assumes independent developments in Mesopotamia, Egypt, India and China, with the native American cultures emerging somewhat later. He also invokes the possibility of a climatic factor, adopting the suggestion that the flooding of the Mediterranean basin by the sea may have driven populations into new environments where agriculture was possible.[30] Within a comparatively short period of time the first great empires of the ancient world had emerged.

Wells's account of the ancient empires follows a conventional path, but he interpolates substantial chapters on the civilizations of India, China and Islam. He also acknowledges that Chinese art, science and industry were for a long

---

[27] I have used both the first book version of *The Outline of History* (1920), which follows the original monthly parts, and the definitive edition of 1923, which keeps the text description of the invasion of the land but drops the chapter title. See chap. 4, pp. 15–18 of the 1920 edition and pp. 11–14 for the equivalent text in the 1923 version. All references below are to the latter.
[28] *The Outline of History*, p. 26.   [29] Ibid., pp. 69 and 72.   [30] Ibid., p. 56.

time superior to those of the West, and asks the question: why did Chinese civilization eventually fall behind? His answer invokes an element of conservatism that affects all cultures and is challenged only rarely:

The superior intellectual initiative, the liberal enterprise, the experimental disposition that is supposed to be characteristic of the Western mind, is manifest in the history of that mind only during certain phases and under exceptional circumstances. For the rest, the Western world displays itself as traditional and conservative as China.[31]

The story of how European culture escaped this inertia forms a major theme in the rest of the *Outline*. Wells insists that the emergence of modern science and its link to technology was not something that could have been predicted on the basis of existing trends. Europe had always been a backwater, and the great empires of the East were the dominant sources of what progress there had been. It was its exposure to a new environment that transformed European society, an environment that focused attention on travel by sea rather than on land. The subtitle of a chapter on the Renaissance is 'Land Ways Give Way to Sea Ways'. Command of the sea, first by sail and later by steam, led Europe to world dominance, but it also created a culture that recognized the value of a rational study of nature and the application of the resulting knowledge to technical and industrial innovation. The unpredictable nature of this revolution is explicitly compared to one of the major steps in evolution:

Now, just as in the Mesozoic Age, while the great reptiles lorded it over the earth, there were developing in odd out-of-the-way corners those hairy mammals and feathered birds who were finally to supersede that tremendous fauna by another far more versatile and capable, so in the limited territories of Western Europe in the Middle Ages, while the Mongolian monarchies dominated the world from the Danube to the Pacific and from the Arctic Seas to Madras and Morocco and the Nile, the fundamental lines of a new and harder and more efficient type of human community were being laid down.[32]

Wells notes how the possibilities of applied science were recognized by Roger Bacon, probably 'of more significance to mankind than any monarch of his time', and then highlights the contribution of Francis Bacon and the founders of the Royal Society. The significance of the new initiatives in science and technology is emphasized by quoting Gibbon's argument that – unlike the Roman Empire – modern Europe cannot be overrun by barbarians 'since, before they can conquer, they must cease to be barbarous'. Roger Bacon's predictions for new technologies were then fulfilled in the nineteenth century, including the creation of new power sources and means of communication. Wells argues that we need to distinguish two parallel revolutions, the conventionally recognized Industrial Revolution, which would probably have occurred anyway as factories

[31] Ibid., p. 300.   [32] Ibid., p. 492.

were developed, and the 'mechanical revolution', in which the sciences were applied to create new power sources including steam and electricity and which drove the Industrial Revolution much further and faster. When combined they ushered in 'a fresh phase of history' which created 'a gigantic material framework' with unlimited potential. More had been done in a century than all that passed between the Old Stone Age and the first civilizations.[33]

This material is embedded in extensive coverage of the social and political development of Europe since the rise of Christianity. Wells was no simple-minded progressionist, and he tells this story at length to emphasize two points. On the one hand he wants his readers to see that the rise of Europe was not an inevitable step in the way civilization is predetermined to advance, but was the consequence of a unique set of circumstances that was unlikely to have been paralleled within any other culture. On the other, he wants to highlight the disastrous consequences of unrestrained industrialization and the rise of imperialism (encouraged by the misuse of Darwinian rhetoric), which led to the calamity of the Great War. His argument is that all the potential advantages of the new science-based technologies have been wasted by the continued activity of precisely the conservative political and social attitudes he had highlighted in his account of China. He identifies the only way forward as the creation of a World Government promoting a single global religion and a more equitable division of the benefits of material prosperity. He notes that Japan and, to a lesser extent, China were now modernizing themselves by adopting the Western model for the application of technology.

Wells wanted the scientific method to be applied to the study and control of society as well as nature. His vision contains elements of utopianism, but he also realized that the opportunities offered by the expansion of scientific knowledge would ensure that the future could not be static:

Gathered together at last under the leadership of man, the student-teacher of the universe, unified, disciplined, armed with the secret powers of the atom, and with knowledge yet beyond dreaming, Life, for ever dying to be born afresh, for ever young and eager, will presently stand upon this earth as upon a footstool, and stretch out its realm amidst the stars.[34]

In this concluding passage he passes beyond the history of progress and reminds his readers that he has already acquired a reputation through his speculations – fictional and non-fictional – about the future.

The Invention of Invention

Wells was by no means the only thinker at this time promoting the claim that the emergence of modern science-based technology represented a major event

---

[33] Ibid., pp. 389–90 and 488–92.   [34] Ibid., p. 590.

in human progress, in effect ushering a new age. Another rationalist, the science writer John Langdon-Davies – who also speculated about the future – wrote his *Man and His Universe* to argue that the emergence of the scientific world-view had finally overcome the barrier presented by ancient superstitions.[35]

From a very different perspective Alfred North Whitehead stressed the significance of science-based innovations and the 'invention of invention' in his *Science and the Modern World* of 1926. He respected the cultural achievements of non-Western civilizations, including the Chinese, but thought that their abilities to generate a scientific way of thinking were limited. He used a counterfactual approach to suggest that had the Persians conquered the ancient Greeks there would have been no suitable foundation for later Europeans to build on. The creation of modern science was a truly ground-breaking revolution, made possible by a unique combination of philosophical developments. Unlike Wells, Whitehead did not stress the role of applied science in the seventeenth century, but he saw the eventual linkage of science with technology and invention as equally transformative:

The greatest invention of the nineteenth century was the invention of the method of invention. A new method entered into life. In order to understand our epoch, we can neglect all the details of change, such as railways, telegraphs, radios, spinning machines, synthetic dyes. We must concentrate on the method itself; that is the real novelty, which has broken up the foundations of the old civilisation. The prophecy of Francis Bacon has now been fulfilled; and man, who at times dreamt of himself as a little lower than the angels, has submitted to become the servant and the minister of nature. It still remains to be seen whether the same actor can play both parts.

He went on to argue that 'In the past human life was lived in a bullock cart; in the future it will be lived in an aeroplane; and the change of speed amounts to a difference in quality.'[36]

For Whitehead, the rise of science had philosophically harmful consequences, whatever its practical benefits. Science had encouraged the creation of mechanistic materialism, and his own 'process philosophy' contributed to efforts now being made to find an alternative. His approach was sympathetic to evolutionism, although he distrusted the conventionally materialistic interpretation of Darwinism and favoured the Bergsonian model of creative evolution. His vision of the future opened up by the revolution in science and technology was certainly an open-ended one:

Modern science has imposed on humanity the necessity for wandering. Its progressive thought and its progressive technology make the transition through time, from

---

[35] See Langdon-Davies, *Man and His Universe* and *A Short History of the Future*.

[36] Whitehead, *Science and the Modern World*, pp. 136 and 137; on the Persians and the Greeks see p. 8. For details of the reaction against materialism in this period see my *Reconciling Science and Religion*.

generation to generation, a true migration into uncharted seas of adventure. The very benefit of wandering is that it is dangerous and needs skill to avert evils. We must expect, therefore, that the future will disclose dangers. It is the business of the future to be dangerous; and it is among the merits of science that it equips the future for its duties.[37]

Among the dangers was the threat of military technology; here at least Whitehead and Wells were in agreement, however different their views on philosophy.

Immediately after World War II, Arnold Toynbee revived Wells's focus on the West's invention of a global travel network based on sail, steam and now aviation as a key step in cultural history. Toynbee's historiography based on the rise and fall of civilizations had always emphasized the response to challenge as the driving force of progress, but he now seems to have accepted that this particular response had opened up wider prospects. In a lecture entitled 'The Unification of the World' he insisted that the West's triumph was by no means the continuation of a previous trend in history:

The agent of this revolutionary change in the affairs of men might have been any one of the divers parochial societies that were on the map when the revolution was put in hand, but the particular parochial society that has actually done the deed is the one that, of all of them, was the most unlikely candidate.[38]

Forced by geography to focus on travel by sea rather than land, the previously marginalized lands of Western Europe had created a global network that united humanity at a level never achieved by any other culture. This was a technological revolution that allowed the West to dominate the world. Like Whitehead, Toynbee feared that the emphasis on utility had led the West to throw off its religious heritage. He had a much greater appreciation of the contributions made by other cultures in this area, however, and insisted that the West now needed to learn from them to fill the vacancy left by the decline of Christianity. Otherwise we might well use our technology to destroy ourselves.

Whitehead's *Science in the Modern World* contributed to a debate within a new discipline that had emerged to study the history of science. It was hardly surprising that with growing awareness of the impact of science, scholars would begin to appreciate the need to understand how this new force for change had emerged. The Belgian historian George Sarton found the journal *Isis* in 1912 to publish material on the topic, and this was adopted by the History of Science Society, founded in America in 1924. In the middle decades of the century the field began to gain recognition as an academic specialization

---

[37] *Science and the Modern World*, p. 291.
[38] Toynbee, 'The Unification of the World and the Change in Historical Perspective', in his *Civilization on Trial*, pp. 62–96, quotation from p. 65.

(although its relationship to the more established areas of history remains somewhat uneasy).

An important contribution to the consolidation of the field was Herbert Butterfield's *The Origins of Modern Science* of 1949. In a chapter entitled 'The Place of the Scientific Revolution in the History of Western Civilisation' Butterfield argued that the revolution that began in the seventeenth century 'represents one of the great episodes in human experience, which ought to be placed ... amongst the epic adventures that have helped to make the human race what it is'. He insisted that the revolution must be regarded 'as a creative product of the West – depending on a complicated set of conditions which existed only in western Europe, depending partly also perhaps on a certain dynamic quality in the life and history of this half of the continent'.[39] Here was an academic historian giving credence to the position sketched out by Wells: progress was intermittent and depended on contingent circumstances, but the last outburst of all had produced the conditions for a whole new phase.

The question of why the scientific revolution occurred in Europe and not elsewhere has been much debated by historians of science, with rival views aired on the relationship between science and capitalism and on the indirect contributions of other cultures. There has been a particular focus on the 'great divergence' when the West's economic power began to expand far beyond that of China. This is linked to the 'Needham question' arising from the work of Joseph Needham on Chinese science and technology. For much of the time this was ahead of anything available in the West, until the latter pulled away rapidly during the early modern period. It is widely agreed that the new scientific and industrial techniques introduced then were made possible by a unique combination of social, economic and cultural developments, just as Wells had argued. Mokyr's introduction to the modern version of this form of progressionism also includes explicit analogies with the Darwinian theory of evolution and the sudden explosion of new types in the fossil record.[40]

Fictions of the Past

The new interpretation of the rise of the West had obvious implications not only for the idea of progress but also for the way historians think about the past. If the Scientific and Industrial Revolutions were not inevitable steps in social evolution but the products of a contingent set of circumstances that affected Europe, the whole model of history as the unfolding of an inevitable

---

[39] Butterfield, *The Origins of Modern Science*, p. 163. For some additional comments on the creation of the history of science as a discipline see Bowler and Morus, *Making Modern Science*, chap. 1.
[40] See Mokyr's works cited above, and also Pomeranz, *The Great Divergence*.

sequence was threatened. Episodes of real progress – increasingly understood as improvements in our control of nature – were rare and unpredictable because they were made possible only by unique combinations of events. Even if the modern wave of inventiveness is seen as a move to a more intense level of activity, it is still just another product of the complex interplay of historical agencies.

An immediate consequence of this transformation of attitudes towards invention was a shift in the balance between history and prediction in the presentations of the idea of progress. If history no longer offers evidence of a coherent progressive trend, trying to work out the range of opportunities that might open up in the future becomes more important. By the same token, the present situation must be the outcome of a contingent sequence of events and it becomes possible to imagine that, given a slightly different set of circumstances at some point in the past, history might have been deflected into a different channel and we would now be living in a very different world. It might even be a world in which the Scientific Revolution had not taken place in Europe, or had not taken place at all. It was no accident that the twentieth century witnessed the emergence of an interest in counterfactual history. A few academic historians began to explore the possibility that things might have turned out differently, and novelists had a field day imagining alternative worlds in which history had turned out very differently.

In 1894 Henry Adams told the American Historical Association that a true science of history would entail the discovery of a law that 'must fix with mathematical certainty the pattern which human society has got to follow'. He thought this would be impossible and warned that in any case, if the law offered no hope of improvements its discovery would do more harm than good.[41] Wells's decision to write *The Outline of History* was not driven by any hope of discovering such a law, but by the belief that his readers needed to understand the complex sequence of social and political events that together had created the modern world. Progress was not inevitable, because the emergence of modern science was a product of precisely the same interactions as those that had generated the rivalries of the European powers. Wells came to believe that another catastrophic war would be needed to destroy the old political system and open the way to the establishment of a rationally ordered society. History was necessary for understanding the current situation but was no guide to the future, so its value for an overall philosophy of progress was limited.

Not surprisingly, people began to ask: did it have to be this way? At some point in the past a crucial event whose outcome turned on a knife-edge might

---

[41] Adams, 'The Tendency of History', in his *The Degradation of the Democratic Dogma*, pp. 125–33, quotation from p. 129.

have come out differently, initiating a cascade of consequences that led to a world different from the one we know. The possibility was relevant to the idea of progress because conditions might be better or worse in the alternative world – and the fact that one could envisage both possibilities fitted in with the view that history is open-ended and unpredictable. The possibility of alternative worlds is unthinkable to anyone convinced that history is governed by rigid trends or laws. Even if there are broad social or cultural trends there may be crucial turning points where the chain of causation could be deflected, the most obvious case being military battles whose outcomes are sometimes determined by quite trivial events.

A few historians now sought to identify such turning points and imagine the consequences of an alternative outcome. In 1907 the *Westminster Gazette* offered a prize for an essay on the subject 'If Napoleon had won the Battle of Waterloo' – a possibility that could easily be imagined if Marshal Blücher and the Prussians had been delayed by a few hours. The prize was won by the historian G. M. Trevelyan, whose essay was later included in a 1932 collection entitled *If It Had Happened Otherwise* along with a chapter by Winston Churchill, himself no mean historian, exploring the consequences of the Confederacy winning the battle of Gettysburg.[42] From the mid-century onwards the exploration of counterfactual histories became a favourite theme with novelists, who often constructed their alternative worlds in considerable detail as backdrops to exciting stories. The possibility that the Allies might have lost World War II was explored in novels such as Philip K. Dick's *The Man in the High Castle*.

The novelist has to imagine the alternative world in some detail in order to be convincing. Academic historians are more interested in identifying potential turning points, following the extended development of the alternative world-line only in outline to suggest the possibilities of how things might have been different. Their ideas are collected in volumes such as Robert Cowley's *What If?* and Niall Ferguson's *Virtual History*. Some historians dismiss the technique as a waste of time, but others think it is a valid way of highlighting the contingency of events. Ferguson's introduction to his volume is a detailed account of the debate between the two positions. Counterfactualism has even made its way into the history of science, as in my own *Darwin Deleted*.[43]

If our vision of the past now seems less stable, the future looks equally uncertain. The new model of progress sees the Scientific and Industrial

---

[42] Trevelyan's essay was first reprinted in his *Clio: A Muse*, pp. 184–200; see also Squire, ed., *If It Had Happened Otherwise*.

[43] Ferguson, introduction to Ferguson, ed., *Virtual History*, pp. 1–90; for a similar collection of counterfactuals see Cowley, ed. *What If?* More generally see Geoffrey Hawthorn, *Plausible Worlds* and my own *Darwin Deleted*, which includes a discussion of how the technique has entered the history of science.

Revolutions as contingent events, but recognizes that they have ushered in a new phase of social development in which the rate of change is massively accelerated. For Wells and many of his contemporaries it was clear that there could be no prospect of creating a static utopia. Discovery and invention were unstoppable forces that would continue to generate new technologies capable of transforming the social order, and it was impossible to predict what new inventions would appear and which of them would actually succeed in gaining an impact. The emergence of science fiction as a popular genre of literature brought home the possibility that there might be many possible future worlds. Wells was perhaps the most active exponent of these new visions, but he was also clear that the open-endedness of the future rested on insights gained from the new and less deterministic interpretation of the past steps in progress.

10   Towards an Uncertain Future

As the twentieth century unfolded the concept of progress was increasingly defined in terms of technological development. By the end of the century Neil Postman could argue that traditional notions of culture itself had been replaced by the worship of technology.[1] As a result the most active debates about progress were increasingly focused not on the past but on the present and the future. It became commonplace to welcome (or complain about) the accelerating rate of progress. New technologies opened up opportunities for transforming industry and everyday life so rapidly that excitement was often overlaid with anxiety. Speculating about these possibilities became a regular feature of popular-science literature and science-fiction writers. The experts disagreed over which innovations were likely to be successful and over what their potential effects on society might be. If future progress was to be driven by innovation, the exact shape of the future became unpredictable. Society would change in ways that the experts might try to predict but which were all too often the unanticipated by-products of what at first seemed to be an obvious improvement.

The very idea of what constituted progress had diverged from what had been accepted in previous centuries. Some, including technophiles such as H. G. Wells, were aware that the model they were now promoting bore a closer resemblance to the Darwinian theory of branching evolution than to the old chain of being. The new approaches to history meshed with the sense of accelerating innovation to generate a more open-ended world-view. There were certainly some differences between the image of progress in the past and the sense of onrushing innovation in the present. Most obvious was the actual rate of change, which had accelerated to unprecedented levels. In prehistoric times episodes of innovation were widely spaced, much as the major breakthroughs in the history of life on earth were. Following the 'invention of invention', innovations began to come thick and fast and were at first concentrated in a single culture that dominated the whole world.

---

[1] Postman, *Technopoly*.

The impact of innovation was now expressed in a different way. Instead of occurring as isolated events in widely spaced cultures, innovations were flooding into the social life of a single culture. The multiplicity of innovations was affecting a single community, more like the mutations that occur in a single biological population than the tree of cultural diversity. Divergence was no longer a matter of separating cultures, but was at first seen as a range of possibilities opening up for a single global society. This appearance of unification now looks increasingly illusory. Globalization has certainly taken place, but it has not been complete, and the West has increasingly faced rival power blocs which are willing to take on board the new interplay of science and technology and use it in different ways. In this case the model of the tree of life remains relevant.

The impression of competing innovations struggling to dominate the market within a single community highlights the Darwinian aspects of the new model of progress. At all levels, innovations are for all practical purposes unpredictable, whatever the efforts of futurologists to anticipate them. The most influential new technologies are often those that were not predicted by the experts. Progress is represented by the sum total of the impacts of those innovations that succeed in the struggle for existence. Just as no one could have predicted that the West would be catapulted to global dominance by the creation of a new approach to applied science, no one can predict the inventions that will be made – or their effects. All too often it is the unanticipated consequences of new technologies that turn out to have the greater impact. New devices and techniques pioneered because they seem to offer benefits and opportunities may have harmful consequences when applied on too wide a scale. Those who doubt that the rush to innovate really does generate improvements as measured by traditional moral standards can dismiss the whole idea of progress as an illusion.

All of these implications of the new situation were already beginning to be recognized in the early decades of the century, just as the Darwinian synthesis began to transform evolutionism in biology. Thinkers such as Bergson expressed the new way of thinking in ways that linked ideas about evolution across the board, in philosophy, in history and in predicting what might come. They ushered in a world that took the open-endedness of change for granted when contemplating both the past and the future. This awareness has only expanded as the flood of technological innovations has continued to transform society.

Invention as Creation

The belief that the modern world was being reshaped by the relentless pressure of invention had become widespread by the early years of the twentieth century. These were what Philip Blom has called the 'vertigo years' when people saw their world being transformed around them by an array of new technologies including electric power, cinema, the telephone and radio, motor vehicles and aircraft. It all happened so quickly that it seemed inevitable that there was more to come. As Roxanne Panchasi suggests in the case of France, this created a 'culture of anticipation', and the same is true for other nations, especially the United States. Much interest was aroused by writers who tried to predict what technologies might be invented and what their implications for society might be. All this came at the same time as our conception of the universe was being unsettled by new theories such as relativity and atomic physics – the latter with its own promise of new energies to exploit.[2]

This was a vision of human creativity with significant parallels with the latest developments in evolution theory, encapsulated in the enthusiasm of Darwinians such as Huxley for Bergson's philosophy of 'creative evolution'. Natural selection working on a constant stream of genetic mutations was seen as a materialistic equivalent of human inventiveness, making available a variety of new structures, some of which turned out to be of greater significance than a mere response to the environment. This is why evolution could sometimes produce a breakthrough that opened up an entirely new level of development. The multiple potentials of innovation, like the genetic mutations in a population, are winnowed out by natural selection. As the technical journalist George Sutherland noted in his 1901 prediction of future developments, most inventions don't actually work in practice, and those that do take some time to perfect.[3] Invention can be seen as the equivalent of genetic mutation, both providing a basis for the Bergsonian notion of creativity.

While biological mutation is essentially undirected, the production of innovations in science and technology is by definition goal-directed, although the goals depend on the interests of those doing the innovating. As Sutherland noted, many of their ideas don't work in the real world, and this is where a process analogous to natural selection sifts out those that really do have the potential to succeed. Inevitably, the whole process now works at a much faster rate: it is as though, in biological evolution, the majority of genetic mutations tended to confer some potential benefit. Human creativity had been immensely slow through the early phases of our species's history, but now it had been pushed into a new level of activity. Where innovators had once been few and far between there was now a dedicated class of people devoted to technical

---

[2] Blom, *The Vertigo Years* and Panchasi, *Future Tense*. See also Corn and Horrigan, eds., *Yesterday's Tomorrows*; Corn, ed., *Imagining Tomorrow*; Segal, *Technological Utopianism in American Culture*; and my own *A History of the Future*.

[3] Sutherland, *Twentieth-Century Inventions*, p. xii.

invention. And where most people had been tended to be wary of proposed changes to their lives there was now a general willingness to take on board what the inventors produced.

Bergson's emphasis on the creativity of evolution recognized that there was no main line of development. Evolution is not being drawn towards a single goal ahead: it is driven upwards by a blind push from behind.[4] He drew an explicit comparison between evolution and human inventiveness, acknowledging that our social life is shaped by the products – although it often takes time to catch up. We should be called *Homo faber*, not *Homo sapiens*, because our intelligence is expressed most clearly in the creation of an indefinitely variable array of tools.[5] Bergson was seen as a major contributor to the new philosophy of nature that was replacing classical physics and challenging determinism. He became an international celebrity, and *Creative Evolution* was widely discussed in the English-speaking world long before it was translated in 1911. By linking evolutionary progress with human creativity he helped to establish a new version of progressionism in which it was impossible to predict what might come next.

Bergson's influence was recognized by Antonio Aliotta of the University of Padua, a leading voice in the reaction against the mechanistic view of nature. He emphasized that evolution was now seen as 'a perennial becoming, an incessant renewal of forms which cannot be foreseen, and which cannot therefore be subject to the rigid necessity of determinism'. The leading British advocate of Bergson's thought, H. Wildon Carr, also stressed that it was intended to provide an alternative to the traditional belief that history worked according to a preconceived plan. Humanity was not preordained: we were merely the highest product so far of the line of development that worked through the promotion of intelligence.[6] The implications of this new vision of evolution were now beginning to influence thinking in philosophy as well as in biology.

Progress and Pragmatism

The link between the Darwinian view of evolution and human inventiveness was also explored by the American thinkers who founded the philosophy of pragmatism. They appreciated that the Darwinian theory provided the basis for a new world-view in which both the natural and the human worlds could be

---

[4] Bergson, *Creative Evolution*, p. 103. On Bergson's philosophy of evolution see for instance Grosz, *The Nick of Time*, part 3, and on his link to the new Darwinism Herring, 'Great Is Darwin and Bergson Is His Prophet'.

[5] Bergson, *Creative Evolution*, pp. 138–9.

[6] Aliotta, *The Idealistic Reaction against Science*, p. xxi; Carr, *Henri Bergson*, esp. pp. 87–91.

seen to operate not according to some preordained pattern of development but by trial and error. Biological evolution and human thought advanced through the creation of new ways to deal with problems, success in the real world determining which were of real value.

As early as 1880, William James's essay on 'great men' attacked the Spencerian view that evolution was determined by law to emphasize the individual creativity that would become a centrepiece of his psychology. The link to Darwinism was explicit in the first sentence: 'A remarkable parallel, which I think has never been noticed, obtains between the facts of social evolution on the one hand, and of zoological evolution as expounded by Mr. Darwin on the other.'[7] James insisted that the appearance of individuals whose creative insights could direct the future course of social evolution was equivalent to the indefinite variation on which natural selection acts. In each case the actual appearance of the innovation might be impossible to explain, but what mattered was that it fitted the needs of the time and shaped the future. The result would restrict future possibilities without predetermining what must happen. History, not the environment, determines how things turn out. James would later become an enthusiastic supporter of Bergson's philosophy of creative evolution.

John Dewey's 1909 lecture 'The Influence of Darwin on Philosophy' also stressed the breakdown of the old finalism and the emergence of a 'Darwinian genetic and experimental logic' in which changes arise from attempts to solve practical problems. In his contribution to a 1917 volume on pragmatism entitled *Creative Intelligence*, he declared that obstacles are 'stimuli to variation, to novel response, and are hence the occasion of progress'. In another essay addressing the theme of progress directly, he argued that the previous generation had been deceived about the inevitability of progress and that it is by no means automatic: 'It is not a wholesale matter, but a retail job, to be contracted for and executed in sections.' Problems are dealt with one by one, not in accordance with some general law. Dewey accepted that the opportunities for improving the world had expanded enormously in modern times. Applied science had now begun to produce social changes far beyond those achieved by the great leaders of the past, giving us at last the means to direct progress in our chosen direction.[8]

---

[7] James, 'Great Men, Great Thoughts, and the Environment', reprinted in James, *The Will to Believe*, pp. 216–54, quotation from p. 216. See Wiener, *Evolution and the Founders of Pragmatism* and Marcell, *Progress and Pragmatism*.

[8] Dewey, 'The Influence of Darwin on Philosophy', in his *The Influence of Darwin on Philosophy*, pp. 1–19, quotation from p. 18; Dewey, 'The Need for a Recovery of Philosophy', in Dewey et al., *Creative Intelligence*, pp. 3–69, quotation from p. 12; Dewey, 'Progress', in his *Characters and Events*, vol. 2, pp. 820–30, quotation from p. 823.

Prophets of Progress

In the 'vertigo years' at the start of the new century there was a growing recognition of the fact that everyday life was now being transformed by the advent of technologies such as the cinema and radio. Despite concerns about the military applications of new innovations, there was still an air of optimism about the potential for life to be improved. Economists and social thinkers hoped that the benefits of the Industrial Revolution could be extended to all by the application of scientific planning. There was a sense that the establishment of a machine-based economy had created the potential for a world of peace and plenty for all, but that the potential had been thwarted by the profit motive that drove the early industrialists. One last push to extend the scientific method into economic and social planning would complete the process, eliminating the inequalities that led to war and ushering in a materialist utopia. But there was also a realization that the hope of creating a tranquil utopia ignored the power of technical innovation to disrupt the social order. The future would be uncertain and unpredictable because no one could foresee what would be invented, or what the long-term consequences would be.

It was in America that this sense of open-ended optimism retained its greatest hold. The American dream was being reinterpreted as a search for material prosperity rather than personal fulfilment, but prosperity now came in many different forms. Historians dealing with the most recent developments in American society recognized this point at the time. In the words of Henry F. May: 'Progress was no longer a universal single movement, but wherever one looked things were getting better. People no longer needed the old fixities; they could do better without them, and few doubted, at bottom, that they knew what better meant.'[9] Charles A. Beard popularized the view that the open-endedness of the sequence of events gives us the freedom to reform the world, presenting his readers with an image of the United States as the vehicle by which progress would best be achieved. Beard edited two books on the idea of progress: *Whither Mankind?* of 1928 collected generally pessimistic views, while *Towards Civilization* two years later included the more optimistic evaluations of scientists and inventors.

In 1927 Charles and Mary Beard began publication of *The Rise of American Civilization*, a survey dismissed by academic historians but read by large numbers of ordinary citizens. The conclusion suggested that people were increasingly optimistic. They had faith in democracy but also:

a faith in the efficacy of that new and mysterious instrument of the modern mind, the 'invention of invention', moving from one technological triumph to another ...

---

[9] May, *The End of American Innocence*, p. 141. On redefining the American dream see Churchwell, *Behold, America*.

effecting an ever wider distribution of the blessings of civilization – health, security, material goods, knowledge, leisure and aesthetic appreciation, and through the cumulative forces of intellectual and artistic reactions, conjuring from the vasty deep of the nameless and unknown creative imagination of the noblest order, subduing physical things to the empire of the spirit – doubting not the capacity of the Power that had summoned into being all patterns of the past and present, living and dead, to fulfill its endless destiny.[10]

Here the idea that progress driven by creativity had entered a new and more productive phase was applied in a specifically American context. Even after the depression and the outbreak of another war, the Beards could still wax lyrical over the possibilities of progress in 'a partially open and dynamic world' where individual and collective acts could have a real effect.[11]

The sense of what constituted a better future was increasingly being defined not by intellectuals pontificating about their ideal world, but by what ordinary people thought would make a better and more exciting life. Some of these expectations were utilitarian in the most mundane way, although one can hardly complain when those living hard lives see how their lot could become more comfortable. But an element of vision remained: people also saw the possibility of realizing things they could once have only dreamed of, like flying, talking to loved ones around the world and even going to the moon. Inventors were driven by the hope of making these dreams a reality.

The diversity of new inventions made it increasingly unlikely that planners could anticipate them and reorganize society to absorb their effects. The hope of a planned, static utopia evaporated in response to the scope of material benefits coming forward. All too often, planning had to deal with the unanticipated consequences of the new inventions. As Sean Johnston has shown, the American engineers who led the technocratic movement of the 1930s became increasingly concerned not with planning utopia but with devising a 'technological fix' for the problems that had emerged as the unintended consequences of previous innovations. Technology offered unlimited progress, but it was not progress towards a fixed goal, and its impacts could not be predicted by the planners and would have to be managed on an ad hoc basis.[12] A new and less structured version of the idea of progress was emerging, popularized by writers such as H. G. Wells who were aware of the unpredictability of technological innovation and used fictional accounts of future worlds to drive home this message.

---

[10] Beard and Beard, *The Rise of American Civilization*, vol. 2, p. 831. In addition to *Towards Civilization* Charles A. Beard also provided an introduction to Bury's *The Idea of Progress* which stresses the opportunities provided by the plethora of new technologies.
[11] Beard and Beard, *The American Spirit*, p. 674.   [12] Johnston, *Techno-fixers*.

Commentators were impressed by the fact that the rate of change seemed to be accelerating. In 1909 Henry Adams suggested that the mechanical phase of progress which had started around 1600 had now been transformed into the electrical phase, which would soon give way to the ethereal. With tongue in cheek he argued that this would last only until 1921 'and bring thought to the limit of its possibilities'.[13] No one seriously thought that any such limit to human inventiveness would be reached, although Adams's claim might be seen as an anticipation of the prophecies of modern technophiles who anticipate a transhuman phase of existence.

It was Wells who came to symbolize this sense of future possibilities. He was the best-known source of new – and often contradictory – predictions about what would happen as new technologies emerged. He addressed this theme directly in a 1902 lecture published as *The Discovery of the Future* and in his book *Anticipations: Of the Reaction of Mechanical and Scientific Progress on Human Life and Thought*. Novels such as *When the Sleeper Wakes* got the message across to a wider audience. He was no simple-minded progressionist, though, being well aware of the potential for technological innovation to threaten the social order and undermine its potential benefits.

After the Great War the darker side of the coin became more apparent, especially in Europe, and historians sometimes present the inter-war years as a 'morbid age' obsessed with the expectation that civilization was collapsing. But for every pessimistic moralist and literary figure there was an enthusiastic technophile who hoped that the benefits would outweigh any potential dangers. A number of commentators and science writers predicted what might be invented next in popular books, magazines and newspapers. The publisher Kegan Paul issued a series of little books under the rubric 'Today and Tomorrow' with titles by scientists such as J. B. S. Haldane and J. D. Bernal. Magazines including *Popular Mechanics* in America, *Armchair Science* in Britain, *Kosmos* in Germany and *Science et vie* in France described the latest developments and often speculated about the future. An array of books by scientists, engineers and popular-science writers presented the public with a variety of predicted futures, all based on the assumption that new technologies would transform life for the better. Some of these predictions were fairly straightforward extensions of existing technologies, such as television and long-range air travel. Others were far more imaginative, including wireless broadcasting of power and artificial foods.[14]

---

[13] Adams, 'The Rule of Phase Applied to History', in his *The Degradation of the Democratic Dogma*, pp. 267–311, quotation from p. 308.

[14] For details of these developments see my *A History of the Future*. On the pessimistic aspects of inter-war life and thought see for instance Overy, *The Morbid Age*.

Wells's emphasis on the future rather than the past marks a crucial turning point in the framework of debate about progress: history was less important now that everyone knew that life was being relentlessly transformed by technology. His experience as a prophet confirmed how unpredictable things had become. Anticipating the impact of a technology that had just been introduced was hard enough; predicting the scientific discoveries that made new technologies possible was virtually impossible. He also knew how difficult it was to anticipate the long-term impacts. Although introduced in the expectation that they would make life better, innovations had a tendency to generate problems when they were applied on a large scale. In 1932 Wells gave a BBC radio talk entitled 'Wanted – Professors of Foresight!' He complained that although there were thousands of historians working in the universities there was no one professionally employed to think about future technical innovations and plan for their reception.

Wells and his fellow commentators worked in a totally unofficial capacity; it would not be until the 1950s that organizations devoted to futurology such as the RAND Corporation would be formed. In the absence of foresight, chaos often ensued. Wells gave as an example the introduction of the motor car, now causing confusion and carnage on roads that had never been designed for this kind of traffic. If those in charge of the country had been warned that when motor vehicles became faster and cheaper a whole new kind of road system would be needed, they could have planned for this in advance. Anticipating the future was never going to be easy, because it was difficult to imagine what the impact of an embryonic technology might be, especially when one could not be sure which innovations would actually catch on.

Examples of these uncertainties abound in the early twentieth century. Marconi pioneered radio as a new form of telegraphy to pass messages between individuals, not realizing its potential for broadcasting to wide audiences. When broadcasting came in, some saw it as a benefit to education and global harmony while others saw the potential for commercialism and propaganda. Almost everyone assumed that the principles underlying radio and the cinema would be combined to give what came to be called television, and this prediction was realized by the late 1930s (although the technical problems were more complex than the prophets had claimed). Aviation was also hailed for its potential to improve worldwide interactions while the pessimists warned about its military applications. Long-haul airlines did emerge – again in the 1930s – but those who thought that soon everyone would fly their own light aircraft (perhaps a helicopter) were doomed to disappointment. There were predictions that agriculture would soon be replaced by the production of food by purely chemical means, which turned out to be another red herring.

Fictions of the Future

Wells saw that imaginative fiction was the most effective medium through which the reading public could be warned of the potential changes. Fictional accounts of futuristic technologies had become popular in the late nineteenth century, most obviously in the writings of Jules Verne. Wells now exploited the potential of the emerging genre that became known as science fiction, using it as a vehicle to get his sociological messages across to a wider audience. By setting a human-interest story in a projected future he encouraged his readers to appreciate how their own lives might be changed. As early as 1899 his 'A Story of the Days to Come' explored life in a future mega-city, a theme taken further in *When the Sleeper Wakes*.[15]

By the 1920s science fiction had become popular with younger readers enthusiastic about technology. There were important development in the early Soviet Union, but the most influential trend was in America, where Hugo Gernsback pioneered the exploration of the genre in cheap 'pulp' magazines. Like Wells, Gernsback wanted to promote speculation about plausible future technologies, and his magazines frequently interpolated science fiction with fact-based accounts of the latest innovations. Space travel was soon introduced to the general reader – and movie-goer – in series featuring the adventures of Flash Gordon and Buck Rogers. Although dismissed as moonshine by all but a few pioneers of rocket technology, space exploration became a major theme in more serious science fiction in the 1950s and 1960s, often promoted by writers such as Arthur C. Clarke who were also publicizing the latest real-world developments. The prospect of exploring the solar system and beyond offered an exciting and open-ended vision of the future, perhaps not a utopia but an alternative to the more pessimistic view of technology expressed by some intellectuals.[16]

Wearing his hat as a social reformer, Wells was quite capable of imagining what looks like a fairly static future, as in his *A Modern Utopia* of 1905 and later in *Men Like Gods*. These contain hints of advanced technologies but focus on the necessity for creating a fairer society. Superficially, the dream of a

---

[15] 'A Story of the Days to Come' is reprinted in *The Short Stories of H. G. Wells*, pp. 796–877. *When the Sleeper Wakes* was later reissued as *The Sleeper Awakes*; the edition cited in the bibliography has a useful introduction to Wells's futurology by Patrick Parrinder. See also Parrinder's *Shadows of the Future*; Roslynn D. Haynes, *H. G. Wells*; and Huntington, *The Logic of Fantasy*.

[16] On the emergence of science fiction see Aldiss and Wingrove, *Billion Year Spree*; Carter, *The Creation of Tomorrow*; Cheng, *Astounding Wonder*; James and Mendlesohn, eds., *The Cambridge Companion to Science Fiction*; Nicholls, ed., *The Encyclopedia of Science Fiction*; and Stover, *Science Fiction from Wells to Heinlein*. For a collection of essays exploring the significance of fiction for the history of science see Rees and Morus, eds., 'Presenting Futures Past'.

rationally ordered worldwide utopia resembles the end-product imagined by the earlier writers who thought that history was moving towards a culminating point. But Wells knew that the goal could not mark the end of history. A *Modern Utopia* opens with an explicit claim that the future cannot be static:

> The Utopia of a modern dreamer must needs differ in one fundamental aspect from the Nowheres and Utopias men planned before Darwin quickened the thought of the world. Those were all perfect and static States, a balance of happiness won for ever against the forces of unrest and disorder that inhere in things. . . . But the Modern Utopia must not be static but kinetic, must shape not as a permanent state but as a hopeful stage, leading to a long ascent of stages.[17]

The hope of achieving a better society remained, but many of Wells's stories explore more complex futures in which both the benefits and the potential dangers of new technologies are apparent. Even when imagining the creation of a rationally organized World State, he acknowledged that it would offer no rest from change.

Wells engaged directly with the uncertainty of prediction. In *When the Sleeper Wakes* he had suggested that the whole human race would soon be concentrated in a few enclosed mega-cities, only to realize a few years later that improved transport would tend to disperse people out into the suburbs. In *The World Set Free* he anticipated both atomic energy for peaceful uses and the atomic bomb. He was among the first to predict the military applications of aviation in *The War in the Air* of 1908. The catastrophic war predicted in *The Shape of Things to Come* of 1933 was made worse by aircraft dropping poison gas – but it was aviation that allowed the surviving technocrats to dominate the earth and found a rational World Government. The film version *Things to Come* ends with a conflict between the enthusiasts who want to explore the cosmos and conservatives who prefer a quiet life. As the space gun (Wells's tribute to Jules Verne) prepares to fire the young volunteers aloft, the leader of the adventurous faction, Osward Cabal (played by Raymond Massy), points upwards and declaims: 'All the universe or nothing . . . Which shall it be?' The scene fades out to the caption 'WHITHER MANKIND?' Here the hint at the end of *The Outline of History* that the earth is but a footstool from which we shall reach to the stars becomes a vivid expression of the exciting future that technology could open up.[18]

This expectation was explored by later enthusiasts for space travel such as Arthur C. Clarke. There were some who took a more prosaic approach. Olaf

---

[17] Wells, *A Modern Utopia*, p. 1. In the edition cited, Krishnan Kumar's introduction and his appendix 'H. G. Wells and His Critics', pp. 244–61, contain useful information on the text and its reception.

[18] The script of the film is published in Wells, *Things to Come*; see p. 204 for the final scene, and and for more details on the film see Stover, *The Prophetic Soul*.

Stapledon''s *Last and First Men* of 1930 predicted that we might have to move to other planets as the earth became uninhabitable. Isaac Asimov simply transferred humanity, warts and all, into space, imagining the decline and fall of a Galactic Empire in his classic *Foundation* trilogy. But Wells seems to hint at something more, at a destiny of cosmic significance that we can achieve if we have the will. The Martian invaders of his *War of the Worlds* could still terrify the public, as in Orson Welles's radio version of 1938, but astronomers were increasingly sure that the solar system and perhaps even the galaxy were devoid of intelligent life. Perhaps we alone have the ability to extend consciousness to the whole universe. If other civilizations do exist, surely we have a duty to interact with them. For Clarke, the latter possibility seemed to offer a kind of surrogate religion, a new sense of purpose for humanity that would replace the vision of a spiritual paradise as the goal of progress. Materialists too could have a meaningful sense of life's destiny. In the film *2001: A Space Odyssey* the vision far transcends the mere building of space stations and moon colonies.

Future Shock

By the 1970s the public had begun to lose interest in such predictions as it became clear that the moon landings were not a prelude to permanent colonization. Clarke's transcendent vision of contact with higher races faded too, leaving the field to more hard-headed proposals such as Gerard O'Neill's plans for giant self-sustaining space habitats. O'Neill is one of the 'visioneers', to use a term introduced by the historian Patrick McCray to denote the enthusiasts who believe in the potential for technology to open new horizons for humanity to explore. Futurology had now become institutionalized, with organizations such as the RAND Corporation (originally a United States Air Force project) backing up Hermann Kahn and other experts predicting new technologies and their implications. Gerald Feinberg called for a 'Prometheus Agency' to establish the long-range goals. In Britain C. P. Snow's *The Two Cultures* lamented the unwillingness of the governing class to acknowledge the work of the scientists who 'have the future in their bones'. He insisted that ordinary people were only too keen to see new inventions that would improve their lives, and a Labour government proclaimed its willingness to exploit the 'white heat of technology'.[19]

---

[19] McCray, *The Visioneers*. For more on space enthusiasm see Benjamin, *Rocket Dreams*; Kilgrove, *Astrofuturism*; and more recently Kaku, *The Future of Humanity*. See also Feinberg, *The Prometheus Project*; Snow, *The Two Cultures and a Second Look*; and Ortolano, *The Two Cultures Controversy*.

In fact many of the expectations articulated in the 1950s and 1960s were the residue of the unfulfilled prophecies of the inter-war years, including cheap atomic power, synthetic food, better air transport and moon colonies. Many of these have turned out to be illusory and (in the case of atomic power) potentially hazardous, while genuinely important new technologies such as the computer were not predicted. Few at first imagined that the miniaturization of electronics would lead to personal computers and entirely new means of communication such as the internet. It was members of the counterculture such as Stewart Brand who first realized the liberating potential of mass personal communication, although their 'digital utopianism' failed to appreciate that the system could also be misused. The science-fiction writers were more perceptive in their explorations of how these technologies might have both beneficial and harmful effects on our lives.[20]

As the pace of research quickened the future became ever more uncertain. Best-sellers such as Alvin Toffler's *Future Shock* of 1970 exploited a public sense that innovations were now coming so fast that no one could cope. In a world of competing visions, enthusiasts for particular technologies sought to capture the public's attention. McCray's other main example of a visioneer is Eric Drexler, the apostle of a new world of opportunity provided by nanotechnology. The ability to manipulate matter at the level of individual atoms was another innovation that appeared largely unheralded by the experts. Its potential to transform a whole range of industrial and biomedical processes may yet turn out to be as profound as Drexler's *Engines of Creation* of 1990 proclaimed. Its most disturbing implication is the potential to modify biological organisms including the human body and brain.

Transforming Ourselves

Francis Fukuyama's claim that history had ended with the triumph of liberal capitalism was undermined by the continued fragmentation of the world into rival power blocks. He continued to doubt the significance of the emerging conflicts, but he did come to appreciate that there was another force at work that might generate further change. As the new century dawned he suggested that history could not reach a stable conclusion if scientific progress continued to generate innovations capable of reconfiguring society. He now worried about the prospect that new technologies might allow us to reconfigure our bodies and brains in ways that would lift us into an entirely new 'posthuman' frame of existence.[21]

---

[20] See Turner, *From Counterculture to Cyberculture.*   [21] Fukuyama, *Our Posthuman Future.*

The prospect that humanity itself might be transformed in the future had long been contemplated. In the conclusion to *The Shape of Things to Come* Wells imagined that the human race would begin to merge into a single entity:

The body of mankind is now one single organism of 2,500 million persons and the individual differences of every one of these persons is like an exploratory tentacle thrust out to test and learn, to savour life in its fullest and bring in new experiences for the common stock. We are all members of one body.[22]

The result would usher in a new phase in evolution. Wells adds in a footnote that this was what St Paul may have been hinting at before his vision was subverted by formal Christianity, and the prospect does seem to hint at the 'omega point' that Teilhard de Chardin would foresee as the end-point of our spiritual evolution.

Others were thinking along far more intrusive lines. People have been modifying animals and plants by artificial selection since time immemorial, but now the process was being applied far more systematically. In the twentieth century advances in biology, especially but by no means exclusively in the study of heredity, led to the expectation that more radical improvements would soon be possible. New varieties of crops and domesticated animals might improve food supplies and alleviate the threat of famine. Some even hoped for a complete taming of nature, with the elimination of pests and parasites allowing the whole earth to be turned into a park-like paradise. Wells himself pandered to this vision in his *Men Like Gods*, contributing to what the historian Jim Endersby calls the hope of a 'biotopia' to come.[23]

From the start, though, it had been apparent that the same techniques could be applied to the human race itself. The eugenics movement had been founded by Darwin's cousin Francis Galton in the hope of improving the human race, or at least preventing its degeneration. Eugenics advocated a policy of artificial selection in which the 'best' people would be encouraged to have more children while those deemed 'unfit' would be discouraged or prevented from passing on their detrimental characters. The new understanding of heredity that was centred on the emergence of genetics boosted the hopes of the eugenicists in the early twentieth century. Now they had the capacity to identify the good and the bad genes they needed to control. Their efforts led to some of the most horrific consequences of the drive to perfection, culminating in the Nazis' efforts to purify the Aryan race and eliminate 'subhuman' types. Less drastic programmes were in place around the globe, often leading to the forced sterilization of the 'unfit'.

It is less well remembered today that those who saw the environment rather than heredity as the determinant of human nature also developed techniques

---

[22] Wells, *The Shape of Things to Come*, p. 431.
[23] Endersby, 'A Visit to Biotopia'. See also my own *History of the Future*, chap. 10.

intended to improve the race. The Lamarckian biologist Paul Kammerer, best known for his efforts to demonstrate the inheritance of acquired characteristics in the midwife toad, believed that individuals could be rejuvenated by hormones and insisted that the benefits of improved social conditions would become permanently implanted in the race.[24]

Most of these expectations were confined within a conventional ideology that defined the 'best' form of humanity that might be attained. Each social or political group envisaged the perfect human who would embody all the characters it found most attractive. In this respect there was still the sense of a preordained goal to be achieved. In his *Out of the Night* of 1936 the geneticist Hermann Muller even proposed inseminating women with the semen of the most eminent men as recognized at the time.

Once the idea of technical innovation could be applied to biology this image of perfection began to look a bit old-fashioned. The new science could be applied to produce any characters we chose, and in an ever-changing social environment it might be better to aim for types pre-adapted to the roles they would be expected to play. The most notorious pioneer of this experimental vision of the human future was the biologist J. B. S. Haldane, whose *Daedalus* of 1924 proposed that in the future all reproduction would be by ectogenesis, the artificial rearing of embryos in controlled conditions. This would allow the production of human types with different abilities; there was no need to imagine the whole race becoming uniformly perfect. In a later essay Haldane imagined such techniques being used to modify the human species so it could move to other planets, a possibility taken up by science-fiction authors such as Stapledon. J. D. Bernal suggested it would be possible to produce a race adapted to living on the space colonies that would explore the universe. Humanity would divide into two, the adventurous space-farers and the stay-at-homes – a vision that was also hinted at in the concluding scene of Wells's *Things to Come*.[25]

The darker potential of Haldane's suggestion was explored in Aldous Huxley's *Brave New World*, in which the techniques were used to create a stratified and soul-less 'perfect' society. Similar predictions became more acute in the 1960s as developments in biology accelerated following the discovery of the structure of DNA. The eminent French biologist Jean Rostand's *Can Man Be Modified?* of 1959 quoted *Brave New World* but argued that the production of 'test tube babies' could also have beneficial

---

[24] There is a huge literature on eugenics; see for instance Kevles, *In the Name of Eugenics* and Bashford and Levine, eds., *The Oxford Handbook of the History of Eugenics*. For an account of Kammerer and the Lamarckian programme see my *A History of the Future*, pp. 188–92.

[25] See Adams, 'Last Judgement' and Bernal, *The World, the Flesh and the Devil*. On Wells's film see n. 18 above.

uses. Technophiles looked forward to a world in which humanity could be modified for the better, although pessimists still worried about the social and moral issues that would arise. In 1964 Robert Ettinger proposed that the dead should be frozen in the hope that future medical developments would be able to extend their lives. Gordon Rattray Taylor's *The Biological Time Bomb* of 1968 predicted a host of new biological technologies by the end of the century, including personality reconstruction, enhancement of intelligence, cloning and genetic engineering. What had begun as wild speculation now became an immediately threatening reality.[26]

The list of innovations predicted by popular writers such as Taylor gives some idea of how open-ended the future had begun to seem. By the end of the century enthusiasts were promoting an ever-wider array of techniques with even deeper implications. The boundary between science fiction and corporate ambitions became blurred: Marvin Minsky's foreword to Drexler's *Engines of Creation* praised the book as one of the few fact-based surveys of forthcoming developments to match those of the fiction writers. Nanotechnology opened up a whole new level of intrusive control of the body and the brain through the use of minute machines that would repair faults and reverse the aging process. Colin Milburn's *Nanovision* of 2008 identified the science-fiction author Vernor Vinge as the first to warn of a 'technological singularity' that would come in the twenty-first century as the human species was transformed by the new technologies. This had already become the theme of Ray Kurzweil's *The Singularity Is Near: When Humans Transcend Biology*.

The futurologists project a sense that technical progress is now accelerating at such a rate that it will become truly transformative. If the 'invention of invention' that created the interaction between science and technology was a step change that pushed progress into a new phase, the coming singularity would be more abrupt and more dramatic and would move human existence onto a new and more exciting plane. Yet the element of unpredictability that had been introduced by the earlier phase remained because the technophiles still could not agree on which of their projected innovations would actually provide the next leap forward. The concerns expressed in Fukuyama's *Our Posthuman Future* and Harari's *Homo Deus* centre on various possibilities, including the biological transformation of our bodies and brains by nanotechnology and rival projects involving links with non-biological systems that have been raised to new levels of sophistication. We might become cyborgs in which biological organs are supplemented or replaced by artificial ones. The most drastic proposal is that we download copies of our personalities into

---

[26] Rostand, *Can Man Be Modified?*, pp. 80–5; Ettinger, *The Prospect of Immortality*; for Taylor's succinct list of predictions see his *The Biological Time Bomb*, p. 218, and for a wider discussion of this period Turney, *Frankenstein's Footsteps*, chap. 7.

computers, where we would live for ever in a cyberspace with unlimited potential.

Such visions began to seem plausible only when computers gained the capacity to out-think human beings. The fascination with artificial intelligence that began with cybernetics and expanded with the power of modern computers has produced a change in the technophiles' very sense of what it is to be a human personality. In their eyes, we ourselves are nothing more than information-processing machines, so there is no reason why we should not bolt on artificial enhancements or even simply copy the software that defines us into a computer, confident that what continues there is somehow really ourselves. And all this is in addition to the claims that we, or our transformed selves, should move out to explore the cosmos. The enthusiasts all think that their pet innovations will become the basis for a more exciting and fulfilling future, but no one can be sure which, if any, will actually succeed in dominating the market. The pessimists, of course, warn that the whole project represents just the kind of dehumanization that Huxley and others predicted long ago.[27]

On the Bandwagon

In his *Outline of History* H. G. Wells had noted an important feature of the recent past. The West may have pioneered the 'invention of invention' and reaped the benefits of this unique cultural breakthrough, but other cultures were now beginning to adopt the technique and were doing so without necessarily giving up their traditional values. Japan was the first example, already becoming a major power with a modern navy in the early years of the century. Cultures can learn from each other and can adopt just those innovations they think they can use. (This seems an apparent difference from the process of biological evolution, although we now know that genetic material has occasionally been transferred between branches of the tree of life.) As Wells ought to have realized, the logic of Gibbon's claim that the West cannot be conquered by barbarians was flawed in the sense that rival cultures can learn how to become modern in their technologies without losing the cultural traits that make them seem alien to the West. It has slowly become apparent that the globalization of the technological imperative does not necessarily mean the emergence of a New World Order based on Western values.[28]

---

[27] For some more imaginative explorations what might happen see Rosenberg and Harding, eds., *Histories of the Future.*

[28] See for instance Bertram and Chitty, eds., *Has History Ended?* and Horsman and Marshall, *After the Nation State.*

This possibility was anticipated by Arnold Toynbee, whose recognition of the West's unique role in unifying the world did not prevent him from stressing that the other cultures had learnt to copy the methods of science-based industrialization. He argued that – having downgraded its own religion – the West now needed to learn from the wisdom of the other civilizations.[29] In the twenty-first century it has become clear that the globalization of industry is not leading to a complete integration of world cultures. Many nations now have all the trappings that we used to associate with the West's ambitions – including atomic power, space programmes and, more recently, artificial intelligence. It now seems possible that if space exploration goes ahead it may be led by the Chinese – and it is no accident that Chinese science fiction is now appearing in which their role is taken for granted, as in Cixin Liu's best-selling *The Three-Body Problem* and its sequels. Yet social and cultural rivalries continue unabated. If there is an approaching 'singularity' leading to a posthuman future, it may be shaped by non-Western values rather than a continuation of liberal capitalism.

We are at last aware of a lesson that was learnt with difficulty in the course of the twentieth century. It seems bizarre in retrospect that anyone could have believed that Western civilization was superior to all others and hence represented the only way forward. Anthropologists slowly became aware of cultural diversity, and that awareness is now widely disseminated in popular books, films and television programmes. The history of other cultures has also attracted wide attention: witness for instance the global fascination with China's terracotta army. A famous BBC television series of 1969 was based on Kenneth Clark's personal view of Western civilization, but it was recently remade as *Civilizations* (plural) to emphasize the growing need for us to be aware of other cultures. Given this awareness of diversity, should we really be surprised that the other cultures may wish to retain as much as possible of their tradition and history even as they modernize? Perhaps the globalization we are all so well aware of will work at a level which, while by no means superficial, allows coexistence with elements defining ways of life and values that are not shared by the West.

If this is so, the branching tree of cultural evolution will not be pruned down to leave a single branch as the basis for future development. And now that others have taken on board the technological imperative, the Darwinian element of competition and selection will be maintained, along with the unpredictability that comes with the creativity of innovation. The open-endedness that derives from a constant plethora of new ideas and inventions will continue within each culture, and because each provides a different social and cultural

---

[29] Toynbee, 'The Unification of the World and the Change in Historical Perspective', in his *Civilization on Trial*, pp. 83–4.

environment the innovations that succeed may not always be the same. It is at least conceivable that the various power-blocks might move in different directions, with who knows what consequences for the global interaction between them. Conflicts will continue, undermining Wells's hope of achieving a rationally ordered World State and threatening the superiority the West has enjoyed since it pioneered the emergence of the modern economy.[30]

The linear models of progress explored in Part I of this study were created by rival factions within a Western culture that was sure of its expanding power and its place at the head of creation. The theory of evolution emerged originally in a form that provided this vision with a foundation in the natural world based on the temporalized chain of being. That vision was gradually undermined as biology came to grips with the implications of Darwinism: humans were not the goal of creation, and progressive innovations were rare in the history of life. Historians too recognized the significance of innovation in the development of civilizations and the acceleration in the rate of change it has produced. In an age of increased uncertainty the technophiles alone remain convinced that their innovations are generating a better and more fulfilling life for all. They have salvaged the idea of progress by shifting the criterion of improvement to one based on operational sophistication and by conceding that such advances can be made in many different ways. Only now can we appreciate the full implications of this model of divergent progress, as the effects of the 'invention of invention' spread out to the other branches of the tree of cultural and social evolution.

---

[30] On the possible decline of the West see Morris, *Why the West Rules – For Now* and Conway and Oreskes, *The Collapse of Western Civilization*. On the unlikelihood of a peaceful global civilization emerging, see Bobbitt, *The Shield of Achilles*.

11   Epilogue
Where Did It All Go Wrong?

To conclude it may be worth taking a brief look at those who have opposed the idea of progress, if only because their strategies throw light on the changes we have been concerned with. Some critiques apply to all versions of the idea, while others focus on a particular manifestation – and some attacks on one form were endorsed by supporters of another. Doubts about the possibility of genuine, sustained progress intensified in the twentieth century and are still active today, but there has been a significant shift in the nature of the opposition, reflecting the growing popularity of the more open-ended vision of progress. The global pandemic raging as this book is completed adds another layer to the uncertainty of what lies ahead.

The most effective critiques of the early and mid-twentieth century worked by depicting a nightmare vision of the future – yet it was usually a dystopia parodying the static utopias of the early progressionists. Now the hope (or fear) of a perfectly ordered society has faded. There are still optimists who support a general vision of progress brought about by advancing technology and rational social reform. Stephen Pinker's *Enlightenment Now* is a recent example, and its reference back to the Enlightenment project shows that the historical background is still seen as relevant. Pinker hopes for improvement, but there is no longer any suggestion that a perfect society can be achieved. The real technophiles promote their own areas of innovation, and much of the opposition is a reaction to the disadvantages of individual projects. Critics see a variety of unpleasant futures, just as the enthusiasts promote a range of different ways forward. No one foresees a static *Brave New World* or *1984*. There is still plenty of mistrust, but it is diffused among a wide range of critics each driven by fear of a particular form of technology.

Earlier critics saw the idea of progress as the product of an unwillingness to face certain general features of the world we live in. Either human nature is so deeply flawed that we are beyond all but supernatural help or the universe itself is so constructed that nothing meaningful can ever gain traction within it. The belief that human nature is permanently flawed was the traditional perspective of Christianity. Liberal Christians transformed the faith by encouraging the hope that we might regain the paradise lost at the Fall. By the nineteenth century many of those who hoped for a realization of Christian values in society had become openly hostile to the industrialization seen by the progressives as a key component of their programme. In the short term at least, progress had generated the miseries of the slums – and even if this were relieved there was the ever-present threat of hedonism. Only by focusing on the spiritual education of humanity could real progress be achieved, a view that became ever less plausible as the twentieth century advanced.

The parallel concern that material nature itself might present insurmountable obstacles to progress was integral to Thomas Malthus's argument against the hopes of Condorcet and Godwin. His 'principle of population' – first promulgated in 1797 – combined factors from both human and material nature to imply that our tendency to produce too many children collides with the limitations of the food supply to ensure a permanent level of misery in the world. The pessimistic interpretation of the Darwinian universe as an endless chaos of random variation and death owed a great deal to the legacy of Malthus, serving as an alternative to the progressive vision of the Spencerians.

This sense of nature as a system that does not allow cumulative improvement was reinforced by the physicists' claim that according to the laws of thermodynamics the universe must inevitably run down to a 'heat death' when all activity will cease. This cosmic pessimism was eloquently expressed by Bertrand Russell, whose views were cited by W. R. Inge, the notoriously pessimistic dean of St Paul's in London, as evidence that 'Naturalism has severed its alliance with optimism and a belief in progress.'[1] The sense of looming catastrophe was heightened by the fear of war and the expectation that future conflicts would be rendered even more terrible by advances in military technology. Novels predicting a war that would destroy civilization and perhaps even the human race abounded in the late nineteenth and early twentieth centuries, providing a counterweight to the enthusiasts who recognized only the potential benefits of innovation.[2]

A more subtle level of critique acknowledged the potential for superficial progress but argued that the results could never be permanent. Oswald Spengler sought to checkmate those who cited the achievements of the nineteenth century as evidence for a general trend to progress by arguing that such a rise was always followed by a decline to eventual extinction as civilizations endlessly repeated the rhythm of birth and death.[3] There had always been a suspicion that civilization itself was a mistake, offering everyone a superficially more pleasant life but undermining the more fundamental aspects of

---

[1] Russell, 'A Free Man's Worship', in *The Basic Writings of Bertrand Russell*, p. 67; Inge, *The Idea of Progress* and, more optimistically, 'The Future of the Human Race'.

[2] For a brief survey of this literature see my *A History of the Future*, pp. 25–7 and chap. 9.

[3] See Fischer, *History and Prophecy* and Farrenkopf, *Prophet of Decline*.

human behaviour. The fear of dehumanization became central to the Romantic movement's reaction against industrialization and the more savage attacks of figures such as Nietzsche. Spencer's vision of a future race of self-reliant individualists perfectly adapted to a free-enterprise marketplace was equally at variance with the rather anodyne Enlightenment vision of a polite and cultured society.

The answer, as Wells and others insisted, was the creation of a World State run on rational lines. For many – although not for Wells and those who recognized the transforming power of technical innovation – such a state would represent the end-point of social evolution as predicted by Condorcet and other reformers. The Soviet system seemed to offer a blueprint for this move, but the vision of a state run by an elite for the benefit of all soon gave way to fears of a totalitarian nightmare. Religious thinkers, moralists and literary figures saw the rigidly ordered paradise as something that would all too easily be subverted to create a society devoid of individualism and creativity. They parodied the idea unmercifully in works now recognized as the foundation stones of science fiction, because in most cases technology, physical, biological and psychological, was seen as a tool that would be used by the rulers to enslave the masses. Yevgeny Zamyatin's *We* of 1924, George Orwell's *1984* and C. S. Lewis's *Cosmic Trilogy* proclaimed the potential for the creation of a brutal totalitarianism. Aldous Huxley's *Brave New World* was almost more terrifying because the rulers of its World State thought that the hedonistic society they had created fulfilled the materialists' hope of a society in which everyone can be happy. Yet these dystopias all missed the point that Wells had tried to make about technical innovation: they were static societies designed to ensure that no further change could take place. For all their power, they were really echoing the death-knell of the view that history must have a definite end-point.[4]

Wells insisted that the World State would not be a static paradise, because new technologies would allow adventurous spirits to explore the cosmos and extend the reach of human activity. Here the new vision of open-ended progress driven by innovation and backed up by a sense that historical forces allow the possibility of advance without predetermining its direction came into its own. Science-fiction authors such as Arthur C. Clarke would explore the potential for this vision to give humanity a sense of cosmic purpose. Others echoed Huxley's pessimism, imagining futures in which the wider opportunities opened up by technology were subverted by human greed. Fears that technologies often did more harm than good inspired a new generation of dystopian visions, but now they were as diverse as the futures envisaged by the technophiles who had taken control of the idea of progress.

---

[4] For more on this point see my *A History of the Future*, pp. 22–5 and chap. 11.

Much of the current hostility towards technical progress comes from the awareness that innovations often have unintended consequences. Except for the military no one promotes or accepts a new technology expecting it to have harmful effects. Inventors go ahead with their projects convinced that they will be of benefit to all. The pioneers of the internet assumed that putting everyone in touch with the system would produce a more informed and more interactive society, little realizing that there would be many opportunities for it to be misused by criminals and others with dangerous agendas. This was exactly the problem identified by Wells when he called for professors of foresight. His professors would take a deeper look at new technologies to imagine what would happen if they took off in unexpected directions. Anticipating future innovations is hard enough; trying to predict their unintended effects is even harder. One of the first lessons we should have learnt when we realized we were living in an unpredictable world was: expect the unexpected.

History matters here precisely because it teaches us that new technologies don't always produce the benefits they seem to promise. As Wells pointed out, the first motor cars were introduced to replace the horse: no one imagined a world of traffic chaos and accidents due to speeding. Atomic energy was supposed to supply us with electricity too cheap to be worth metering but turned out to be too dangerous to use. The negative consequences are usually pointed out by self-appointed critics, who are at first vilified by those who benefit from the industries concerned but are eventually hailed as the heroes.

In many cases it was the science-fiction writers who stood in for the professors of foresight, predicting both the opportunities and the dangers of new technologies. The potential consequences of a nuclear holocaust were highlighted in novel such as Nevil Shute's *On the Beach* of 1957. John Wyndham's *The Day of the Triffids* was an early attempt to point out the dangers of artificially produced living things. Novels such as William Gibson's *Neuromancer* of 1984 were among the first efforts to suggest that the world of cyberspace might not be as nice a place as its promoters suggested. These authors joined with a range of critics writing popular-science exposés of new technologies to alert us to the potentially harmful consequences of what we were being offered. The technophiles insist that we can always innovate our way out of danger, but even they don't agree on which innovations are most likely to be crucial. The transhuman future predicted by those who expect a whole new phase of existence turns out to be a range of possibilities, any one of which might actually be realized.[5]

---

[5] For more on these issues see for instance Greenfield, *Radical Technologies*; Susskind, *Future Politics*; Mann, *The Wizard and the Prophet*; Hayles, *How We Became Posthuman*; and Graham, *Representations of the Post/human*.

Commentators on our modern predicament mine recent history for examples of technologies that have given rise to both opportunities and problems. In this book I have argued that in order to understand how the current situation emerged we need to look further back in time to chart a transformation in both the idea of progress and our wider sense of how history unfolds. There are parallels between the changing views in evolutionary biology, in archaeology and anthropology, in studies of modern history and in how we think about the future. Darwinism serves as a useful framework for the project because it pioneered some of the new insights as the various areas interacted by trading ideas, analogies and metaphors.

The key transition was from the linear, goal-directed model of progress originating from the temporalized chain of being to the more Darwinian vision of history in which progress took on a different meaning. Instead of being regarded as a development along a scale leading to a goal defined by moral or spiritual values, progress came to be measured in utilitarian terms. Greater sophistication in exploiting the material environment was the key factor, both in the theory of natural selection and in any study of history which recognized technical innovation as the driving force of change. Evolution, whether biological or social, had to be seen as a branching tree of developments which had no predetermined goal, and in which new branches only occasionally advanced to a higher level of organization. Since innovations and their effects are essentially unpredictable the future became open-ended.

The final step in the emergence of the modern idea of progress came when it became clear that the link between scientific discovery and industrial invention had pushed the world into a new phase of development more rapid than anything that had gone before. This was the 'invention of invention', a step still being explored by historians seeking to explain the source of the West's (probably temporary) rise to global prominence. The impact of new technologies could be appreciated in every generation and soon in every decade, yet no one could be sure what the next step would be. The whole idea of progress is rejected by those who dislike technology and fear its impact. But even the technophiles cannot agree on which inventions are most crucial, nor be certain what their long-term impact will be. It should be no surprise that those of us trying to live with the consequences find it hard to be sure whether we should look forward to the future or fear it.

I add this last paragraph as I enter self-isolation amid the escalating pandemic in 2020. There are dire predictions of the devastations that this may wreak on the global economy, raising the question of what effect this episode may have on our thinking about progress. Pessimists will see it as confirmation of their claim that technology has led us into a fools' paradise. Technophiles will expect a rapid solution from modern medicine and may hope that the crisis will teach us to make better use of our resources. It seems certain that we shall become even more aware of the contingency of events. Experts such as Nassim Talib have urged the need to be prepared for the unexpected, and the possibility of a global pandemic was predicted as one such 'black swan' event.[6] The unpredictability of technological innovation is not the only source of contingency, and how we respond to nature's challenges adds another dimension to the uncertainties that those who still believe in progress must allow for. Which is, perhaps, another lesson we can learn from the Darwinian perspective.

---

[6] Talib, *The Black Swan*; see also Kay and King, *Radical Uncertainty*.

Bibliography

Ackerman, Robert. *J. G. Frazer: His Life and Work*. Cambridge: Cambridge University Press, 1987.

Acton, J. E. E. D, Lord. *Essays on Freedom and Power*. Introduction by Gertrude Himmelfarb. Gloucester, MA: Peter Smith, 1972.

Lectures on Modern History. London: Macmillan, 1921.

Selected Writings, vol. 2: *Essays on the Study and Writing of History*. Ed. J. Rufus Fears. Indianapolis, IN: Liberty Classics, 1985.

Adams, Henry. *The Degradation of the Democratic Dogma*. Introduction by Brooks Adams. New York: Macmillan, 1920.

Adams, Mark B., ed. *The Evolution of Theodosius Dobzhansky*. Princeton, NJ: Princeton University Press, 1994.

Adams, Mark B., 'Last Judgement: The Visionary Biology of J. B. S. Haldane'. *Journal of the History of Biology*, 33 (2000): 457–91.

Adams, Robert McC. *The Evolution of Urban Society: Early Mesopotamia and Prehistoric Mexico*. London: Weidenfeld and Nicolson, 1966.

Agassiz, Louis. *Essay on Classification*. London: Longman, Brown, 1869.

'On the Succession and Development of Organized Beings at the Surface of the Terrestial Globe'. *Edinburgh New Philosophical Journal*, 33 (1842): 388–99.

Agassiz, Louis, and A. A. Gould. *Outlines of Comparative Physiology*. Rev. ed. London: H. G. Bohn, 1851.

Ahlers, Rolf. 'The Dialectic in Hegel's Philosophy of History'. In Robert L. Perkins, ed., *History and System: Hegel's Philosophy of History*. Albany: State University of New York Press, 1984, pp. 149–61.

Akin, William E. *Technocracy and the American Dream: The Technocratic Movement, 1900–1941*. Berkeley: University of California Press, 1977.

Aldiss, Brian, and David Wingrove. *Billion Year Spree: The History of Science Fiction*. London: Paladin, 1986.

Aliotta, Antonio. *The Idealistic Reaction against Science*. Trans. Agnes McCaskill. London: Macmillan, 1914.

Al-Khalili, Jim. *What's Next? Even Scientists Can't Predict the Future – or Can They?* London: Profile Books, 2017.

Alter, Stephen G. *Darwinism and the Linguistic Image: Language, Race and Natural Theology in the Nineteenth Century*. Baltimore: Johns Hopkins University Press, 1999.

Allinson, Henry E. 'Teleology and History in Kant: The Critical Foundations of Kant's Philosophy of History'. In Amélie Oksenberg Rorty and James Schmidt, eds.,

Kant's Idea for a Universal History with a Cosmopolitan Aim: A Critical Guide. Cambridge: Cambridge University Press, 2009, pp. 22–45.

Ameriks, Karl, 'The Purposive Development of Human Capacities'. In Amélie Oksenberg Rorty and James Schmidt, eds., *Kant's Idea for a Universal History with a Cosmopolitan Aim: A Critical Guide.* Cambridge: Cambridge University Press, 2009, pp. 46–67.

Anderson, Lorin. *Charles Bonnet and the Order of the Known.* Dordrecht: D. Reidel, 1982.

Appel, Toby A. *The Cuvier–Geoffroy Debate: French Biology in the Decades before Darwin.* Oxford: Oxford University Press, 1987.

'Henri de Blainville and the Animal Series: A Nineteenth-Century Chain of Being'. *Journal of the History of Biology,* 13 (1980): 291–319.

Ardrey, Robert. *The Territorial Imperative: A Personal Inquiry into the Animal Origins of Property and Nations.* New York: Athenaeum, 1966.

Argyll, George Douglas Campbell, Duke of. *Primeval Man: An Examination of Some Recent Speculations.* London: Alexander Strahan, 1869.

Arkright, Frank. *The ABC of Technocracy.* London: Hamish Hamilton, 1933.

Arnold, Thomas. *Introductory Lectures on Modern History Delivered in Lent Term 1842.* Oxford: John Henry Parker, 1842.

Babbage, Charles. *The Ninth Bridgewater Treatise: A Fragment.* 2nd ed. Reprinted. London: Cass, 1968 [London, 1838].

Bacon, Francis. *The Advancement of Learning.* Ed. G. W. Kitchin. London: Dent, 1973 [1605].

The Novum Organon: Or a True Guide to the Interpretation of Nature. Trans. G. W. Kitchin. Oxford: Oxford University Press, 1855.

Badash, Lawrence. 'The Completeness of Nineteenth-Century Science'. *Isis,* 63 (1972): 48–58.

Baer, Karl Ernst von. *Über Entwickelungsgeschichte der Thiere: Beobachtung und Reflexion; Erster Theil.* Reprinted. Brussels: Culture et Civilization, 1967 [Königsberg, 1828].

Baillie, John. *The Belief in Progress.* London: Geoffrey Cumberlege and Oxford University Press, 1950.

Baker, Keith Michael. *Condorcet: From Natural Philosophy to Social Mathematics.* Chicago: University of Chicago Press, 1975.

Baldwin, James Mark. *Development and Evolution: Including Psychophysical Evolution, Evolution by Orthoplasy, and the Theory of Genetic Modes.* New York: Macmillan, 1902.

Bannister, Robert C. *Social Darwinism: Science and Myth in Anglo-American Social Thought.* Philadelphia: Temple University Press, 1979.

Barkan, Elazar. *The Retreat of Scientific Racism: Changing Concepts of Race in Britain and the United States between the World Wars.* Cambridge: Cambridge University Press, 1992.

Barnard, F. M. *Herder on Nationality, Humanity, and History.* Montreal: McGill-Queen's University Press, 2003.

Bashford, Alison, and Philippa Levine, eds. *The Oxford Handbook of the History of Eugenics.* Oxford: Oxford University Press, 2010.

Batchelor, John. *H. G. Wells.* Cambridge: Cambridge University Press, 1984.

Bauer, Ludwig. *War Again Tomorrow.* Trans. W. Harsfell Carter. London: Faber and Faber, 1932.

Beard, Charles A., ed. *Towards Civilization*. London: Longman Green, 1930.

Beard, Charles A., *Whither Mankind? A Panorama of Modern Civilization*. London: Longman Green, 1928.

Beard, Charles A., and Mary P. Beard. *The American Spirit: A Study of the Idea of Civilization in the United States*. New York: Macmillan, 1942.

The Rise of American Civilization. New ed. 2 vols. in 1. London; Jonathan Cape, n.d.

Beatty, John. 'Replaying Life's Tape'. *Journal of Philosophy*, 103 (2006): 336–62.

Becker, Carl. L. *The Heavenly City of the Eighteenth-Century Philosophers*. New Haven, CT: Yale University Press, 1966.

Bellamy, Edward. *Looking Backward: 2000–1887*. Reprinted with introduction by Cecelia Tichi. Harmondsworth: Penguin, 1982 [1888].

Benjamin, Marina. *Rocket Dreams: How the Space Age Shaped Our Vision of a World Beyond*. London: Vintage, 2004.

Bergson, Henri. *Creative Evolution*. Trans. Arthur Mitchell. New York: Henry Holt, 1911.

Bernal, J. D. *The World, the Flesh and the Devil: An Enquiry into the Future of the Three Enemies of the Rational Soul*. London: Kegan Paul, 1929.

Bertram, Christopher, and Andrew Chilty, eds. *Has History Ended? Fukuyama, Marx, Modernity*. Aldershot: Avebury, 1994.

Billingham, John, ed. *Life in the Universe*. Cambridge, MA: MIT Press, 1981.

Blinderman, Charles. *The Piltdown Inquest*. Buffalo, NY: Prometheus Books, 1986.

Blom, Philip. *The Vertigo Years: Change and Culture in the West, 1900–1914*. London: Weidenfeld and Nicolson, 2008.

Blumenbach, Johann Friedrich. *On the Natural Varieties of Mankind*. Trans. Thomas Bendyshe. Reprinted. New York, Bergman, 1969 [1865].

Boas, Franz. *Race, Language, and Culture*. New York: Macmillan, 1940.

Bobbitt, Philip. *The Shield of Achilles: War, Peace and the Course of History*. London: Allen Lane, 2002.

Bober, M. M. *Karl Marx's Interpretation of History*. Cambridge, MA: Harvard University Press, 1948.

Bois, Jean-Pierre. *L'abbé de Saint-Pierre: entre classicisme et lumières*. Ceyzérieu: Champ Vallon, 2017.

Bonnet, Charles. *Considérations sur les corps organizés*. 2 vols. Amsterdam: Marc-Michel Rey, 1762.

Contemplation de la nature. 2 vols. Amsterdam: Marc-Michel Rey, 1764.

La plaingénésie philosophique: ou idées sur l'état passé et sur l'état future des êtres vivans. 2 vols. Geneva: Clause Philibert et Barthelemi Chirol, 1769.

Oeuvres d'histoire naturelle et de philosophie. 19 vols. Neuchâtel: S. Fauche, 1779.

Boule, Marcellin. *Fossil Men: Elements of Human Palaeontology*. Edinburgh: Oliver and Boyd, 1923.

Bowler, Peter J. 'American Paleontology and the Reception of Darwinism'. *Studies in History and Philosophy of Biological and Biomedical Sciences*, 66 (2017): 3–7.

'Bonnet and Buffon: Theories of Generation and the Problem of Species'. *Journal of the History of Biology*, 6 (1973): 259–81.

Charles Darwin: The Man and His Influence. Oxford: Basil Blackwell. Reprinted. Cambridge: Cambridge University Press, 1990.

Darwin Deleted: Imagining a World without Darwin. Chicago: University of Chicago Press, 2013.

The Eclipse of Darwinism: Anti-Darwinian Evolution Theories in the Decades around 1900. Baltimore: Johns Hopkins University Press, 1983.

'Evolutionism in the Enlightenment'. *History of Science*, 12 (1974): 159–83.

Evolution: The History of an Idea. 25th anniversary ed. Berkeley: University of California Press, 2009.

Fossils and Progress: Paleontology and the Idea of Progressive Evolution in the Nineteenth Century. New York: Science History Publications, 1976.

'Herbert Spencer and Lamarckism'. In Mark Francis and Michael Taylor, eds., *Herbert Spencer: Legacies*. London: Routledge, 2015, pp. 203–21.

A History of the Future: Predicting the Future from H. G. Wells to Isaac Asimov. Cambridge: Cambridge University Press, 2017.

The Invention of Progress: The Victorians and the Past. Oxford: Basil Blackwell, 1989.

Life's Splendid Drama: Evolutionary Biology and the Reconstruction of Life's Ancestry, 1860–1940. Chicago: University of Chicago Press, 1996.

The Mendelian Revolution: The Emergence of Hereditarian Concepts in Modern Science and Society. London: Athlone, and Baltimore: Johns Hopkins University Press, 1989.

Monkey Trials and Gorilla Sermons: Evolution and Christianity from Darwin to Intelligent Design. Cambridge, MA: Harvard University Press, 2007.

The Non-Darwinian Revolution: Reinterpreting a Historical Myth. Baltimore: Johns Hopkins University Press, 1988.

'Philosophy, Instinct, Intuition: What Motivates a Scientist in Search of a Theory?' *Biology and Philosophy*, 15 (2000): 93–101.

'Preformation and Pre-existence in the Seventeenth Century: A Brief Analysis'. *Journal of the History of Biology*, 4 (1971): 221–44.

Reconciling Science and Religion: The Debates in Early Twentieth-Century Britain. Chicago: University of Chicago Press, 2001.

Theories of Human Evolution: A Century of Debate, 1844–1944. Baltimore: Johns Hopkins University Press; Oxford: Basil Blackwell, 1986.

Bowler, Peter J., and Iwan Rhys Morus, *Making Modern Science: A Historical Survey*. Chicago: University of Chicago Press, 2005. 2nd ed. 2020.

Boyd, Robert, and Peter J. Richerson. *Culture and the Evolutionary Process*. Chicago: University of Chicago Press, 1985.

Brace, C. Loring. 'The Fate of the "Classic" Neanderthals: A Study in Hominid Catastrophism'. *Current Anthropology*, 5 (1964): 3–43.

Brin, David. *Startide Rising*. New York: Bantam, 1983.

Broca, Paul. *On the Phenomena of Hybridity in the Genus Homo*. London: Longmans, Green for the Anthropological Society, 1864.

Bronn, Heinrich Georg. *Essai d'une réponse a la question de prix proposée en 1850 par l'Académie des sciences ... savoir: – étudier les lois de la distribution des corps organisés fossiles ...* Paris: Académie des Sciences, 1861.

Broom, Robert. *The Coming of Man: Was It Accident or Design?* London: H. F. and G. Witherby, 1933.

Finding the Missing Link. London: Watts, 1950.

The Mammal-Like Reptiles of South Africa. London: H. and F. Witherby, 1932.

Brown, David. *Walter Scott and the Historical Imagination*. London: Routledge, 1979.

Browne, Janet. *Charles Darwin: The Power of Place*. London: Jonathan Cape, 2002.

Charles Darwin: Voyaging. London: Jonathan Cape, 1995.

Buckle, Henry. *History of Civilization in England*. 3 vols. London: Longman Green, 1903 [1857].

Buckley, Arabella. *Winners in Life's Race*. London: Edward Stanford, 1882.

Buffon, Georges Louis Leclerc, comte de. *The Epochs of Nature*. Trans. Jan Zalasiewicz, Anne-Sophie Milon and Mateuz Zalasiewicz. Chicago: University of Chicago Press, 2018.

    Histoire naturelle, générale et particulière. 15 vols. Paris: Imprimerie Royale, 1749–67.

    Histoire naturelle: supplément. 7 vols. Paris: Imprimerie Royale, 1774–89.

Bunsen, Christian Charles Josias. *Christianity and Mankind: Their Beginnings and Prospects*. 7 vols. London: Longmans, Green, 1854.

Burkhardt, Richard W., Jr. *The Spirit of System: Lamarck and Evolutionary Biology*. Cambridge, MA, and London: Harvard University Press, 1977.

Burnet, Thomas. *Archaeologiae philosophicae; Or, the Ancient Doctrine Concerning the Origin of Things*. Trans. Mr Foxton. London: E. Curll, 1729.

    Archaeologiae philosophicae: sive doctrina antiqua de rerum originibus, libri duo. London: G. Kettelby, 1692.

    The Sacred Theory of the Earth. 2nd ed. Reprinted with introduction by Basil Willey. London: Centaur Press, 1965 [1691].

Burrow, J. W. *Evolution and Society: A Study in Victorian Social Theory*. Cambridge: Cambridge University Press, 1966.

Bury, J. B. *The Idea of Progress: An Inquiry into Its Growth and Origins*. Reprinted. New York: Dover, 1955 [1920].

    Selected Essays of J. B. Bury. Ed. Howard Temperley. Cambridge: Cambridge University Press, 1930.

Butler, Judith. *Subjects of Desire: Hegelian Reflections in Twentieth-Century France*. New York: Columbia University Press, 1999 [1987].

Butterfield, Herbert. *The Origins of Modern Science: 1300–1800*. London: G. Bell, 1949.

Bynum, William F. 'The Great Chain of Being after Forty Years: An Appraisal'. *History of Science*, 13 (1975): 1–28.

Campbell, Reginald John. *The New Theology*. London: Chapman and Hall, 1907.

Carr, E. H. *What Is History?* New introduction by Richard Evans. London: Palgrave Macmillan, 2001 [1961].

Carr, H. Wildon. *Henri Bergson: The Philosophy of Change*. London: T. C. and E. C. Jack, 1911.

Carter, Paul A. *The Creation of Tomorrow: Fifty Years of Magazine Science Fiction*. New York: Columbia University Press, 1977.

Cavalli-Sforza, L. L., and M. W. Feldman. *Cultural Transmission and Evolution: A Quantitative Approach*. Princeton, NJ: Princeton University Press, 1981.

Chamberlain, Alexander Francis. *The Child: A Study in Evolution*. London: Walter Scott, 1900.

Chamberlin, J. Edward, and Gilman, Sander L. eds. *Degeneration: The Dark Side of History*. New York: Columbia University Press, 1985.

Chambers, Robert. *Explanations: A Sequel to the Vestiges of the Natural History of Creation*. 2nd ed. London, 1846.

    Vestiges of the Natural History of Creation. Reprinted with introduction by Sir Gavin De Beer. Leicester: Leicester University Press, 1969 [London, 1844].

Vestiges of the Natural History of Creation and Other Evolutionary Writings. Ed. James Secord. Chicago: University of Chicago Press, 1994.

Cheng, John. *Astounding Wonder: Imagining Science and Science Fiction in Interwar America*. Philadelphia: University of Pennsylvania Press, 2012.

Childe, V. Gordon. *Man Makes Himself*. 4th ed. London: Fontana, 1965 [1936].

What Happened in History. New ed. Harmondsworth: Penguin, 1982 [1942].

Churchwell, Sarah, *Behold, America: A History of America First and the American Dream*. London: Bloomsbury, 2018.

Cixin Liu. *The Three-Body Problem*. London: Head of Zeus, 2015.

Claeys, Gregory. *Searching for Utopia: The History of an Idea*. London: Thames and Hudson, 2011.

Clark, Robert T. *Herder: His Life and Thought*. Berkeley: University of California Press, 1955.

Clarke, I. F., ed. *British Future Fiction*. 8 vols. London: Pickering and Chatto, 2001.

Cleland, John. 'Terminal Forms of Life'. *Journal of Anatomy and Physiology*, 18 (1884): 345–62.

Clive, John. *Thomas Babington Macaulay: The Shaping of a Historian*. London: Secker and Warburg, 1973.

Cohen, G. A. *Karl Marx's Theory of History: A Defence*. Oxford: Clarendon Press, 1978.

Coleman, William. *Georges Cuvier, Zoologist: A Study in the History of Evolution Theory*. Cambridge, MA: Harvard University Press, 1964.

Colp, Ralph, Jr. 'The Myth of the Darwin–Marx Letter'. *History of Political Economy*, 14 (1982): 461–82.

Comte, Auguste. *August Comte and Positivism: The Essential Writings*. Ed. Gertrud Lenzer. New York: Harper & Row, 1975.

Condorcet, Jean-Antoine-Nicolas Caritat, marquis de. *Condorcet: Political Writings*. Ed. Steven Lukes and Nadia Urbinati. Cambridge: Cambridge University Press, 2012.

Condorcet, Jean-Antoine-Nicolas Caritat, *Esquisse d'un tableau historique des progrès de l'esprit humain*. Paris: Vrin, 1970 [1795].

Conklin, Edwin Grant. *The Direction of Human Evolution*. London: Humphrey Milford/Oxford University Press, 1921.

Conway, Erik M., and Naomi Oreskes. *The Collapse of Western Civilization: A View from the Future*. New York: Columbia University Press, 2014.

Conway Morris, Simon. *The Crucible of Creation: The Burgess Shale and the Rise of Animals*. Oxford: Oxford University Press, 1998.

Life's Solution: Inevitable Humans in a Lonely Universe. Cambridge: Cambridge University Press, 2006.

Cooper, Barry. *The End of History: An Essay on Modern Hegelianism*. Toronto: University of Toronto Press, 1984.

Cooter, Roger. *The Cultural Meaning of Popular Science: Phrenology and the Organization of Consent in Nineteenth-Century Britain*. Cambridge: Cambridge University Press, 1985.

Cope, Edward Drinker. *The Origin of the Fittest: Essays on Evolution*. London: Macmillan, 1887.

The Primary Factors of Organic Evolution. Chicago, 1896.

The Theology of Evolution. Philadelphia: Arnold, 1887.

Corn, Joseph J., ed. *Imagining Tomorrow: History, Technology and the American Future*. Cambridge, MA: MIT Press, 1986.

Corn, Joseph J., and Brian Horrigan, eds. *Yesterday's Tomorrows: Past Visions of the American Future*. Baltimore: Johns Hopkins University Press, 1984.

Corren, Michael. *The Invisible Man: The Life and Liberties of H. G. Wells*. New York: Atheneum, 1993.

Corsi, Pietro. *The Age of Lamarck: Evolutionary Theories in France, 1790–1830*. Berkeley: University of California Press, 1988.

Cowley, Robert, ed. *What If? The World's Foremost Military Historians Imagine What Might Have Been*. London: Macmillan, 1999.

Cowley, Robert, *More What If? Eminent Historians Imagine What Might Have Been*. London: Macmillan, 2002.

Cramb, J. A. *The Origins and Destiny of Imperial Britain*. London: John Murray, 1915. *Reflections on the Origin and Destiny of Imperial Britain*. London: Macmillan, 1900.

Cravens, Hamilton. *The Triumph of Evolution: American Scientists and the Heredity Environment Controversy, 1900–1941*. Philadelphia: University of Pennsylvania Press, 1978.

Crowe, Michael J. *The Extraterrestrial Life Debate, 1750–1900: The Idea of the Plurality of Worlds from Kant to Lowell*. Cambridge: Cambridge University Press, 1986.

Culler, A. Dwight. *The Victorian Mirror of History*. New Haven, CT: Yale University Press, 1985.

Cuvier, Georges. *Recherches sur les ossemens fossiles de quadrupèdes* .... 4 vols. Reprinted. Brussels: Culture et Civilisation, 1969 [Paris, 1812].

Dale, Eric Michael. *Hegel, the End of History, and the Future*. Cambridge: Cambridge University Press, 2014.

Daniel, Glyn. *A Hundred and Fifty Years of Archaeology*. London: Duckworth, 1975.

Darwin, Charles Robert. *Charles Darwin's Notebooks (1836–1844)*. Ed. Paul H. Barrett et al. Cambridge: Cambridge University Press, 1987.

The Correspondence of Charles Darwin. Ed. Frederick Burkhardt and Sydney Smith. Cambridge: Cambridge University Press, 1984– (publication ongoing).

The Descent of Man and Selection in Relation to Sex. London: John Murray, 1871. 2 vols. 2nd ed. London: John Murray, 1874.

On the Origin of Species by Means of Natural Selection: Or the Preservation of Favoured Races in the Struggle for Life. London: John Murray, 1859. Reprinted with introduction by Ernst Mayr. Cambridge, MA: Harvard University Press, 1964.

Darwin, Erasmus. *The Temple of Nature*. London: J. Johnson, 1803.

Zoonomia: Or the Laws of Organic Life. 2 vols. Reprinted. New York: AMS Press, 1974 [London, 1794–96].

Daudin, Henri. *Études d'histoire des sciences naturelles*: Vol. 1: *De Linné à Jussieu: méthodes de la classification et l'idée de série en botanique et en zoologie*. Vol. 2: *Cuvier et Lamarck: les classes zoologiques et l'idée de série animale*. Paris: Alcan, 1926.

Davies, G. L. *The Earth in Decay: A History of British Geomorphology, 1578 to 1878*. New York: American Elsevier, 1969.

DeArney, Michael H., and James Good, eds. *The St. Louis Hegelians*. 3 vols. Bristol: Thoemmes, 2001.

De Beer, Gavin. *Embryology and Evolution*. Oxford: Oxford University Press, 1930.

De Mortillet, Gabriel. *Le préhistorique: L'antiquité de l'homme*. Paris: C. Reinwald, 1883.

'Promenades préhistoriques à l'Exposition Universelle'. *Matériaux pour l'histoire positive et philosophique de l'homme*, 3 (1867): 181–368.

Desmond, Adrian. *Archetypes and Ancestors: Palaeontology in Victorian London, 1850–1875*. London: Blond and Briggs, 1982. Reprinted. Chicago: University of Chicago Press, 1984.

Huxley: Evolution's High Priest. London: Michael Joseph, 1997.

Huxley: The Devil's Disciple. London: Michael Joseph, 1994.

The Politics of Evolution: Morphology, Medicine and Reform in Radical London. Chicago: University of Chicago Press, 1989.

Desmond, Adrian and Moore, James R. *Darwin*. London: Michael Joseph, 1991.

Darwin's Sacred Cause: Race, Slavery and the Quest for Human Origins. London: Allen Lane, 2009.

Dewey, John. *Characters and Events: Popular Essays in Social and Political Philosophy*. New York: Octagon Books, 1970 [1929].

The Influence of Darwin on Philosophy: And Other Essays in Contemporary Thought. Reprinted. Bloomington: Indiana University Press, 1965 [1910].

et al. *Creative Intelligence: Essays in the Pragmatic Attitude*. New York: Octagon Books, 1970 [1917].

Dick, Steven J. *The Biological Universe: The Twentieth-Century Extraterrestrial Life Debate and the Limits of Science*. Cambridge: Cambridge University Press, 1996.

Diderot, Denis. *Eléments de physiologie*. Ed. Jean Meyer. Paris: Marcel Didier, 1964.

Oeuvres philosophiques. Ed. Paul Verniere. Paris: Classiques Garnier, 1964.

Di Gregorio, Mario A. *From Here to Eternity: Ernst Haeckel and Scientific Faith*. Göttingen: Vandenhoek and Ruprecht, 2005.

T. H. Huxley's Place in Natural Science. New Haven, CT: Yale University Press, 1984.

Dobzhansky, Theodosius. *The Biology of Ultimate Concern*. London: Fontana, 1971 [1967].

Genetics and the Origin of Species. New York: Columbia University Press, 1937.

Mankind Evolving: The Evolution of the Human Species. Toronto: Bantam Books, 1970 [1962].

Dohrn, Anton. 'The Origin of Vertebrates and the Principle of Succession of Functions'. Trans. Michael T. Ghiselin. *History and Philosophy of the Life Sciences*, 16 (1993): 1–98.

Drexler, K. Eric. *Engines of Creation*. London: Fourth Estate, 1996 [1990].

Drouet, Joseph. *L'abbé de Saint-Pierre: l'homme et l'oeuvre*. Paris: Honoré Champion, 1912.

Drummond, Henry. *The Ascent of Man*. New York: James Pott, 1904 [1894].

Eamon, William. 'From the Secrets of Nature to Public Knowledge'. In David C. Lindberg and Robert S. Westman, eds., *Reappraisals of the Scientific Revolution*. Cambridge: Cambridge University Press, 1990, pp. 333–65.

Easton, Lloyd D. 'Hegelianism in Nineteenth-Century Ohio'. *Journal of the History of Ideas*, 23 (1962): 355–78.

The Ohio Hegelians: John B. Stallo, Peter Kaufmann, Moncure Conway, and August Willich, with Key Writings. Athens: Ohio University Press, 1966.

Eiseley, Loren. *Darwin's Century: Evolution and the Men Who Discovered It*. New York: Doubleday Anchor, 1958.

The Immense Journey. London: Gollancz, 1958.

Ekirch, Arthur A., Jr. *The Idea of Progress in America, 1815–1860*. New York: Columbia University Press, 1944.

Ellul, Jacques. *The Technological Culture*. Introduction by Robert K. Merton. Reprinted. New York: Vintage, n.d. [1964].

Endersby, Jim. 'A Visit to Biotopia: Genre, Genetics and Gardening in the Early Twentieth Century'. *British Journal for the History of Science*, 51 (2018): 423–56.

Engels, Friedrich. *Herr Eugen Dühring's Revolution in Science [Anti-Dühring]*. London: Martin Laurence, n.d.

The Origins of the Family, Private Property and the State. London: Penguin Classics, 2010.

Ettinger, Robert C. W. *The Prospect of Immortality*. London: Sidgwick and Jackson, 1965 [1964].

Fara, Patricia. *Erasmus Darwin: Sex, Science and Serendipity*. Oxford: Oxford University Press, 2012.

Farrar, Frederic W. *Families of Speech: Four Lectures Delivered before the Royal Institution of Great Britain in March 1869*. New ed. London: Longmans, Green, 1873.

Farrenkopf, John. *Prophet of Decline: Spengler on World History and Politics*. Baton Rouge: Louisiana State University Press, 2001.

Fay, Margaret A. 'Did Marx Offer to Dedicate *Capital* to Darwin?' *Journal of the History of Ideas*, 39 (1978): 133–46.

Feinberg, Gerald. *The Prometheus Project: Mankind's Search for Long-Range Goals*. New York: Doubleday, 1969.

Ferguson, Adam. *An Essay on the History of Civil Society, 1767*. Ed. Duncan Forbes. Edinburgh: Edinburgh University Press, 1966.

Ferguson, Niall, ed. *Virtual History: Alternatives and Counterfactuals*. London: Picador, 1997.

Ferraro, Joseph. *Freedom and Determinism in History according to Marx and Engels*. New York: Monthly Review Press, 1992.

Feuer, Lewis S. 'Is the Darwin–Marx Correspondence Authentic?' *Annals of Science*, 32 (1975): 1–12.

Fichman, Martin, *An Elusive Victorian: The Evolution of Alfred Russel Wallace*. Chicago: University of Chicago Press, 2004.

Fichte, Johann Gottlieb. *The Destination of Man*. Trans. Mrs Percy Sinnett. London: Chapman Brothers, 1846.

The Popular Works of Johann Gottlieb Fichte. Trans. William Smith. 2 vols. London: Chapman and Hall, 1849.

Figes, Orlando. *A People's Tragedy: The Russian Revolution, 1891–1924*. London: Pimlico, 1997.

Filene, Edward A. *The Way Out: A Business-Man Looks at the World*. London: Routledge, 1925.

Fisch, Max Harold. 'Introduction'. In Gianbattista Vico, *The New Science of Gianbattista Vico*. Trans. Thomas Goddard Bergin and Max Harold Fisch. Ithaca, NY: Cornell University Press, 1970, pp. xxi–liii.

Fischer, Klaus P. *History and Prophecy: Oswald Spengler and the Decline of the West.* Durham, NC: Moore International, 1977.

Fisher, R. A. *The Genetical Theory of Natural Selection.* Oxford: Clarendon Press, 1930.

Fiske, John. *Darwinism and Other Essays.* London: Macmillan, 1879.

  The Destiny of Man Viewed in the Light of His Origin. Boston: Houghton Mifflin, 1884.

  Outlines of Cosmic Philosophy: Based on the Doctrine of Evolution. 2 vols. London: Macmillan, 1874.

Fontenelle, Bernard le Bovier de. *Entretiens sur la pluralité des mondes / Digression sur les anciens et les modernes.* Ed. Robert Shackleton. Oxford: Clarendon Press, 1955.

Forbes, Duncan. *The Liberal Anglican Idea of History.* Cambridge: Cambridge University Press, 1952.

Foster, William Trufant, and Waddell Catchings. *The Road to Plenty.* London: Pitman, 1929.

Foucault, Michel. *Les mots et les choses: une archéologie des sciences humaines.* Paris: Gallimard, 1966.

  The Order of Things: The Archaeology of the Human Sciences. New York: Pantheon Books, 1970.

Francis, Mark. *Herbert Spencer and the Invention of Modern Life.* Stocksfield: Acumen, 2007.

Francis, Mark, and Michael Taylor, eds. *Herbert Spencer: Legacies.* London: Routledge, 2015.

Frangsmyr, Tore, ed. *Linnaeus: The Man and His Work.* Berkeley: University of California Press, 1984.

Frankel, Charles. *The Faith of Reason: The Idea of Progress in the French Enlightenment.* New York: Columbia University Press, 1948.

Frazer, James George. *The Golden Bough: A Study of Magic and Religion.* Abridged ed. London: Macmillan, 1924 [1890].

  Psyche's Task: A Discourse Concerning the Influence of Superstition on the Growth of Institutions. London: Macmillan, 1913.

Freud, Sigmund. *Civilization and Its Discontents.* Reprinted in *The Standard Edition of the Complete Psychological Works of Sigmund Freud.* London: Hogarth Press and the Institute of Psychoanalysis, 1953–74, vol. 21, pp. 59–145 [1930].

  'Five Lectures on Psycho-analysis'. Reprinted in *The Standard Edition of the Complete Psychological Works of Sigmund Freud.* London: Hogarth Press and the Institute of Psychoanalysis, 1953–74, vol. 11, pp. 3–55 [1910].

Fukuyama, Francis. *The End of History and the Last Man.* London: Hamish Hamilton, 1992.

  Our Posthuman Future: Consequences of the Biotechnology Revolution. London: Profile Books, 2002.

Gabor, Dennis. *Inventing the Future.* London: Secker and Warburg, 1963.

Gandy, D. Ross. *Marx and History: From Primitive Society to the Communist Future.* Austin: University of Texas Press, 1979.

Gascoigne, Robert M. 'Julian Huxley and Biological Progress'. *Journal of the History of Biology*, 24 (1991): 433–55.

Gates, R. Ruggles. *Human Ancestry from a Genetical Point of View*. Cambridge, MA: Harvard University Press, 1948.

Gay, Peter. *The Enlightenment: An Interpretation; The Rise of Modern Paganism*. New York: Vintage, 1966.

    The Party of Humanity: Essays in the French Enlightenment. New York: Norton, 1971.

Gibbon, Edward. *The History of the Decline and Fall of the Roman Empire*. New ed. 12 vols. London: W. Allason, 1820.

Gilbert, Felix. *History: Politics or Culture? Reflections on Ranke and Burkhardt*. Princeton, NJ: Princeton University Press, 1990.

Gillispie, Charles Coulston. 'Lamarck and Darwin in the History of Science'. In Bentley Glass, Owsei Temkin and William Strauss, Jr., eds., *Forerunners of Darwin, 1745–1859*. Baltimore: Johns Hopkins University Press, 1959, pp. 265–91.

Ginsberg, Morris. *The Idea of Progress: A Revaluation*. Reprinted. Westport, CT: Greenwood Press, 1972 [1953].

Glanvill, Joseph. *The Vanity of Dogmatizing: The Three 'Versions'*. Introduction by Stephen Metcalf. Hove, Sussex: Harvester Press, 1970 [1661].

Glass, Bentley. 'The Germination of the Biological Species Concept'. In Bentley Glass, Owsei Temkin and William Strauss, Jr., eds., *Forerunners of Darwin, 1745–1859*. Baltimore: Johns Hopkins University Press, 1959, pp. 30–48.

    'Heredity and Variation in the Eighteenth-Century Concept of the Species'. In Bentley Glass, Owsei Temkin and William Strauss, Jr., eds., *Forerunners of Darwin, 1745–1859*. Baltimore: Johns Hopkins University Press, 1959, pp. 144–72.

    'Maupertuis, Pioneer of Genetics'. In Bentley Glass, Owsei Temkin and William Strauss, Jr., eds. *Forerunners of Darwin, 1745–1859*. Baltimore: Johns Hopkins University Press, 1959, pp. 51–83.

Glass, Bentley, Owsei Temkin and William Strauss, Jr., eds. *Forerunners of Darwin, 1745–1859*. Baltimore: Johns Hopkins University Press, 1959.

Gliboff, Sander. *H. G. Bronn, Ernst Haeckel, and the Origins of German Darwinism: A Study in Translation and Transformation*. Cambridge, MA: MIT Press, 2008.

Gobinau, Arthur de. *The Inequality of Human Races*. Trans. Adrian Collins. New York: G. P. Putnam's Sons, 1915.

Godwin, William. *Enquiry Concerning Political Justice and Its Influence on Morals and Happiness*. Ed. Isaac Kramnick. London: Penguin, 2015 [1793].

Goetzmann, William H., ed. *The American Hegelians: An Intellectual Episode in the History of Western America*. New York: Knopf, 1973.

Goldsmith, Oliver. *History of the Natural World*. Foreword by Gerald Durrell. London: Studio Editions, 1990.

Good, James A., ed. *The Ohio Hegelians*. 3 vols. Bristol: Thoemmes, 2005.

Gould, Stephen Jay. *The Mismeasure of Man*. New York: Norton, 1981.

    Ontogeny and Phylogeny. Cambridge, MA: Harvard University Press, 1977.

    Wonderful Life: The Burgess Shale and the Nature of History. London: Hutchinson Radius, 1989.

Gould, Stephen J., and Eldredge, Niles. 'Punctuated Equilibria: The Tempo and Mode of Evolution Reconsidered'. *Paleobiology*, 3 (1977): 115–51.

Graham, Elaine L. *Representations of the Post/human: Monsters, Aliens and Others in Popular Culture*. Manchester: Manchester University Press, 2002.

Grayson, Donald K. *The Establishment of Human Antiquity*. New York: Academic Press, 1983.

Greene, John C. 'Biology and Social Theory in the Nineteenth Century: Auguste Comte and Herbert Spencer'. In M. Clagett, ed., *Critical Problems in the History of Science*. Madison: University of Wisconsin Press, pp. 419–66.

'The Interaction of Science and World View in Sir Julian Huxley's Evolutionary Biology'. *Journal of the History of Biology*, 23 (1990): 39–55.

Greenfield, Adam. *Radical Technologies: The Design of Everyday Life*. London: Verso, 2018.

Gregory, William King. *Our Face from Fish to Man: A Portrait Gallery of Our Ancient Ancestors and Kinfolk, Together with a Concise History of Our Best Features*. New York: G. P. Putnam's Sons, 1929.

Grosz, Elizabeth. *The Nick of Time: Politics, Evolution, and the Untimely*. Durham, NC: Duke University Press, 2004.

Haeckel, Ernst. *The Evolution of Man: A Popular Exposition of the Principal Points of Human Ontogeny and Phylogeny*. New York: Appleton, 1879.

The History of Creation: Or the Development of the Earth and Its Inhabitants by the Action of Natural Causes. A Popular Exposition of the Doctrine of Evolution in General and of That of Darwin, Goethe and Lamarck in Particular. 2 vols. New York: Appleton, 1876.

The Last Link: Our Present Knowledge of the Descent of Man. London: A. and C. Black, 1898.

Haldane, J. B. S. *The Causes of Evolution*. London: Longmans, 1932.

Daedalus: Or Science and the Future. London: Kegan Paul, 1924.

Heredity and Politics. London: Allen and Unwin, 1938.

The Inequality of Man and Other Essays. London: Chatto and Windus, 1932.

Haldane, J. B. S., and Julian Huxley. *Animal Biology*. Oxford: Clarendon Press, 1927.

Haldar, Hiralal. *Neo-Hegelianism*. London: Heath Cranton, 1927.

Halévie, Elie. *The Growth of Philosophic Radicalism*. Trans. Mary Morris. Boston: Beacon Press, 1955.

Hall, G. Stanley. *Adolescence: Its Psychology and Its Relation to Physiology, Anthropology, Sociology, Sex, Crime, Religion and Education*. 2 vols. New York: Appleton, 1904.

Haller, John S., Jr. *Outcasts from Evolution: Scientific Attitudes of Racial Inferiority, 1859-1900*. Urbana: University of Illinois Press, 1975.

Hammond, Michael. 'Anthropology as a Weapon of Social Combat in Late Nineteenth-Century France'. *Journal of the History of the Behavioral Sciences*, 16 (1980): 118–32.

'The Expulsion of the Neanderthals from Human Ancestry: Marcellin Boule and the Social Context of Scientific Research'. *Social Studies of Science*, 12 (1982): 1–36.

Harari, Yuval Noah. *Homo Deus: A Brief History of Tomorrow*. London: Vintage, 2017.

Sapiens: A Brief History of Humankind. London: Vintage, 2011.

Haraway, Donna. *Primate Visions: Gender, Race, and Nature in the World of Modern Science*. New York: Routledge, 1989.

Hard, Michael, and Andrew Jameson, eds. *The Intellectual Appropriation of Technology: Discourses on Modernity, 1900–1939*. Cambridge, MA: MIT Press, 1998.

Harris, Marvin. *The Rise of Anthropological Theory: A History of Theories of Culture*. New York: Thomas Y. Crowell, 1968.

Harrison, Harry. *West of Eden*. New York: Bantam Books, 1984.

Harrison, Peter. *The Bible, Protestantism and the Rise of Natural Science*. Cambridge: Cambridge University Press, 1998.

Hartley, David. *Observations on Man: His Frame, His Duty and His Expectations*. 2nd ed. 2 vols. Reprinted. Washington, DC: Woodstock Books, 1998 [London, 1791].

Hartmann, Eduard von. *Philosophy of the Unconscious: Speculative Results according to the Inductive Method of Physical Science*. London: Kegan Paul, 1931.

Harvey, Joy. 'Evolutionism Transformed: Positivists and Materialists in the Société d'Anthropologie de Paris from Second Empire to Third Republic'. In D. Oldroyd and I. Langham, eds., *The Wider Domain of Evolutionary Thought*. Dordrecht: Reidel, 1983, pp. 261–310.

Hatch, Elvin. *Theories of Man and Culture*. New York: Columbia University Press, 1973.

Hatch, Ronald B. 'Joseph Priestley: An Addition to Hartley's *Observations*'. *Journal of the History of Ideas*, 36 (1975): 548–50.

Hattersley, C. Marshall. *This Age of Plenty – Its Problems and Their Solution*. London: Pitman, 1929.

Hawthorn, Geoffrey. *Plausible Worlds: Possibilities and Understanding in the Social Sciences*. Cambridge: Cambridge University Press, 1995 [1991].

Hayles, N. Katherine. *How We Became Posthuman: Virtual Bodies in Cybernetics, Literature and Informatics*. Chicago: University of Chicago Press, 1999.

Haynes, Rosslyn D. *H. G. Wells: Discoverer of the Future*. London: Macmillan, 1980.

Hegel, G. W. F. *Lectures on the Philosophy of History*. Trans. Joseph Sibrae. London: George Bell, 1881.

Lectures on the Philosophy of World History: Introduction. Reason in History. Trans. H. B. Nisbet. Cambridge: Cambridge University Press, 1975.

Lectures on the Philosophy of World History: Manuscripts of the Introduction and the Lectures of 1822–3. Ed. Robert F. Brown and Peter C. Hodgson. Oxford: Clarendon Press, 1996.

Herbert, Frank. *Dune*. New York: Ace Books, 1965.

Herder, Johann Gottfried von. *Herder: Philosophical Writings*. Ed. Michael N. Foster. Cambridge: Cambridge University Press, 2002.

J. G. Herder on Social and Political Culture. Trans. F. M. Barnard. Cambridge: Cambridge University Press, 1969.

Reflections on the Philosophy of the History of Mankind. Trans. T. O. Churchill, ed. Frank E. Manuel. Chicago: University of Chicago Press, 1968.

Herring, Emily. 'Great Is Darwin and Bergson Is His Prophet: Julian Huxley's Other Evolutionary Synthesis'. *Annals of Science*, 75 (2018): 40–54.

Hillegas. Mark R. 'Victorian "Extraterrestrials"'. In Jerome H. Buckley, ed., *The Worlds of Victorian Fiction*. Cambridge, MA: Harvard University Press, 1975, pp. 391–414.

Hodge, M. J. S. 'Lamarck's Science of Living Bodies'. *British Journal for the History of Science*, 5 (1971): 323–52.

'The Universal Gestation of Nature: Chambers' *Vestiges* and *Explanations*'. *Journal of the History of Biology*, 6 (1972): 127–52.

Hodgson, Peter C. *Shapes of Freedom: Hegel's Philosophy of World History in Theological Perspective*. Oxford: Oxford University Press, 2012.

Hofstadter, Richard. *Social Darwinism in American Thought*. Rev. ed. New York: George Braziller, 1959.

Hooke, Robert. *Micrographia: Or Some Physiological Descriptions of Minute Bodies Made by Magnifying Glasses, with Observations and Inquiries Thereupon*. Facsimile reprint. Lincolnwood, IL: Science Heritage, 1987 [1665].

Hooton, Earnest Albert. *Up from the Ape*. London: Allen and Unwin, 1931.

Horsman, Matthew, and Andrew Marshall. *After the Nation State: Citizens, Tribalism and the New World Order*. London: Harper Collins, 1994.

Huntington, John. *The Logic of Fantasy: H. G. Wells and Science Fiction*. New York: Columbia University Press, 1982.

Huxley, Aldous. *Brave New World*. Reprinted. Harmondsworth: Penguin, 1955 [1932].

Huxley, Julian S. *Essays of a Biologist*. Harmondsworth: Penguin, 1939 [1923].

Evolution: The Modern Synthesis. London: Allen and Unwin, 1942.

Memories. London: Allen and Unwin, 1970.

The Humanist Frame. London: Allen and Unwin, 1961.

The Uniqueness of Man. London: Chatto and Windus, 1941.

Huxley, Thomas Henry. *Collected Essays*, vol. 8: *Discourses: Biological and Geological*. London: Macmillan, 1894.

Man's Place in Nature. London: Williams and Norgate, 1863.

Iggers, Georg G. *The Cult of Authority: The Political Philosophy of the Saint-Simonians: A Chapter in the Intellectual History of Authoritarianism*. The Hague: Martinus Nijhoff, 1954.

Inge, William Ralph. 'The Future of the Human Race'. *Proceedings of the Royal Institution of Great Britain*, 26 (1929–31): 494–515.

The Idea of Progress. Oxford: Clarendon Press, 1920.

James, David. *Fichte's Republic: Idealism, History, and Nationalism*. Cambridge: Cambridge University Press, 2015.

James, Edward, and Farah Mendlesohn, eds. *The Cambridge Companion to Science Fiction*. Cambridge: Cambridge University Press, 2003.

James, William. *The Will to Believe and Other Essays in Popular Philosophy*. New ed. New York: Longmans Green, 1912 [1896].

Jenkins, Bill. 'The Platypus in Edinburgh: Robert Jameson, Robert Knox and the Place of the *Ornithorhynchus* in Nature, 1821–24'. *Annals of Science*, 73 (2016): 425–41.

Jerónimo, Helena M., José Louís Garcia and Carl Micham, eds. *Jacques Ellul and the Technological Society in the 21st Century*. Dordrecht: Springer, 2013.

Johnston, Sean F. *Techno-fixers: Origins and Implications of Technological Faith*. London: McGill-Queen's University Press, 2020.

Jones, Frederic Wood. 'The Origin of Man'. In Arthur Dendy, ed., *Animal Life and Human Progress*. London: Constable, 1919, pp. 99–131.

Jones, Greta. *Social Darwinism and English Thought: The Interaction between Biological and Social Theory*. London: Harvester Press, 1980.

Social Hygiene in Twentieth-Century Britain. London: Croom Helm, 1986.

Jones, Richard Foster. *Ancients and Moderns: A Study of the Rise of the Scientific Movement in Seventeenth-Century England.* 2nd ed. Berkeley and Los Angeles: University of California Press, 1961.

Jordanova, L. *Lamarck.* Oxford: Oxford University Press, 1984.

Jussieu, Anton Laurent de. *Genera plantarum.* Reprinted with introduction by Frans A. Stafleu. Weinheim: J. Cramer, 1964 [1789].

Kaku, Michio. *The Future of Humanity: Terraforming Mars, Interstellar Travel, Immortality and Our Destiny beyond Earth.* London: Allen Lane, 2018.

Kant, Immanuel. *Universal Natural History and Theory of the Heavens.* Introduction by Milton K. Munitz. Ann Arbor: University of Michigan Press, 1969 [1755].

Kay, John, and Mervin King, *Radical Uncertainty: Decision-Making beyond the Numbers.* London: Hachette, 2020.

Keith, Sir Arthur. *The Antiquity of Man.* London: Williams and Norgate, 1915.

The Construction of Man's Family Tree. London: Watts, 1934.

A New Theory of Human Evolution. New York: Philosophical Library, 1949.

Kevles, Daniel. *In the Name of Eugenics: Genetics and the Uses of Human Heredity.* New York: Knopf, 1985.

Kilgrove, De Witt Douglas. *Astrofuturism: Science, Race, and Visions of Utopia in Space.* Philadelphia: University of Pennsylvania Press, 2003.

Kingsley, Charles. *Hypatia: Or New Foes with an Old Face.* London: Macmillan, 1895 [1853].

The Roman and the Teuton: A Series of Lectures Delivered before the University of Cambridge. Cambridge and London: Macmillan, 1864.

Westward Ho! London: Macmillan, 1917.

Knox, Robert. *The Races of Man: A Philosophical Enquiry into the Influence of Race on the Destiny of Nations.* 2nd ed. London: Henry Renshaw, 1862 [1850].

Kojève, Alexandre. *Introduction to the Reading of Hegel: Lectures on the Phenomenology of Spirit Assembled by Raymond Queneau.* Ed. Alan Bloom. New York: Basic Books, 1969.

Kroeber, A. L. 'The Superorganic'. *American Anthropologist,* n.s.19 (1917): 163–213.

Kuper, Adam. 'The Development of Lewis Henry Morgan's Evolutionism'. *Journal for the History of the Behavioral Sciences,* 21 (1985): 3–21.

The Invention of Primitive Society: Transformations of an Illusion. London: Routledge, 1988.

Kurzweil, Ray. *The Singularity Is Near: When Humans Transcend Biology.* London: Duckworth, 2005.

Lamarck, Jean-Baptiste Pierre Antoine de Monet, chevalier de. *Histoire naturelle des animaux sans vertèbres.* 6 vols. Reprinted. Brussels: Culture et Civilisation, 1969 [Paris, 1815–22].

Zoological Philosophy. Trans. Hugh Elliot. Reprinted. New York, Hafner, 1963 [London, 1914].

La Mettrie, Julien Offray de. *L'homme machine.* Ed. Aram Vartanian. Princeton, NJ: Princeton University Press, 1960.

Oeuvres philosophiques. 2 vols. Berlin, 1774.

Langdon-Davies, John. *Man and His Universe.* London: Watts, 1937.

A Short History of the Future. London: Routledge, 1936.

Langham, Ian. *The Building of British Social Anthropology: W. H. R. Rivers and His Cambridge Disciples in the Development of Kinship Studies*. Dordrecht: Reidel, 1981.

Lankester, E. Ray. *Degeneration: A Chapter in Darwinism*. London: Macmillan, 1880.
   Great and Small Things. London: Methuen, 1923.
   The Kingdom of Man. London: Constable, 1907.
   'Notes on the Embryology and Classification of the Animal Kingdom: Comprising a Revision of Speculations Relative to the Origin and Significance of the Germ Layers'. *Quarterly Journal of Microscopical Science*, 17 (1877): 399–454.

Larson, James L. *Reason and Experience: The Representation of Natural Order in the Work of Carl von Linné*. Berkeley: University of California Press, 1971.

Laurent, Goulven. ed. *Jean-Baptiste Lamarck: 1744–1828*. Paris: Comité des Travaux Historiques et Scientifiques, 1997.

Layton, Robert. *Introduction to Theory in Anthropology*. Cambridge: Cambridge University Press, 1997.

Leakey, Louis S. B. *Adam's Ancestors: An Up-to-Date Outline of What Is Known about the Origin of Man*. 3rd ed. London: Methuen, 1934.

LeConte, Joseph. *Evolution: Its Nature, Its Evidences and Its Relation to Religious Thought*. 2nd ed. New York: Appleton, 1899.

Le Gros Clark, Wilfrid, *Early Forerunners of Man: A Morphological Study of the Evolutionary Origin of the Primates*. London: Ballière, Tindall and Cox, 1934.

Lehmann, William C. *John Millar of Glasgow, 1735–1801: His Life and Thought and His Contributions to Social Analysis*. Cambridge: Cambridge University Press, 1960.

Leigh, John. *Voltaire: A Sense of History*. Oxford: Voltaire Foundation, 2004.

Lenoir, Timothy. 'Generational Factors in the Origin *of Romantische Naturphilosophie*'. *Journal of the History of Biology*, 11 (1978): 57–100.
   The Strategy of Life: Teleology and Mechanics in Nineteenth-Century German Biology. Dordrecht: D. Reidel, 1982.

Leslie, Margaret. 'Mysticism Misunderstood: David Hartley and the Idea of Progress'. *Journal of the History of Ideas*, 33 (1972): 625–32.

Levit, Gregory S., Uwe Hossfeldt and Lornart Olsen. 'From the "Modern Synthesis" to Cybernetics: Ivan Ivanovich Smallhausen (1884–1963) and His Research Program for a Synthesis of Evolutionary and Developmental Biology'. *Journal of Experimental Zoology*, 306B (2006): 89–106.
   'The Integration of Darwinism and Evolutionary Morphology: Alexej Nikolajevich Sewetrtzoff (1866–1976) and the Developmental Basis of Evolutionary Change'. *Journal of Experimental Zoology*, 302B (2004): 343–54.

Lewis, C. S. *The Cosmic Trilogy: Out of the Silent Planet; Perelandra; That Hideous Strength*. London: Bodley Head, 1969.

Lightman, Bernard. *The Origins of Agnosticism: Victorian Unbelief and the Limits of Knowledge*. Baltimore: Johns Hopkins University Press, 1987.
   Victorian Popularizers of Science: Designing Nature for New Audiences. Chicago: University of Chicago Press, 2007.

Lightman, Bernard, ed. *Victorian Science in Context*. Chicago: University of Chicago Press, 1997.

Linnaeus, Carolus (Karl von Linné). *Philosophia botanica*. Reprinted. Codicote, Hants.: Wheldon and Wesley, 1966 [1751].

Systema naturae. Reprinted. Ed. M. S. J. Engel-Ledeboer and H. Engel. Nieuwkoop:
    B. De Graff, 1964 [1735].
Lodge, Sir Oliver. *Evolution and Creation*. London: Hodder and Stoughton, 1926.
The Making of Man: A Study in Evolution. London: People's Library, 1929 [1924].
London, Jack. *The Iron Heel*. Preface by Anatole France. London: Mills and Boon,
    1932 [1907].
Losos, Jonathan. *Improbable Destinies: How Predicable Is Evolution?* London: Allen
    Lane, 2017.
Lovejoy, Arthur O. 'The Argument for Organic Evolution before the Origin of Species,
    1830–1858'. In Bentley Glass, Owsei Temkin and William Strauss, Jr., eds.,
    Forerunners of Darwin, 1745–1859. Baltimore: Johns Hopkins University Press,
    1959, pp. 356–414.
'Buffon and the Problem of Species'. In Bentley Glass, Owsei Temkin and William
    Strauss, Jr., eds., *Forerunners of Darwin, 1745–1859*. Baltimore: Johns Hopkins
    University Press, 1959, pp. 84–113.
The Great Chain of Being: A Study in the History of an Idea. Reprinted. New York:
    Harper, 1960 [1936].
'Herder: Progressionism without Transformism'. In Bentley Glass, Owsei Temkin
    and William Strauss, Jr., eds., *Forerunners of Darwin, 1745–1859*. Baltimore:
    Johns Hopkins University Press, 1959, pp. 207–21.
Lubbock, John. *The Origin of Civilization and the Primitive Condition of Man: Mental
    and Social Conditions of Savages*. London: Longmans, Green, 1870.
*Prehistoric Times: As Illustrated by Ancient Remains, and the Manners and Customs
    of Modern Savages*. London: Williams and Norgate, 1865.
Lull, Richard Swann. *Organic Evolution: A Textbook*. New York: Macmillan, 1917.
Lurie, Edward. *Louis Agassiz: A Life in Science*. Chicago: University of Chicago Press,
    1960.
Lyell, Charles. *Geological Evidences of the Antiquity of Man: With Remarks on
    Theories of the Origin of Species by Variation*. London: John Murray, 1863.
*Principles of Geology: Being an Attempt to Explain the Former Changes of the
    Earth's Surface by Reference to Causes Now in Operation*. 3 vols. Reprinted with
    introduction by M. J. S. Rudwick. Chicago: University of Chicago Press, 1990–1
    [London, 1830–3].
Lynskey, Winifred. 'Goldsmith and the Chain of Being'. *Journal of the History of
    Ideas*, 6 (1945): 363–74.
Macaulay, Thomas Babington. *Critical and Historical Essays*. New ed. London:
    Longmans, Green, 1908.
McCarney, Joseph. *Hegel on History*. London: Routledge, 2000.
McCray, W. Patrick. *The Visioneers: How a Group of Elite Scientists Pursued Space
    Colonies, Nanotechnologies and a Limitless Future*. Princeton, NJ: Princeton
    University Press, 2013.
McGee, John Edwin. *A Crusade for Humanity: The History of Organized Positivism in
    England*. London: Watts, 1931.
Mackintosh, R. *Hegel and Hegelianism*. Edinburgh: T. and T. Clark, 1903.
Malthus, Thomas Robert. *Population: The First Essay*. Ann Arbor: University of
    Michigan Press, 1959 [1797].
Mann, Charles C. *The Wizard and the Prophet: Two Groundbreaking Scientists and
    Their Conflicting Visions of the Future of Our Planet*. London: Picador, 2018.

Manuel, Frank E. *The New World of Henri de Saint-Simon*. Cambridge, MA: Harvard University Press, 1956.
  The Prophets of Paris. Cambridge, MA: Harvard University Press, 1962.
Marcell, David W. *Progress and Pragmatism: James, Dewey, Beard, and the American Idea of Progress*. Westport, CT: Greenwood Press, 1974.
Marsden, Ben, and Crosbie Smith. *Engineering Empires: A Cultural History of Technology in Nineteenth-Century Britain*. London: Palgrave Macmillan, 2007.
Marsh, Othniel Charles. *Introduction and Succession of Vertebrate Life in America*. New York: Appleton, 1877.
Marx, George, ed. *Bioastronomy – The Next Step*. Dordrecht: Kluwer, 1988.
Marx, Karl. *Capital: A Critique of Political Economy*. 3 vols. New York: International Publishers, 1967.
Marx, Karl, and Friedrich Engels. *The Communist Manifesto*. Trans. Samuel Moore, introduction by A. J. P. Taylor. Harmondsworth: Penguin, 1967.
  Marx and Engels: Basic Writings on Politics and Philosophy. Trans. and ed. Lewis S. Feuer. New York: Doubleday, 1959.
Mason, Frances, ed. *Creation by Evolution*. New York: Macmillan, 1928.
Mason, Frances, *The Great Design: Order and Progress in Nature*. Introduction by Arthur Thomson. London: Duckworth, 1934.
Matthew. William Diller. 'Life on Other Worlds'. *Science*, 53 (1921): 239–41.
  Outline and General Principles of the History of Life. Berkeley: University of California Press, 1928.
Maupertuis, Pierre Louis Moreau de. *The Earthly Venus*. Trans. Simone Brangier Boas. New York: Johnson Reprint Corporation, 1968.
  Oeuvres. 3 vols. Lyons: Jean-Marie Bruyset, 1766.
May, Henry F. *The End of American Innocence: The First Years of Our Own Time, 1912–1917*. Oxford: Oxford University Press, 1979 [1959].
Mayr, Ernst. *The Growth of Biological Thought: Diversity, Evolution and Inheritance*. Cambridge, MA: Harvard University Press, 1982.
  'The Search for Extraterrestrial Intelligence'. In Ben Zuckerman and Michael H. Hart, eds., *Extraterrestrials: Where Are They?* 2nd ed. Cambridge: Cambridge University Press, 1995 [1982], pp. 152–6.
  Systematics and the Origin of Species. New York: Columbia University Press, 1942.
  Toward a New Philosophy of Biology: Observations of an Evolutionist. Cambridge, MA: Harvard University Press, 1988.
Mayr, Ernst, and William B. Provine, eds. *The Evolutionary Synthesis: Perspectives on the Unification of Biology*. Cambridge, MA: Harvard University Press, 1980.
Meijer, Miriam Claude. *Race and Aesthetics in the Anthropology of Petrus Camper (1722–1789)*. Amsterdam: Rodopi, 1999.
Mercier, Louis-Sébastien. *L'an deux mille cent quarante: rêve s'il en fut jamais*. Reprinted. Paris: Éditions Ducros, 1971 [1771].
Mesouli, Alex. *Cultural Evolution: How Darwinian Theory Can Explain Human Culture and Synthesize the Social Sciences*. Chicago: University of Chicago Press, 2011.
Milburn, Colin. *Nanovision: Engineering the Future*. Durham, NC: Duke University Press, 2008.
Mill, James. *The History of British India*. 4th edn. 6 vols. London: James Madden, 1848.

Millar, Ronald. *The Piltdown Men: A Case of Archaeological Fraud*. London: Victor Gollancz, 1972.

Miller, Hugh. *The Testimony of the Rocks*. Edinburgh: William P. Nimmo, 1870 [1841].

Mivart, St George Jackson. *The Genesis of Species*. New York: Appleton, 1871.

Mokyr, Joel. *Culture and Growth: The Origins of the Modern Economy*. Princeton, NJ: Princeton University Press, 2016.

 The Gifts of Athena: Historical Origins of the Knowledge Economy. Princeton, NJ: Princeton University Press, 2002.

Montesquieu, Charles Louis Secondat, baron de. *The Spirit of the Laws*. Ed. Anne M. Cohler. Cambridge: Cambridge University Press, 1989.

Moore, James R. *The Post-Darwinian Controversies: A Study of the Protestant Struggle to Come to Terms with Darwin in Great Britain and America, 1870–1900*. New York: Cambridge University Press, 1979.

Morgan, Lewis Henry. *Ancient Society: Or Researches into the Lines of Human Progress from Savagery through Barbarism to Civilization*. Reprinted. Cambridge, MA: Harvard University Press, 1964 [1877]. Also reprinted Chicago: Charles H. Kerr, 1909.

Morris, Ian. *Why the West Rules – For Now*. New York: Farrar, Straus and Giroux, 2010.

Morss, J. R. *The Biologizing of Childhood: Developmental Psychology and the Darwinian Myth*. Hove: Laurence Erlbaum, 1990.

Müller, Fritz. *Facts and Arguments for Darwin*. London: John Murray, 1869.

Muller, H. J. *Out of the Night: A Biologist's View of the Future*. London: Gollancz, 1936.

Müller, Max. *Lectures on the Science of Language Delivered at the Royal Institution of Great Britain in April, May and June, 1861*. London: Longmans, Green, 1861.

Murphy, Terence D. 'Jean-Baptiste Robinet: The Career of a Man of Letters'. *Studies in Voltaire and the Eighteenth Century*, 150 (1976): 183–250.

Muschinske, David. 'The Nonwhite as Child: G. Stanley Hall on the Education of Nonwhite Peoples'. *Journal of the History of the Behavioural Sciences*, 13 (1977): 328–36.

Needham, Joseph. *History Is on Our Side: A Contribution to Political Religion and Scientific Faith*. London: Allen and Unwin, 1946.

 Time, the Refreshing River: Essays and Addresses, 1932–1942. London: Allen and Unwin, 1943.

Nicholls, Peter, ed. *The Encyclopedia of Science Fiction*. London: Granada, 1974.

Nisbet, Robert. *History of the Idea of Progress*. London: Heinemann, 1980.

Nordau, Max. *Degeneration*. 2nd ed. London: Heinemann, 1895.

Nyhart, Lynn K. *Biology Takes Form: Animal Morphology and the German Universities, 1800–1900*. Chicago: University of Chicago Press, 1995.

O'Brien, George Dennis. *Hegel on Reason and History: A Contemporary Interpretation*. Chicago: University of Chicago Press, 1975.

O'Duffy, Eimar. *Life and Money: Being a Critical Examination of the Principles and Practices of Orthodox Economics with an Outline of the Principles and Proposals of Social Credit*. London: Pitman, 1932.

Ogilvie, Brian W. *The Science of Describing: Natural History in the Renaissance*. Chicago: University of Chicago Press, 2006.

Oken, Lorenz. *Elements of Physicophilosophy*. Trans. Alfred Tulk. London: Ray Society, 1847.

Ortolano, Guy. *The Two Cultures Controversy: Science, Literature and Cultural Politics in Postwar Britain*. Cambridge: Cambridge University Press, 2009.

Osborn, Henry Fairfield. *The Age of Mammals in Europe, Asia and North America*. New York: Macmillan, 1910.

   Cope: Master Naturalist: The Life and Writings of Edward Drinker Cope. Princeton, NJ: Princeton University Press, 1931.

   Man Rises to Parnassus: Critical Epochs in the Prehistory of Man. Princeton, NJ: Princeton University Press, 1928.

   Men of the Old Stone Age: Their Environment, Life and Art. 3rd ed. London: G. Bell, 1927.

   The Origin and Evolution of Life on the Theory of Action, Reaction and Interaction of Energy. New York: Scribners, 1917.

Ospovat, Dov. 'The Influence of Karl Ernst von Baer's Embryology, 1828–1859: A Reappraisal in Light of Richard Owen and William B. Carpenter's "Paleontological Application of von Baer's Law"'. *Journal of the History of Biology*, 9 (1976): 1–28.

Overy, Richard. *The Morbid Age: Britain and the Crisis of Civilization*. London: Allen Lane, 2009.

Owen, Richard. *The Anatomy of the Vertebrates*. 3 vols. Reprinted. New York: AMS Press, 1973 [London, 1866–8].

   On the Archetype and Homologies of the Vertebrate Skeleton. London: J. van Voorst, 1848.

   On the Nature of Limbs: A Discourse. Preface by Brian K. Hall, introduction by Ron Amundson. Reprinted. Chicago: University of Chicago Press, 2007 [1849].

   Palaeontology: Or a Systematic Study of Extinct Animals and Their Geological Relations. Edinburgh: A. and C. Black, 1860.

Pancaldi, Giuliano. 'Darwin's Technology of Life'. *Isis*, 110 (2019): 680–700.

Panchasi, Roxanne. *Future Tense: The Culture of Anticipation in France between the Wars*. Ithaca, NY: Cornell University Press, 2009.

Parrinder, Patrick. *Shadows of the Future: H. G. Wells, Science Fiction and Prophecy*. Liverpool: Liverpool University Press, 1995.

Pavuk, Alexander. 'Biologist Edwin Grant Conklin and the Idea of a Religious Direction of Human Evolution in the Early 1920s'. *Annals of Science*, 74 (2017): 64–82.

Perkins, Robert L., ed., *History and System: Hegel's Philosophy of History*. Albany: State University of New York Press, 1984.

Pick, Daniel. *Faces of Degeneration: Aspects of European Cultural Disorder, 1848–1918*. Cambridge: Cambridge University Press, 1989.

Pickering, Mary. *Auguste Comte: An Intellectual Biography*. 3 vols. Cambridge: Cambridge University Press, 1993–2009.

Pietsch, Theodore W. *Trees of Life: A Visual History of Evolution*. Baltimore: Johns Hopkins University Press, 2012.

Pinker, Stephen. *Enlightenment Now: The Case for Science, Reason and Progress*. London: Allen Lane, 2018.

Pinto-Correia, Clara. *The Ovary of Eve: Egg and Sperm in Preformation*. Chicago: University of Chicago Press, 1997.

Poliakov, Walter N. *Mastering Power Production: The Industrial, Economical and Social Problems Involved and Their Solution*. London: Cecil Palmer, 1922.

Pollard, Sydney. *The Idea of Progress: History and Society.* Reprinted.
    Harmondsworth: Penguin, 1971.
Pomeranz, Kenneth. *The Great Divergence: China, Europe, and the Making of the
    World Economy* (Princeton, NJ: Princeton University Press, 2000).
Pope, Alexander. *The Works of Alexander Pope, Esq.* 8 vols. London: J. and
    P. Knapton, 1751.
Popper, Karl. *The Logic of Scientific Discovery.* London: Hutchinson, 1959.
    The Open Society and Its Enemies. 2 vols. 4th ed. Princeton, NJ: Princeton
        University Press, 1962.
    The Poverty of Historicism. London: Routledge and Kegan Paul, 1957.
Porter, Roy. *The Enlightenment.* 2nd ed. Basingstoke: Macmillan, 2000 [1990].
Postman, Neil. *Technopoly: The Surrender of Culture to Technology.* New York:
    Vintage, 1993.
Pouchet, Georges. *The Plurality of the Human Race.* London: Longmans, Green for the
    Anthropological Society, 1864.
Priestley, Joseph. *An Essay on the First Principles of Government; and on the Nature of
    Political, Civil, and Religious Liberty.* London: J. Dodsley, T. Cadell and
    J. Johnson, 1768.
    Hartley's Theory of the Human Mind. London: J. Johnson, 1775.
    *Lectures on History and General Policy; to Which Is Prefixed an Essay on a
        Course of Liberal Education for Civil and Active Life.* Birmingham: J. Johnson,
        1788.
Pritchard, James Cowles. *Researches into the Physical History of Man.* Reprinted with
    introduction by George W. Stocking, Jr.Chicago: University of Chicago Press,
    1973 [1813].
Provine, William B. *The Origins of Theoretical Population Genetics.* Chicago:
    University of Chicago Press, 1971.
Rader, Melvin. *Marx's Interpretation of History.* New York: Oxford University Press,
    1979
Rainger, Ronald. *An Agenda for Antiquity: Henry Fairfield Osborn and Vertebrate
    Paleontology at the American Museum of Natural History.* Tuscaloosa: University
    of Alabama Press, 1991.
Ranke, Leopold von. *The Theory and Practice of History.* Introduction by Georg
    G. Iggers. London: Routledge, 2011.
Rasmussen, Nicolas. 'The Decline of Recapitulationism in Early Twentieth-Century
    Biology'. *Journal of the History of Biology*, 24 (1991): 51–89.
Ratzel, Friedrich. *The History of Mankind.* Trans. A. J. Butler, introduction by E. B.
    Tylor. 3 vols. London: Macmillan, 1896.
Raven, Charles E. *The Creator Spirit: A Survey of Christian Doctrine in the Light of
    Biology, Psychology and Mysticism.* London: Martin Hopkinson, 1927.
    John Ray, Naturalist: His Life and Work. Cambridge: Cambridge University Press,
        1942.
    Teilhard de Chardin: Scientist and Seer. London: Collins, 1962.
Ray, John. *The Wisdom of God Manifested in the Works of the Creation.* 2nd ed.
    London: Samuel Smith, 1692 [1691].
Reade, Winwood. *The Martyrdom of Man.* 9th ed. London: Trübner, 1884 [1872].
Rees, Amanda, and Iwan Morus, eds. 'Presenting Futures Past: Science Fiction and the
    History of Science'. Special issue, *Osiris*, 34 (2019).

Regal, Brian. *Henry Fairfield Osborn: Race and the Search for the Origins of Man.* Aldershot: Ashgate, 2002.

Regis, Edward, Jr., ed. *Extraterrestrials: Science and Alien Intelligence.* Cambridge: Cambridge University Press, 1985.

Rehbock, Philip F. *The Philosophical Naturalists: Themes in Early Nineteenth-Century British Biology.* Madison: University of Wisconsin Press, 1983.

Richards, Robert J. *Darwin and the Emergence of Evolutionary Theories of Mind and Behavior.* Chicago: University of Chicago Press, 1987.

The Meaning of Evolution: The Morphological Construction and Ideological Reconstruction of Darwin's Theory. Chicago: University of Chicago Press, 1992.

The Romantic Conception of Life: Science and Philosophy in the Age of Goethe. Chicago: University of Chicago Press, 2002.

The Tragic Sense of Life: Ernst Haeckel and the Struggle over Evolutionary Thought. Chicago: University of Chicago Press, 2008.

Richards, Robert J., and Michael Ruse. *Debating Darwin.* Chicago: University of Chicago Press, 2016.

Richardson, Robert D., Jr., *Emerson: The Mind on Fire.* Berkeley: University of California Press, 1995.

Ritvo, Licille B. *Darwin's Influence on Freud.* New Haven, CT: Yale University Press, 1990.

Rivers, W. H. R. 'President's Address, Anthropology Section'. In *Report of the British Association for the Advancement of Science, 1911.* London: John Murray, 1912, 490–9.

Robertson, James Burton. 'Memoir of the Literary Life of Frederick von Schlegel'. In Friedrich von Schlegel, *The Philosophy of History in a Course of Lectures Delivered in Vienna.* London: Bell and Daldy, 1873, pp. 1–64.

Robinet, Jean-Baptiste. *De la nature.* 4 vols. Amsterdam: E. van Harrevelt, 1761–6.

Roger, Jacques. *Buffon: A Life in Natural History.* Trans. Sarah L. Bonnefoi, ed. L. Pierce Williams. Ithaca, NY: Cornell University Press, 1997.

Les sciences de la vie dans la pensée française du XVIIIe siècle. Paris: Armand Collin, 1971.

The Life Sciences in Eighteenth-Century French Thought. Trans. Robert Ellrich, ed. Keith R. Benson. Stanford, CA: Stanford University Press, 1998.

Romanes, George John. *Darwin and after Darwin: An Exposition of the Darwinian Theory and a Discussion of Post-Darwinian Problems.* 3 vols. London: Longmans, Green, 1892–7.

Mental Evolution in Animals. London: Kegan Paul, 1883.

Mental Evolution in Man. London: Kegan Paul, 1888.

Romer, Alfred Sherwood. *Man and the Vertebrates.* Chicago: University of Chicago Press, 1933.

Vertebrate Paleontology. Chicago: University of Chicago Press, 1933.

Rorty, Amélie Oksenberg, and James Schmidt, eds. *Kant's Idea for a Universal History with a Cosmopolitan Aim: A Critical Guide.* Cambridge: Cambridge University Press, 2009.

Rosenberg, Daniel, and Susan Harding, eds. *Histories of the Future.* Durham, NC: Duke University Press, 2005.

Rostand, Jean. *Can Man Be Modified?* Trans. Jonathan Grifffin. London: Secker and Warburg, 1959.

Roth, Michael S. *Knowing and History: Appropriations of Hegel in Twentieth-Century France*. Ithaca, NY: Cornell University Press, 1988.

Rudwick, Martin J. S. *Bursting the Limits of Time: The Reconstruction of Geohistory in the Age of Revolution*. Chicago: University of Chicago Press, 2005.

    Earth's Deep History: How It Was Discovered and Why It Matters. Chicago: University of Chicago Press, 2014.

    The Meaning of Fossils: Episodes in the History of Paleontology. 2nd ed. New York: Science History Publications, 1976.

    Worlds before Adam: The Reconstruction of Geohistory in the Age of Reform. Chicago: University of Chicago Press, 2008.

Rupke, Nicolaas A. 'Neither Creation nor Evolution: The Third Way in Mid-Nineteenth Century Thought about the Origin of Species'. *Annals of the History and Philosophy of Biology*, 10 (2005): 143–72.

    Richard Owen: Victorian Naturalist. New Haven, CT: Yale University Press, 1994.

Ruse, Michael. *Monad to Man: The Concept of Progress in Evolutionary Biology*. Cambridge, MA: Harvard University Press, 1996.

Russell, Bertrand. *The Basic Writings of Bertrand Russell*. London: Allen and Unwin, 1961.

Russell, Bertrand, and Dora Russell, *The Prospects of Industrial Civilization*. London: Allen and Unwin, 1923.

Russell, Dale A. 'Speculations on the Evolution of Intelligence in Multicellular Organisms'. In John Billingham, ed., *Life in the Universe*. Cambridge, MA: MIT Press, 1981, pp. 259–75.

Russell, E. S. *Form and Function: A Contribution to the History of Animal Morphology*. London: John Murray, 1916.

Sagan, Carl. *Cosmos*. London: Macdonald, 1980.

    The Dragons of Eden: Speculations on the Evolution of Human Intelligence. New York: Random House, 1977.

Sahlins, Marshall D., 'Evolution, Specific and General'. In Marshall D. Sahlins and Elman R. Service, eds., *Evolution and Culture*. Ann Arbor: University of Michigan Press, 1960, pp. 12–44.

Sahlins, Marshall D., and Elman R. Service, eds. *Evolution and Culture*. Ann Arbor: University of Michigan Press, 1960.

Saint-Pierre, Charles Irenée Castel, abbé de. *Ouvrajes de morale et de politique*. 16 vols. Rotterdam: Jean Daniel Beman, 1737.

Saint-Simon, Claude-Henri de Rouvray, comte de. *The Doctrine of Saint-Simon: An Exposition: First Year, 1828–1829*. Ed. and trans. Georg G. Iggers. New York: Schocken, 1972.

    Henri Saint-Simon (1760–1825): Selected Writings on Science, Industry and Social Organization. Ed. and trans. Keith Taylor. London: Croom Helm, 1975.

Sampson, R. V. *Progress in the Age of Reason*. Cambridge, MA: Harvard University Press, 1956.

Schiller, Joseph. 'L'échelle des êtres et la série chez Lamarck'. In Joseph Schiller, ed., *Collque international 'Lamarck' tenue au Museum national d'histoire naturelle*. Paris: Blanchard, 1971, pp. 87–103.

Schiller, Joseph, ed. *Collque international 'Lamarck' tenue au Museum national d'histoire naturelle*. Paris: Blanchard, 1971.

Schlegel, Friedrich von. *The Philosophy of History in a Course of Lectures Delivered in Vienna*. Trans. with a memoir of the author by James Burton Robertson. London: Bell and Daldy, 1873.

Schneider, Jeremy Roger. 'The First Mite: Insect Genealogy in Hooke's *Micrographia*'. *Annals of Science*, 75 (2018): 165–200.

Schofield, Robert E., *The Enlightened Joseph Priestley: A Study of his Life and Work from 1773 to 1804*. University Park: Pennsylvania State University Press, 2004.

The Enlightenment of Joseph Priestley: A Study of his Life and Work from 1733 to 1773. University Park: Pennsylvania State University Press, 1997.

Schwalbe, Gustav. 'The Descent of Man'. In A. C. Seward., ed., *Darwin and Modern Science*. Cambridge: Cambridge University Press, 1909, pp. 112–36.

Die Vorgeschichte des Menschen. Brunswick: Friedrich Vieweg, 1904.

Secord, James A. *Victorian Sensation: The Extraordinary Publication, Reception and Secret Authorship of* Vestiges of the Natural History of Creation. Chicago: University of Chicago Press, 2001.

Segal, Howard P. *Technological Utopianism in American Culture*. Chicago: University of Chicago Press, 1985.

Shklovskii, I. S., and Carl Sagan. *Intelligent Life in the Universe*. San Francisco: Holden-Day, 1966.

Shuttleworth, Sally. *The Mind of the Child: Child Development in Literature, Science, and Medicine, 1840–1900*. Oxford: Oxford University Press, 2010.

Simon, W. M. *European Positivism in the Nineteenth Century: An Essay in Intellectual History*. Ithaca, NY: Cornell University Press, 1963.

Simpson, George Gaylord. *The Meaning of Evolution: A Study of the History of Life and of Its Significance for Man*. Rev. ed. New York: Bantam Books, 1971 [1949].

Tempo and Mode in Evolution. New York: Columbia University Press, 1944.

This View of Life: The World of an Evolutionist. New York: Harcourt, Brace and World, 1964.

Sinclair, Upton. *The Industrial Republic: A Study of the America of Ten Years Hence*. London: Heinemann, 1907.

Sinnerbrink, Robert. *Understanding Hegelianism*. Stocksfield: Acumen, 2007.

Skelton, Matthew. 'The Paratext of Everything: Constructing and Marketing H. G. Wells's The Outline of History'. *Book History*, 4 (2000): 237–75.

Sloan, Philip R. 'John Locke, John Ray, and the Problem of the Natural System'. *Journal of the History of Biology*, 5 (1972): 1–53.

Slobodin, Richard. *W. H. R. Rivers*. New York: Columbia University Press, 1978.

Smellie, William. *The Philosophy of Natural History*. 2 vols. Edinburgh: The Heirs of Charles Elliot, 1790–99.

Smith, Adam. *Lectures on Jurisprudence*. Ed. R. L. Meek. Oxford: Clarendon Press, 1978.

The Wealth of Nations. 2 vols. Reprinted. London: Everyman, 1910.

Smith, Grafton Elliot. *The Evolution of Man: Essays*. Oxford: Oxford University Press, 1924.

Human History. London: Jonathan Cape, 1930.

Smith, Roger. *The Fontana History of the Human Sciences*. London: Fontana, 1997.

Smocovitis, Vassiliki Betty. *Unifying Biology: The Evolutionary Synthesis and Evolutionary Biology*. Princeton, NJ: Princeton University Press, 1996.

Snow, C. P. *The Two Cultures and a Second Look*. Cambridge: Cambridge University Press, 1969.

Soddy, Frederick, *The Interpretation of Radium*. London: John Murray, 1909.

Wealth, Virtual Wealth and Debt. London: Allen and Unwin, 1926.

Sollas, William J. *Ancient Hunters and Their Modern Representatives*. London: Macmillan, 1911.

Soule, George. *The Coming American Revolution*. London: Routledge, 1934.

Spadaforda, David. *The Idea of Progress in Eighteenth-Century Britain*. New Haven, CT: Yale University Press, 1990.

Spencer, Frank. *Piltdown: A Scientific Forgery*. London: Oxford University Press, 1990.

Spencer, Herbert. *Essays Scientific, Political and Speculative*. 3 vols. London: Williams and Norgate, 1883.

First Principles of a New Philosophy. London: Williams and Norgate, 1862.

Principles of Biology. 2 vols. London: Williams and Norgate, 1864.

Principles of Psychology. 2 vols. London: Longman, Brown, 1855.

Principles of Sociology. 3 vols. London: Williams and Norgate, 1876–96.

Social Statics: Or the Conditions Essential to Human Happiness Specified, and One of Them Adopted. London: John Chapman, 1851.

Squire, J. C., ed. *If It Had Happened Otherwise*. London: Sidgwick and Jackson, 1972 [1932].

Stanley, Arthur Penrhyn. *Whether States, Like Individuals, after a Certain Period of Maturity, Inevitably Tend to Decay. A Prize Essay Read in the Sheldonian Theatre, Oxford, July 1, 1840*. Oxford: J. Vincent, 1840.

Stanton, William. *The Leopard's Spots: Scientific Attitudes toward Race in America, 1815–1859*. Chicago: Phoenix Books, 1960.

Stapledon, Olaf. *Last and First Men*. London: Millennium, 1999 [1930].

Stepan, Nancy. *The Idea of Race in Science: Great Britain, 1800–1960*. London: Macmillan, 1982.

Stephen, Sir Leslie. *History of English Thought in the Eighteenth Century*. New preface by Crane Brinton. 2 vols. New York: Harcourt Brace and World, 1962 [1876].

Stephens, Lester G. *Joseph LeConte: Gentle Prophet of Evolution*. Baton Rouge: Louisiana State University Press, 1982.

Steward, Julian H. *Theory of Culture Change: The Methodology of Multilinear Evolution*. Urbana: University of Illinois Press, 1955.

Stirling, James Hutcheson. *Hegel not Haeckel: 'The Riddle of the Universe' Solved*. Birmingham: C. Combridge, 1906.

Stocking, George W., Jr. *After Tylor: British Social Anthropology, 1888–1951*. London: Athlone, 1996.

Stocking, George W., *Race, Culture and Evolution*. New York: Free Press, 1968.

Victorian Anthropology. New York: Free Press, 1987.

Stover, Leon. *The Prophetic Soul: A Reading of H. G. Wells's* Things to Come *Together with His Film Treatment* Whither Mankind? *and the Postproduction Script*. Jefferson, NC: McFarland, 1987.

Science Fiction from Wells to Heinlein. Jefferson, NC: McFarland, 2002.

Sulloway, Frank J. *Freud, Biologist of the Mind: Beyond the Psychoanalytic Legend*. London: Burnett Books, 1979.

Sully, James. *Studies of Childhood*. New ed. London: Longmans, Green, 1903 [1895].

Suriano, Dominique. *L'abbé de Saint-Pierre ou les infortunes de la raison*. Paris: l'Harmattan, 2005.

Susskind, Jamie. *Future Politics: Living Together in a World Transformed by Technology*. Oxford: Oxford University Press, 2018.

Sutherland, George. *Twentieth-Century Inventions: A Forecast*. London: Longmans Green, 1901.

Swetlitz, Marc. 'Julian Huxley and the End of Evolution'. *Journal of the History of Biology*, 28 (1995): 181–217.

Talib, Nassim, *The Black Swan: The Impact of the Highly Improbable*. New York: Random House, 2007.

Taylor, Gordon Rattray. *The Biological Time Bomb*. London: Panther, 1969 [1968].

Taylor, Michael. *The Philosophy of Herbert Spencer*. London: Continuum, 2007.

Teilhard de Chardin, Pierre. *The Phenomenon of Man*. Trans. Bernard Wall, introduction by Julian Huxley. London: Collins, 1959.

Thompson, John M. *A Vision Unfulfilled: Russia and the Soviet Union in the Twentieth Century*. Lexington, MA: D. C. Heath, 1996.

Thomson, J. Arthur. *The Gospel of Evolution*. London: George Newnes, n.d.
  'Introduction'. In Frances Mason, ed., *The Great Design: Order and Progress in Nature*. London: Duckworth, 1934, pp. 11–16.
  The System of Animate Nature. 2 vols. London: Williams and Norgate, 1920.

Thucydides. *The History of the Peloponnesian War. With Notes Chiefly Historical and Geographical by Thomas Arnold*. 3 vols. Oxford: S. Collingwood, 1830–5.

Toews, John Edward. *Hegelianism: The Path toward Dialectical Humanism, 1805–1841*. Cambridge: Cambridge University Press, 1980.

Toffler, Alvin. *Future Shock*. London: Bodley Head, 1970.

Toynbee, Arnold. *Civilization on Trial*. New York: Oxford University Press, 1948.
  A Study of History: Abridgement of Volumes 1–6. Ed. D. C. Somervell. Oxford: Oxford University Press, 1946.
  A Study of History, vol. 12: *Reconsiderations*. Oxford: Oxford University Press, 1961.

Trautmann, Thomas R. *Lewis Henry Morgan and the Invention of Kinship*. Berkeley: University of California Press, 1987.

Trenn, Thaddeus J. 'The Central Role of Energy in Soddy's Holistic and Critical Approach to Nuclear Science, Economics, and Social Responsibility'. *British Journal for the History of Science*, 12 (1979): 261–76.

Trevelyan, G. M. *Clio: A Muse and Other Essays Literary and Pedestrian*. London: Longmans Green, 1913.

Trigger, Bruce G. *A History of Archaeological Thought*. Cambridge: Cambridge University Press, 1989.

Turgot, Anne Robert Jacques. *Turgot on Progress, Sociology and Economics*. Trans. Ronald L. Meek. Cambridge: Cambridge University Press, 1973.

Turner, Fred. *From Counterculture to Cyberculture: Stewart Brand, the Whole Earth Network, and the Rise of Digital Utopianism*. Chicago: University of Chicago Press, 2006.

Turney, John. *Frankenstein's Footsteps: Science, Genetics and Popular Culture*. New Haven, CT: Yale University Press, 1998.

Tuveson, Lee. *Millennium and Utopia: A Study in the Background to the Idea of Progress*. Berkeley: University of California Press, 1949.

Tylor, Edward B. *Anthropology: An Introduction to the Study of Man and Civilization*. London: Macmillan, 1881.

Primitive Culture: Researches into the Development of Mythology, Philosophy, Religion, Art and Custom. 2 vols. London: John Murray, 1870.

Researches into the Early History of Mankind and the Development of Civilization. London: John Murray, 1865.

Uglow, Jenny. *The Lunar Men: The Friends Who Made the Future*. London: Faber and Faber, 2002.

Urwick, Lyndall. *Management of Tomorrow*. London: Nisbet, 1931.

Van Doren, Charles L. *The Idea of Progress*. New York: F. A. Praeger, 1967.

Van Riper, A. Bowdoin. *Men among the Mammoths: Victorian Science and the Discovery of Human Prehistory*. Chicago: University of Chicago Press, 1993.

Van Wyhe, John. *Phrenology and the Origins of Victorian Scientific Naturalism*. Aldershot: Ashgate, 2004.

Veblen, Thorstein. *The Engineers and the Price System*. Introduction by Daniel Bell. New York: Harcourt, Brace and World, 1963 [1921].

Vico, Gianbattista. *The New Science of Gianbattista Vico*. Trans. Thomas Goddard Bergin and Max Harold Fisch. Ithaca, NY: Cornell University Press, 1970.

Vogt, Carl. *Lectures on Man: His Place in Creation, and in the History of the Earth*. Ed. James Hunt. London: Longmans, Green for the Anthropological Society, 1864.

Voltaire, François-Marie Arouet de. *Oeuvres complètes de Voltaire*. Oxford: Voltaire Foundation, vol. 22 (2009); vol. 27 (2016); vol. 59 (1969).

Oeuvres historiques. Ed. René Pomeu. Paris: Gallimard, 1957.

Philosophical Dictionary. Trans. and ed. Theodore Besterman. Harmondsworth: Penguin, 1971.

The Philosophy of History. Reprint. New York: Citadel Press, 1966 [1766].

Wagar, W. Warrren. *Good Tidings: The Belief in Progress from Darwin to Marcuse*. Bloomington: Indiana University Press, 1972.

The Idea of Progress since the Renaissance. New York: Wiley, 1969.

The Next Three Futures: Paradigms of Things to Come. New York: Greenwood Press, 1991.

A Short History of the Future. 2nd ed. London: Adamantine Press, 1992.

Terminal Visions: The Literature of Last Things. Bloomington: Indiana University Press, 1982.

Wagner, Peter. *A Sociology of Modernity: Liberty and Discipline*. London: Routledge, 1994.

Walker, John. *History, Spirit and Experience: Hegel's Conception of the Historical Task of Philosophy in His Age*. Frankfurt: Peter Lang, 1995.

Wallace, Alfred Russel. *Darwinism: An Exposition of the Theory of Natural Selection*. London: Macmillan, 1889.

Man's Place in the Universe. 4th ed. London: Chapman and Hall, 1904.

The World of Life: A Manifestation of Creative Power, Directive Mind and Ultimate Purpose. London: G. Bell, 1911.

Ward, Lester Frank. 'The Relation of Sociology to Anthropology'. *American Anthropologist*, 8 (1895): 241–56.

Waters, C. Kenneth, and Albert Van Helden, eds. *Julian Huxley: Biologist and Statesman of Science*. Houston, TX: Rice University Press, 1992.

Weiner, J. S. *The Piltdown Hoax*. Oxford: Oxford University Press, 1955.

Weismann, August. *The Germ Plasm: A Theory of Heredity*. Trans. W. Newton Parker and Harriet Ronfeldt. London: Scott, 1893.

Wells, Herbert George. *Anticipations: Of the Reaction of Mechanical and Scientific Progress upon Human Life and Thought*. New ed. With author's specially written introduction. London: Chapman and Hall, 1914.

The Discovery of the Future: A Lecture Delivered at the Royal Institution on January 24, 1902. London: T. Fisher Unwin, 1902.

The First Men in the Moon. London: Gollancz, 2017 [1901].

Men Like Gods. London: Cassell, 1923.

A Modern Utopia. Reprinted. Ed., introduction and appendix by Krishnan Kumar, London: Dent, 1997 [1905].

The Outline of History: Being a Plain History of Life and Mankind. 2 vols. London: Newnes, 1920. Definitive ed. 1 vol. London: Cassell, 1924.

The Shape of Things to Come: The Ultimate Revolution. London: Hutchinson, 1933.

The Short Stories of H. G. Wells. London: Benn, 1926.

The Sleeper Awakes. Ed. and introduction by Patrick Parrinder. London: Penguin Classics, 2005 [1899].

Things to Come: A Critical Text of the 1935 London First Edition. Ed. Leon Stover. Jefferson, NC: McFarland, 2007.

'Wanted – Professors of Foresight!' *The Listener*, 8 (1932): 729–30.

The War of the Worlds. Harmondsworth: Penguin, 1946 [1898].

The World Set Free: A Story of Mankind. London: Macmillan, 1914.

Wells, H. G., Julian Huxley and G. P. Wells. *The Science of Life*. London: Newnes, 1931.

West, Anthony. *H. G. Wells: Aspects of a Life*. London: Hutchinson, 1984.

Whewell, William. *Philosophy of the Inductive Sciences*. New ed. 2 vols. London: John Parker, 1847.

White, Charles. *An Account of the Regular Gradation of Nature in Man, and in Different Animals and Vegetables; and from the Former to the Latter*. London: C. Dilly, 1799.

White, L. A. *The Science of Culture: A Study of Man*. New York: Farrar, 1949.

White, Leslie. 'Foreword'. In Marshall D. Sahlins and Elman R. Service, eds., *Evolution and Culture*. Ann Arbor: University of Michigan Press, 1960, pp. v–xii.

Whitehead, Alfred North. *Science and the Modern World*. Cambridge: Cambridge University Press, 1926.

Whitman, Charles Otis. 'Charles Bonnet's Theory of Evolution'. *Biological Lectures Delivered at the Marine Biological Laboratory of Woods Hole* (1894): 225–40.

'The Palingenesia and the Germ Doctrine of Bonnet'. *Biological Lectures Delivered at the Marine Biological Laboratory of Woods Hole* (1894): 241–72.

Wiener, Philip P. *Evolution and the Founders of Pragmatism*. Cambridge, MA: Harvard University Press, 1949.

Wilkins, Burleigh Taylor. *Hegel's Philosophy of History*. Ithaca, NY: Cornell University Press, 1974.

Wilson, Edward O. *On Human Nature*. Cambridge, MA: Harvard University Press, 1978.

Sociobiology: The New Synthesis. Cambridge, MA: Harvard University Press, 1975.

Wright, Sewall. *Evolution and the Genetics of Populations.* 4 vols. Chicago: University of Chicago Press, 1968–78.

Young, A. P. *Forward from Chaos.* London: Nisbet, 1933.

Young, Robert M. *Mind, Brain and Adaptation in the Nineteenth Century: Cerebral Localization and Its Biological Context from Gall to Ferrier.* Oxford: Clarendon Press, 1970.

Zamyatin, Yevgeny. *We.* Trans. and introduction by Clarence Brown. London: Penguin, 1993 [1924].

Zuckerman, Ben, and Michael H. Hart, eds., *Extraterrestrials: Where Are They?* 2nd ed. Cambridge: Cambridge University Press, 1995 [1982].

Index